T0298012

From Quarks to the Universe

Eleftherios N. Economou

From Quarks to the Universe

A Short Physics Course

Second Edition

Eleftherios N. Economou
FORTH, IESL
University of Crete
Iraklion
Greece

ISBN 978-3-319-20653-0 ISBN 978-3-319-20654-7 (eBook)
DOI 10.1007/978-3-319-20654-7

Library of Congress Control Number: 2015954584

Springer Cham Heidelberg New York Dordrecht London

Printed on acid-free paper

Springer International Publishing AG Switzerland is part of Springer Science+Business Media
(www.springer.com)

There is, of course, an immense liberating role of Science at a central existential level. It is what Aristotle was saying about "θαυμάζειν". Science humanizes us, liberating us from our animal instincts, just because it makes us wondering and at the same time desiring to explain...Yet it shows us our limits and our mortality... Thus Science is something immeasurably precious....Science can help us approach anew the real poetic and mythical dimension of human existence.

C. Castoriadis

The Castoriadis quote is from the Castoriadis and Evangelopoulos book, *Philosophy and Science,* (Editions Eurasia, Athens, 2010)

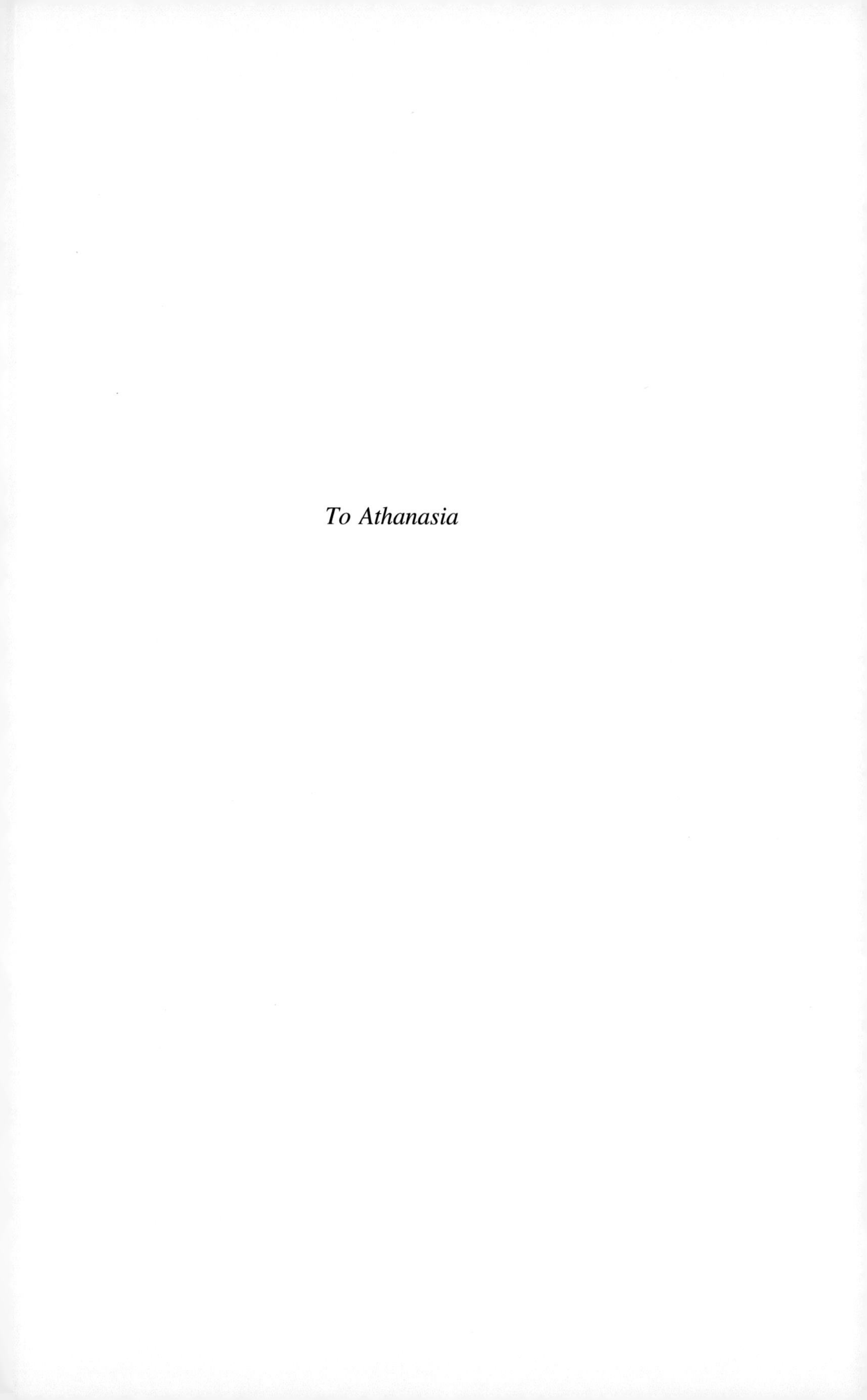

To Athanasia

Preface to the Second Edition

This book in this second edition has been enlarged (its size now is more than twice that of the first edition) and has been enriched in order to also serve as a senior undergraduate textbook; nevertheless, it retains its main feature of deriving most of the basic formulae governing the behavior of the various structures of the physical word by applying “a little thinking” and employing dimensional considerations.

Explicitly, in each chapter, besides more background information, new sections have been added: One of them includes a summary of the main relevant formulae; another contains many multiple choice questions/statements (their correct answers are given at the end of the book). Finally, there are two more sections in every chapter involving solved and unsolved problems respectively.

Moreover, six new appendices have been added in this new edition: In two of them a summary of the subjects of Electrodynamics of Continuous Media and of Thermodynamics and Statistical Mechanics is presented. These two appendices together with the last three, presenting a list of the required background concepts, formulae, and numbers, make the book to a large degree self-contained. In another new appendix a few basic concepts regarding semiconductor physics are introduced.

As I mentioned before, this book in this second edition may well serve a senior undergraduate course: The students in such a course will be asked to wrap up their basic knowledge and reasoning and apply them to *derive* and *understand* the basic features of the physical world. Of course, as it was stated in the preface to the first edition, graduate students, research scientists, physics teachers and others may find this book intellectually stimulating and entertaining.

I would like thank again my colleague, Prof. V. Charmandaris, for reading the entire text of this second edition and for making many useful suggestions. Of course, whatever misprints or misrepresentations remained are my own responsibility only. I am also grateful to Ms. Maria Dimitriadi for her invaluable help in bringing my manuscript to its final form.

Iraklion — Eleftherios N. Economou

Preface to the First Edition

This short book grew out of lectures presented to different audiences (physics students, physicists, material scientists, engineers) and on various occasions (colloquia and seminars in physics and other departments, conferences, special events). The main purpose of these lectures and, obviously, of the present book is to show that basic formulae concerning the various structures of the *physical world* pop out quickly, if some *basic ideas*, the *universal physical constants*, and *dimensional considerations* are exploited. Of course, as R. Feynman pointed out, "a little thinking has to be applied too".

The basic ideas include the three cornerstones of science, namely the atomic idea, the wave-particle duality, and the minimization of free energy as the necessary and sufficient condition for equilibrium (these are presented in Chaps. 2, 3, and 4 respectively). These fundamental ideas exhibit their worth when accompanied by the values of the physical constants: the universal ones, $\hbar$, c, the coupling constants of the four interactions, G, e, g_w, g_s and the masses of the elementary particles, m_p, m_n, m_e, m_w, ... An important consequence of the atomic idea is that the relevant (for each case) physical constants will appear in the quantities characterizing the various structures of the world either microscopic or *macroscopic*. Combining this last observation—often overlooked—with dimensional analysis, presented in Chap. 5, and "a little thinking", one can obtain, in several cases, an amazing short-cut derivation of formulae concerning the various structures of Nature from the smallest (baryons and mesons) to the whole Universe, as shown Chaps. 6–13. In each one of these 8 chapters, in parallel with a demonstration of the method just outlined, a *condensed* (sometimes too condensed) introduction to the relevant subject matter together with a few physical remarks are presented.

I must admit that the main fronts on which our scientific horizons are widened, namely the *small*, the *large*, and the *complex* could not be treated even remotely adequately in this short book. Actually the *complex*, as represented by the living matter, was too complex for our simple method; so it was left out completely (however, see the epilogue). The *large* (cosmology) and the *small* (elementary particles) tend to converge to a unified subject (the snake in Fig. 1.1, p.2, is biting its tail) fed with novel observational data from special instruments mounted usually on satellites, and boosted by high experimental expectations from the Large Hadron Collider. Nevertheless, in these fields there are several open fundamental questions concerning conditions well beyond our present or near future experimental capabilities. This vacuum of confirmed knowledge is filled with new intriguing, imaginative ideas and novel proposed theories (such as supersymmetry, string theory, M-theory, see reference [P1]) which, if established, will radically change our world view. In spite of the wider interest in these ideas and theories and their high intellectual value, I decided for several reasons to restrict myself in the present book to experimentally or observationally tested ideas and theories.

The intended readers of this book are senior undergraduate or graduate students in Physics, Engineering, Applied Mathematics, Chemistry, and Material Science. They may find the book a useful supplement to their courses as a concise overall picture of the physical world. Research physicists, physics teachers, and other scientists may also find this short book intellectually stimulating and entertaining. The required background is no more than a *working* familiarity with the Science/Engineering material taught in the first University year.

I am deeply indebted to my colleague, Prof. V. Charmandaris, for his encouragement during the writing of this book and for reading my entire manuscript and making many useful suggestions. Of course, whatever misprints or misrepresentations remained are my own responsibility only. I am also grateful to Ms. Maria Dimitriadi for her invaluable help in bringing my manuscript to its final form.

Iraklion
January 2011

Eleftherios N. Economou

Contents

Chapter 1
Introduction: The World According to Physics

Ah, but a man's reach should exceed his grasp.
Or what's a heaven for?

R. Browning.

Abstract In this introductory chapter the subject matter as well as the methodology of Physics are briefly presented together with a list of the basic equilibrium structures of matter. The main properties of the latter can be obtained by minimizing the internal energy and are expressed in terms of universal physical constants $\hbar, c, e, m_e, m_p \approx m_n \approx u, G$.

1.1 The Nature of Physics

Modern-era Physics started as Natural Philosophy. As this former name implies, Physics is built (and continues to be developed) around the age-old, yet ever-present questions:

- *What the World and its parts are made of? How?*
- *Is there a hidden underlying simplicity in its immense complexity and diversity?*

The first question implies that the subject matter of Physics is the World, both natural and man-made; from its smallest constituent to the whole Universe. In this sense, Physics tends to encompass the other natural sciences (such as Chemistry and Biology) and even Engineering, while at the same time serves also as their foundation. What allows Physics to have this foundational role is its characteristic methodology. The latter is precise and quantitative, yet capable of abstraction (therefore mathematical). It is based on observations and well controlled experiments both as sources of ideas as well as tests for falsification or tentative confirmations of newly proposed and –even– established theories. Moreover, as the second question suggests, the methodology of physics requires the formulation of a few fundamental quantitative relations on which everything else is based. These features of the methodology of Physics account for its role as the foundation of

E.N. Economou, *From Quarks to the Universe*,
DOI 10.1007/978-3-319-20654-7_1

every other science and engineering, but explain also its limited penetration into very complex, yet very important, parts of the World (such as the molecular and the biological structures). This leaves plenty of space to more specialized sciences such as Chemistry and Biology and, of course, Engineering.

Over the last 50 years or so Physics is actively concerned over another fundamental, age-old, but much more difficult question which stretches its methodology to the limit:

- *How did the World start, how did it evolve, and where is it going*?

Detailed observational data, such as the recession of distant galaxies at a speed proportional to their distance from Earth, the spectral and angular distribution of the Cosmic Microwave Background Radiation, etc, combined with established physical theories, allowed us to reconstruct roughly some of the main events in the history of the Universe. Naturally, other crucial events, including the emergence of life, remain unknown and they are the subject of on-going research. Subject to current theoretical research is also the development of a successful quantum theory of gravity, which is expected to let us describe in a concise manner the very moment of the genesis of the Universe.

1.2 The Subject Matter of Physics

The subject matter of Physics is summarized in Fig. 1.1 and Table 1.1:

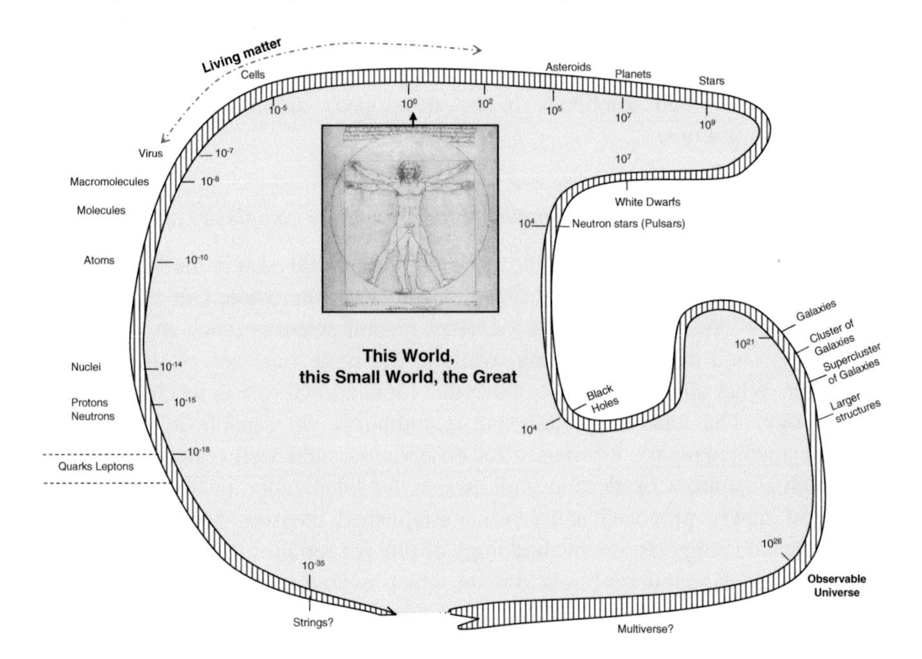

Fig. 1.1 The main structures of matter from the smallest to the largest size (*clockwise*) and the suspected connection of the two extremes (see [1]). The indicated sizes are in meters (see also next page)

Table 1.1 Levels of the structure of matter (see also [2])

Level of the structure of matter	Size (in m)	Constituents	Interaction(s) responsible for the structure
Quarks	$<10^{-18}$	It seems to be elementary	–
Electron	$<10^{-18}$	It seems to be elementary	–
Proton	10^{-15}	u, u, d quarks	Strong, weak, E/M
Neutron	10^{-15}	u, d, d quarks	Strong, weak, E/M
Nuclei	10^{-15}–10^{-14}	Protons, neutrons	Strong, E/M, weak
Atoms	10^{-10}	Nucleus, electrons	E/M
Molecules	$>10^{-10}$	Atoms and/or ions and electrons	E/M
Solids (primitive cell)	$>10^{-10}$	Atoms and/or ions and electrons	E/M
Cells	$\geq 10^{-6}$	Molecules	E/M
Biological entities (e.g., *Homo sapiens*)	10^{-8}–10^{2} (10^{0})	Molecules, cells, tissues, organs, microbes	E/M
Planets	10^{6}–10^{7}	Solids, liquids, gases	E/M, gravitational
Stars, Sun	10^{8}–10^{12},10^{9}	Electrons, nuclei, ions, photons	Gravitational, strong, weak, E/M
White dwarfs	10^{7}	Nuclei, electrons	Gravitational
Neutron stars	10^{4}	Neutrons and some protons and electrons	Gravitational
Astrophysical black holes	10^{4}	?	Gravitational
Galaxies	10^{21}	Stars, ordinary and dark matter, photons, neutrinos	Gravitational
Observable universe	10^{26}	Galaxies, dust, dark matter, dark energy	Gravitational, others?

1.3 Various Branches of Physics

In concluding these introductory remarks regarding the subject matter of Physics, we present below some of the various branches of Physics and their correspondence and/or overlap with more specialized sciences as well as some examples of the impact of Physics on important technologies (Table 1.2):

Table 1.2 Connection of branches of Physics with Technologies and other Sciences and Mathematics

Mathematics	Elementary particle physics Nuclear physics Atomic and molecular physics Condensed matter physics	Chemistry, material science
	Biophysics	Biology
	Geophysics	Geology
	Atmospheric and space physics	Meteorology, global climate
	Astrophysics, cosmology	Astronomy
Technology	E/M waves, lasers	Telecommunications
	Solid state devices Integrated circuits Magnetic devices	Computers
	X-rays γ-rays Magnetic resonance (MRI) Positron annihilation (PET)	Medical technologies

1.4 The Main Points of This Book: Basic Ideas Applied to Equilibrium Structures of Matter[1]

1. Out of the elementary matter-particles presented in Chapter 2, Table 2.1 only the *up quark* (u) and the *down quark* (d) make up the proton consisting of two u's and one d and the neutron consisting of two d's and one u. *Electrons* (e^-) are trapped around nuclei, made of protons and neutrons, to form atoms.
2. The constituents of all composite equilibrium structures are mutually attracted and self-trapped because of one or more of the four interactions presented in Table 2.2.
3. The interactions, which tend to continuously squeeze the composite structures, are counterbalanced by the pressure due to the perpetual motion of the constituents microscopic particles. This motion is of quantum nature and stems from the uncertainty principle aided by the exclusion principle (if more than two fermions (see Sect. 3.3) are involved). In other words, equilibrium of composite systems is established when the squeezing pressure of the interactions is exactly balanced by the expanding pressure of the quantum perpetual motion of the constituent particles. The equality of pressures is a consequence of the general principle of the minimization of the internal energy U (under conditions of negligible external pressure and temperature). Thus:

[1]Section 1.4 summarizes the content of this book. It may be useful for the reader to return to this section at later times.

Equilibrium of composite structures ⇔ Minimum of internal energy U. Internal energy U = Internal potential energy E_P + internal kinetic energy E_K. Internal quantum kinetic energy of N identical fermions:

$$\boxed{E_K = 2.87\frac{\hbar^2 N^{5/3}}{m\,V^{2/3}} = 1.105\frac{\hbar^2 N^{5/3}}{m\,R^2} \quad \text{non-relativistic}}$$

Internal quantum kinetic energy of N identical fermions:

$$\boxed{E_K = 2.32\frac{\hbar\, c\, N^{4/3}}{V^{1/3}} = 1.44\frac{\hbar\, c\, N^{4/3}}{R} \quad \text{extreme-relativistic}}$$

4. The combination of the first and the second law of thermodynamics leads to the following relation: The so-called Gibbs free energy, $G \equiv U + PV - TS$, under conditions of constant pressure and temperature, is always decreasing during the system's path towards equilibrium and reaches its minimum value when equilibrium is established. G reduces to U when PV and TS are negligible.
5. Dimensional analysis is a powerful method for producing physics formulae. It requires the identification of the parameters and/or the universal physical constants on which a quantity X may depend. Then the formula for X is a product (with the same dimensions as X) of appropriate powers of usually three of those parameters/physical constants times a function of their dimensionless combinations.

Next, we apply the general principles presented above to several basic equilibrium structures; the main properties of them are expressed in terms of universal physical constants.

(i) *Nuclei* consist of Z protons and N neutrons, i.e. of $A = Z + N$ nucleons. Both the strong interactions and the Coulomb interactions contribute to the potential energy: $E_P = -\frac{1}{2}AN_{nn}\mathcal{V}_s + \frac{1}{2}\sum_{ij} e^2/r_{ij}$, where N_{nn} is the number of nearest neighbors of a nucleon, $-\mathcal{V}_s$ is the strong interaction between a pair of nearest neighbor nucleons and the double summation in the Coulomb term is over all Z protons. The nuclei's radius R is proportional to the 1/3 power of A and the kinetic energy is the sum of the kinetic energy of the protons and that of the neutrons

$$E_K = 1.105\frac{\hbar^2 Z^{5/3}}{m_p\,R^2} + 1.105\frac{\hbar^2 N^{5/3}}{m_n\,R^2}$$

(ii) *Atoms* consist of a single nucleus (of Z protons) and Z electrons trapped around it by Coulomb interactions: $E_P = -\sum_i Ze^2/r_i + \frac{1}{2}\sum_{ij} e^2/r_{ij}$. The ground state energy of the electrons is found approximately by calculating the single electron atomic orbitals (see Sect. 10.3) and the corresponding energy levels; the lower ones of the latter are fully populated by the electrons, within the restrictions imposed by Pauli's principle, starting from that of lowest energy until all Z electrons are exhausted.

(iii) *Molecules* are consisting of atoms (or cations and anions) of practically unlimited combinations held together by Coulomb interactions. The molecular orbitals are approximately expressed as linear combinations of atomic orbitals or hybridized atomic orbitals.

(iv) In *solids* and *liquids* a huge number of atoms and/or molecules are coming in contact under the action of Coulomb interactions. In metals the main kinetic energy is that of *detached* electrons given approximately by $E_K = 2.87 \frac{\hbar^2 N_e^{5/3}}{m_e V^{2/3}}$, while the potential energy is of Coulomb nature and of the form $E_P = -N_a E_o \gamma / \bar{r}$, where N_a is the total number of atoms, $E_o = e^2/a_B$, $a_B \equiv \hbar^2/m_e e^2$ is the so-called Bohr radius, $\bar{r}$ is connected to the volume by the relation $\frac{4\pi}{3}(\bar{r}\, a_B)^3 = (V/N_a)$, $\gamma \approx 0.56\zeta^{4/3} + 0.9\zeta^2$, and ζ is the valence. For semiconductors and insulators, as for molecules, a linear combination of atomic (hybridized or not) orbitals turns out to be a more convenient way of studying them.

(v) *Planets* are spherical objects of mass $M = N_\nu u$ and radius R, where $N_\nu = A_W N_a \approx 10^{49}$–$10^{55}$ is the total number of nucleons within all nuclei, A_W is the average atomic weight, and $u \equiv \frac{1}{12} m(C^{12})$. In planets both the Coulomb interaction as well as the gravitational one, $E_G = -a_G G M^2/R$, contribute to the potential energy, while the kinetic energy is due mainly to the electrons as in solids and liquids. The spherical shape of moons, planets and stars is a consequence of the long range character of the gravitational interaction which becomes appreciable as a result of the huge mass involved.

(vi) Dead stars are of three types: *White dwarfs* $(M < 1.4 M_S)$, *Neutron stars* $(1.4 M_S \leq M < 3 M_S)$, and *Black Holes* $(3 M_S < M)$, where M_S is the present mass of the Sun. White dwarfs have a radius comparable to that of Earth, i.e. about 100 times smaller than that of their typical previous phase as active stars. Because of this large compression all electrons have been detached from the parent atoms; thus the white dwarf consists of electrons and bare nuclei. The kinetic energy is mainly that of electrons, as in the case of metals, and the potential energy is the gravitational one. Minimization of the total energy gives the radius as a function of the mass: $R = 1.42 \frac{\hbar^2}{G u^2 m_e N_\nu^{1/3}}$, $M = N_\nu u$. If the mass of the white dwarf keeps increasing, the kinetic energy of the electrons tends to the extreme relativistic limit which is of the form A/R, i.e. similar to the gravitational one, $-B/R$. Thus when $B \geq A$ the white dwarf will collapse to a neutron star. Hence, the equality $B = A$ gives the collapse critical value, which is $N_{\nu,cr1} = 0.77 \left(\frac{\hbar c}{G u^2}\right)^{3/2}$ corresponding to $1.4 M_S$. After the collapse, the electrons will be forced within the nuclei, which in turn lose their identity; thus a neutron star consists mainly of neutrons $(N_n \approx 0.934 N_\nu)$ and of a small percentage of protons and electrons $(N_p = N_e \approx 0.066 N_\nu)$. The potential energy in a neutron star is that of gravity as given before. The kinetic energy is mainly that of neutrons and to a very small degree that of protons (both of which are non-relativistic) plus that of electrons (which is extreme relativistic). Minimizing the total energy with respect to R we obtain, $R = 3.16 \left(\frac{\hbar^2}{G m_n^3 N_\nu^{1/3}} \right) \approx 10\,\text{km}$. This radius is smaller than that of a white dwarf by a factor of the order of $m_e/m_n \approx 10^{-3}$. As the mass of a neutron star increases the kinetic energy of both the neutrons and the protons tends to become extreme relativistic, i.e. of the form again A/R. Thus, when $A = B$ the neutron star will collapse to a black hole. The condition $A = B$ gives now $N_{\nu cr2} \approx 1.6\left(\frac{\hbar c}{G m_n^2}\right)^{3/2}$ which corresponds to about $3M_S$.

Both the minimum and the maximum mass of *active stars* can be expressed in terms of physical constants: $N_{\nu,\min} \approx 0.2\left(\frac{u}{m_e}\right)^{3/4}\left(\frac{e^2}{G u^2}\right)^{3/2}$, $N_{\nu,\max} \approx 100\left(\frac{\hbar c}{G u^2}\right)^{3/2}$.

The *Universe*, according to observational data as well as the general theory of relativity, is expanding in the sense that the distance R between two distant points is increasing at a rate $\dot{R}$ proportional to R, $\dot{R}/R = H$, where H is the so-called Hubble constant. The basic equation obeyed by the ratio $\dot{R}/R$ is the following: $\left(\frac{\dot{R}}{R}\right)^2 = \frac{8\pi}{3}\frac{G\varepsilon}{c^2}$, where ε, the total average energy density of the Universe, consists of several contributions: $\varepsilon = \varepsilon_{ph} + \varepsilon_\nu + \varepsilon_b + \varepsilon_{dm} + \varepsilon_{de}$. The first term refers to photons (proportional to $1/R^4$), the second to neutrinos, the third to baryons (proportional to $1/R^3$), the fourth to dark matter (proportional to $1/R^3$), and the fifth one to the dark energy (which seems to be a constant independent of R).

References

1. S.W. Hawking, L. Mlodinow, *The Grand Design* (Bantam Books, NY, 2010)
2. P. Morrison, P. Morrison, *Powers of Ten* (Scientific American Books, NY, 1982)

Part I
Three Key-Ideas and a Short-Cut

Summary

First Idea: The Atomic Structure of the Cosmos
Everything consists of indivisible microscopic particles. As a result the properties of each system depend on the properties and on the motions of these elementary particles and on their interactions.

Second Idea: Everything is a Waveparticle (⇒QM)
A waveparticle either elementary or composite is of particle nature in its very fabric but nevertheless moves as a wave. As a result it is subject to Quantum Laws. Mutual attractive interactions lead to confinement; the latter, according to Quantum laws, produces perpetual motion which prevents collapse and establishes equilibrium.

Third Idea: Equilibrium Corresponds to Minimum Energy
To be more precise, it corresponds to minimum free energy. This minimum implies the equality of the compressive pressure of the interactions with the expansive pressure of the perpetual motion dictated by the Quantum laws, and therefore equilibrium.

The Short-Cut
Physical quantities depend, at least in principle, on a few universal constants and on some parameters. Dimensional considerations and a little thinking may allow us to find what this dependence is.

Chapter 2
The Atomic Idea

If, in some cataclysm, all of scientific knowledge were to be destroyed, and only one sentence passed on to the next generation of creatures, what a statement would contain the most information in the fewest words? I believe it is the atomic hypothesis that all things are made of atoms–little particles that move around in perpetual motion, attracting each other when they are a little distance apart, but repelling upon being squeezed into one another. In that one sentence, there is an enormous amount of information about the world, if just a little imagination and thinking are applied.

R.P. Feynman, *The Feynman Lectures on Physics.*

Abstract In this chapter we introduce the elementary particles from which all things are made of as well as the interactions which bring the particles together. The interactions are transmitted by indivisible particle-like entities. The so-called Feynman diagrams, describing in a vivid pictorial way the various interactions, are also presented.

2.1 Introduction

According to the atomic idea everything is made of indivisible elementary particles (to be called here *m-particles*[1]) which attract each other and are self-trapped without collapsing because they move perpetually. Thus they form, in a hierarchical way, composite stable structures of ever increasing size and complexity.

For example, the matter inside and around us is made from only three kinds of indivisible elementary particles: The *electron*, the so-called *up quark*, and the so-called *down quark*. Two up quarks and one down quark are self-trapped through the strong nuclear interactions to form the *proton*. Two down quarks and one up quark are self-trapped through the strong nuclear interactions to form the *neutron*.

[1]m stands for matter.

E.N. Economou, *From Quarks to the Universe*,
DOI 10.1007/978-3-319-20654-7_2

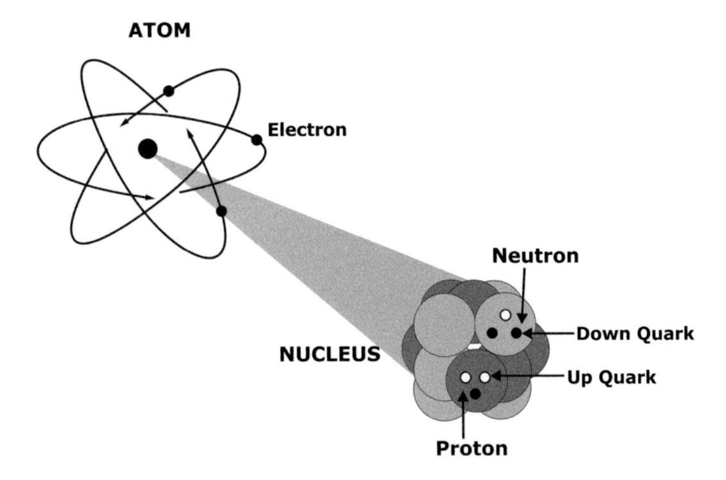

Fig. 2.1 Up and down quarks form protons and neutrons which in turn combine to form atomic nuclei. The latter trap around them electrons to form atoms or ions

Protons and neutrons in various combinations are self-trapped through residual strong nuclear interactions to form the various atomic *nuclei*. The latter attract electrostatically (by Coulomb's law) and trap around them electrons to form *atoms* (when the number of trapped electrons is equal to the number of protons in the nucleus, see Fig. 2.1) or *ions* (when the number of trapped electrons is not equal to the number of protons in the nucleus). Atoms or ions are combined together through electrostatic interactions to form molecules of a huge variety of sizes and shapes. Figure 1.1 and Table 1.1 in the previous chapter show how this hierarchical way of building larger and larger stable structures continues.

To accurately substantiate the atomic idea we must answer in a systematic way the following questions:

(a) What are the various kinds of *m*-particles (see Footnote 1) and what are their properties?
(b) What kind of interactions lead to their mutual attraction?[2]
(c) What counterbalances this attraction and establishes equilibrium?

In this chapter we provide some answers to questions (a) and (b); (c) will be examined in the next chapter.

The atomic idea is the decisive step towards the sought-after underlying simplicity in the World; this is so, because the immense complexity and diversity of the World can be deduced in principle from the *properties* and the *microscopic motions*

---

[2]It turns out that the interactions are actually transmitted through indivisible elementary quantities to be named here (for distinguishing them from the m-particles) *interaction-carrying-particles* (ic-particles). The ic-particles, in addition to mediating forces, are capable of making certain transformations of the m-particles among each other; thus the indestructibility of elementary m-particles (assumed by Demokritos) or ic-particles is of questionable validity.

Table 2.1 The established elementary m-particles are of two types (leptons and quarks) and of three families (1st, 2nd, 3rd) [2]

Leptons					Quarks				
	Symbol, name	Mass $\times c^2$ (MeV)	Electric charge	Spin	Symbol, name	Mass $\times c^2$ (MeV)	Electric charge	Spin	Baryon number
1st	ν_e, e-neutrino	10^{-7}?	0	1/2	u, up quark	1.5–4.5	2/3	1/2	1/3
	e^-, electron	0.511	−1	1/2	d, down quark	5–8.5	−1/3	1/2	1/3
2nd	ν_μ, μ-neutrino	10^{-7}?	0	1/2	c, charm quark	1000–1400	2/3	1/2	1/3
	μ, muon	105.7	−1	1/2	s, strange quark	80–155	−1/3	1/2	1/3
3rd	ν_τ, τ-neutrino	10^{-7}?	0	1/2	t, top quark	174,000	2/3	1/2	1/3
	τ, tau	1777	−1	1/2	b, bottom quark	4000–4500	−1/3	1/2	1/3

Only three of the m-elementary particles (e^-, u, d) are participating in the structure of ordinary matter. Neutrinos are too light and weakly interacting to be bound inside ordinary matter. Particles of the 2nd and 3rd family, having excessive rest energy, are metastable and decay to particles of the 1st family. More exotic particles, which have not been directly observed yet, may account for the so-called dark matter and, possibly, for the so-called dark energy; the existence of these invisible components of the Cosmos is supported by astrophysical and cosmological observations (see also [2]). The neutrino masses are arbitrary guesses

(under the influence of interactions) of all twelve kinds of elementary m-particles presented in Table 2.1. Actually, out of the 12 kinds of elementary particles included in Table 2.1, only the three mentioned before (electrons, up quarks, and down quarks) do participate in the structure of the known ordinary matter. The neutrinos are too light and interact too weakly to be trapped inside the matter. The remaining eight kinds of elementary particles of Table 2.1 are too unstable to be part of existing long-lived structures.

As shown in Table 2.1, all m-particles have spin[3] 1/2. This is an important characteristic property which distinguishes the m-particles from the ic-particles; the latter have integer spin, as shown in Table 2.2. In Table 2.1 each rest mass is multiplied by c^2 to be given in energy units (1 MeV = 10^6 eV = 1.6022×10^{-13} J). The rest mass of each neutrino is not zero, but its value has not been determined yet. The numbers given in Table 2.1 are arbitrary guesses consistent with established upper limit: $\left(m_{\nu_e} + m_{\nu_\mu} + m_{\nu_\tau}\right)c^2 < 0.23$ eV.

It is useful to also introduce here the concept of lepton and baryon number: All leptons, such as electrons and neutrinos, have lepton number 1 and baryon number 0: all quarks have lepton number 0 and baryon number 1/3 so that protons, neutrons and other composite entities consisting of three quarks to have baryon number 1. The composite particles that have baryon number 1 are also called *baryons*.

[3]Spin times the universal constant $\hbar$ ($\hbar = 1.0545716 \times 10^{-34}$ J s) gives the intrinsic angular momentum of each m-particle or ic-particle.

The electric charge e of proton is taken as the unit of electric charge ($e = 1.6021765 \times 10^{-19}$ C).

Returning to the main point of this section, we have to stress that the atomic idea implies that the properties of macroscopic equilibrium matter ought to be expressed, at least in principle, in terms of a few numbers characterizing the elementary particles and their interactions. This is true as long as the considered properties of macroscopic matter do not depend on some uncontrolled initial conditions.

2.2 The Elementary Particles of Matter

We must point out that for each m-particle of Table 2.1 there is an m-antiparticle which is denoted with a bar above the symbol of the corresponding particle. So $\bar{\nu}_e$ is the antiparticle of the e-neutrino and $\bar{d}$ is the antiparticle of the down quark. For historical reasons, the antiparticle of electron is denoted as e^+ and it is called *positron*. Each antiparticle has exactly the same rest mass and spin but exactly the opposite electric charge, the opposite lepton number, and the opposite baryon number of the corresponding particle. Note that antiparticles do not participate in the structure of matter, since a pair of same particle/antiparticle sharing the same volume is annihilated by giving rise to two photons[4] and/or, rarely, a pair of neutrinos/antineutrinos. However, antiparticles can be produced artificially and their properties have been determined experimentally in spite of their short life time.

Quarks have never been observed as isolated free particles; they have been seen always as constituents of composite entities such as baryons (e.g., protons or neutrons) or *mesons*.[5] This is the reason for which the rest mass of the quarks as free particles, especially the lighter ones, is not known accurately.

It must also be pointed out that quarks besides their electric charge, carry another type of charge which will be called here *color-charge*, or c-charge (although there is no relation with the usual color). The c-charge is the emitter and the absorber of the strong nuclear force, the same way that the electric charge is the emitter and the absorber of the electromagnetic (E/M) force (see Table 2.2). In contrast to the electric charge, which is of one type, the c-charge is of three different types termed as *red* (R), *green* (G), and *blue* (B). The opposite charges of these three types of c-charges are called *antired*, *antigreen*, and *antiblue* and are denoted by $\overline{R}$, $\overline{G}$, $\overline{B}$, respectively. Each quark carries an R, or a G, or a B, c-charge. Each antiquark carries an $\overline{R}$, or $\overline{G}$, or $\overline{B}$, c-anticharge. The combination of a pair of quark/antiquark

[4]The medical diagnostic method known as Positron Emission Tomography (PET) is based on the introduction of positron-emitting substances to the body of the patient; the positrons, subsequently, by meeting electrons annihilate giving rise to two photons of opposite direction and the same energy each (equal to the rest mass of electrons, 0.511 MeV).

[5]*Mesons* are composite short-lived entities consisting of one quark and one antiquark. Baryons and mesons are collectively called *hadrons*.

of the form $R\overline{R}$, or $G\overline{G}$, or $B\overline{B}$ has zero c-charge (in other words, can be considered as colorless). Similarly colorless is the combination of three quarks of c-charges R, G, and B respectively or of three antiquarks of c-anticharges $\overline{R}$, $\overline{G}$, $\overline{B}$ respectively. In view of this property the notation R, G, and B for the three types of c-charge is not unreasonable, since it is reminiscent of Newton's disc, where the equal combination of red, green, and blue real colors produces white. Following this analogy with real colors, we could say that an antiblue c-charge, which is equivalent to the RG combination of c-charges, could be presented as a "yellow" c-charge [2].

We should stress once more that out of the twelve elementary *m*-particles of Table 2.1 and the corresponding twelve m-antiparticles, only three (the electron, the up quark, and the down quark) participate in the structure of all ordinary matter surrounding us and in us.

2.3 The Interactions and Their Elementary Interaction-Carrying-Particles

There are four basic established interactions: The gravitational, the electromagnetic (EM), the weak (nuclear), and the strong (nuclear). All forces among *m*-particles are mediated through interactions as follows: An m-particle emits an ic-particle which in turn is absorbed by another m-particle; this causes a force between the two *m*-particles. Sometimes the emission or the absorption of an ic-particle may transform an m-particle to another m-particle. An ic-particle besides disappearing by being absorbed, may disappear by creating a pair of m-particle/m-antiparticle. All these elementary processes are subject to certain conservation laws, meaning that specific quantities must be conserved, i.e., must be the same before and after each of these elementary processes. The electric charge is such a quantity, so that the sum of the electric charges of the particles entering an elementary process must be the same as the sum of the electric charges of the particles coming out of this process. Other conserved quantities, besides the electric charge, are the momentum or the angular momentum, the color-charge, the baryon number, the lepton number (separately for each of the three families[6]). The energy must also be conserved *overall*, meaning that the total energy of the initial particles entering a *physical* process must be the same as the energy of the final particles, after this process is over. (For additional comments concerning energy conservation, see Sect. 6.1). Some interactions, such as the strong one, conserve additional quantities as well. The conservation laws impose severe restrictions regarding which physical processes are allowable and which are forbidden.

In Table 2.2 we present some properties of the four interactions and of their elementary ic-particles.

---

[6]Neutrinos of a given family (see Table 2.1) as they travel after leaving a reaction exhibit a transformation from one family to another; e.g. an e-neutrino may change to a μ-neutrino and vice versa.

Table 2.2 The four basic interactions and their interaction-carrying-particles [2]

Name	Dimensionless strength	Range (m)	Type	ic-particle name-symbol	Rest mass (MeV)	e-charge	Spin	Emitters/absorbers
Gravitational	$a_G \equiv \frac{Gm_p^2}{\hbar c} = 5.9 \times 10^{-39}$	∞	Attractive	Graviton?	0	0	2	All *m*- and ic-particles
Electromagnetic	$\alpha \equiv \frac{e^2}{\hbar c} = \frac{1}{137}$	∞[a]	Attractive or repulsive	Photon, γ	0	0	1	Electrically-charged particles
Weak "nuclear"	$a_w \equiv \frac{g_w^2}{\hbar c}\frac{\sqrt{2}m_p^2}{m_w^2} \approx 10^{-5}$	10^{-18}	$u \rightleftarrows d + \ldots$	Vector Bosons Z^0	91,000	0	1	All *m*-particles, photons, vector bosons
				W^+	80,000	1	1	
				W^-	80,000	−1	1	
Strong "nuclear"	$a_s \equiv \frac{g_s^2}{\hbar c} \approx 1$	10^{-15}	Attractive or repulsive	Gluons, g	0	0	1	Quarks and gluons

[a]E/M interaction being either attractive or repulsive lead to the formation of neutral systems for which their residual E/M interaction becomes short range of the order of a few Angstroms

The physical symbols appearing in the second column of this table are as follows: G is the gravitational constant, m_p is the rest mass of the proton, $\hbar$ is the reduced Planck's constant, "trade mark" of Quantum Mechanics, c is the velocity of light in vacuum, g_w is the strength of the weak interaction, m_w is the mass of the W^+ vector boson, and g_s is the strength of the strong interaction. Numerical values of some of these quantities are given in Table I.1 in Appendix I. The range r_o of each interaction is connected to the rest mass m of the corresponding ic-particle through the formula $r_o = \hbar/m\,c$. The strong interaction to be discussed in Chap. 7, is an apparent exception.

There are several points to be noticed in Table 2.2: the strength of the gravitational interaction is unimaginably weaker than any other interaction. To stress this point take into account that when you lift an object, such as a chair, the residual electric force generated by your muscles overcomes the gravitational force exercised on the chair by the whole Earth. Nevertheless, gravity, in spite of its extreme weakness compared to the other three forces, becomes significant and eventually dominates in systems, such as planets and stars, consisting of a very large number N of m-particles. The reason is that each of these N m-particles interacts gravitationally with every other as a result of gravity's long range character. Moreover, each of the $N(N-1)/2$ contributions to the gravitational self-energy of the system is of the same sign because of gravity's always attractive nature.

Neither the gravitational waves nor their ic-particles, the gravitons, have been detected yet experimentally, because they carry such a minute amount of energy. Finally, one must keep in mind that all particles, both the m- and the ic-ones are emitters and absorbers of the gravitational force, since all of them possess energy and, hence, relativistic mass.

The E/M interaction has also some unique and very important features: First, it dominates the structure of matter from the scale of atoms all the way to that of an asteroid, i.e., it reigns over 15 orders of magnitude from 10^{-10} to 10^5 m. Second, its ic-particle, the photon, is the only one that, besides mediating a force, can also travel free in space carrying over long distances both energy and information. No other ic-particle can do that: Gravitons are too weak to be noticed at least with present day technology; vector bosons decay too fast; and gluons are trapped forever within baryons, mesons, or any other composite quark structure. Third, photons can be easily emitted and detected. These unique features account for the dominant role of photons in the World, including their biological and technological role. Photons deserve further examination; so we shall return to them in Chap. 6.

2.4 Feynman Diagrams

The interactions, besides mediating forces, exhibit other functions as well: They transform particles, they create pairs of particles/antiparticles, etc. All these functions are better presented by the so-called Feynman diagrams which provide a vivid

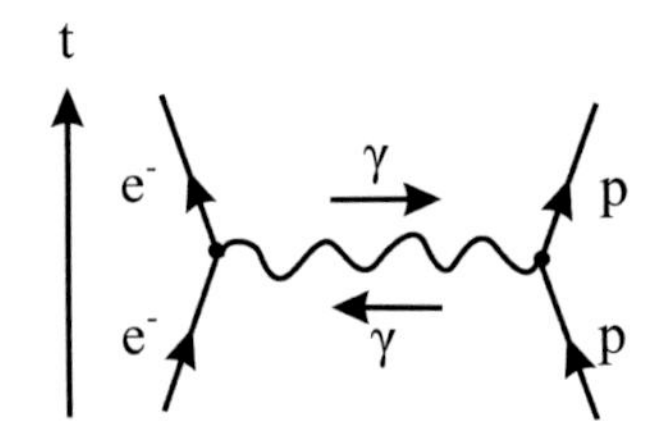

Fig. 2.2 In this simple Feynman diagram the force between two electrically charged particles, an electron and a proton, is represented by the exchange of a photon: the electron emits a photon and the proton absorbs it, or vice versa. This force is associated with the so-called potential energy $\mathcal{V}$, which in the present case is given approximately by the well-known *Coulomb formula*: $\mathcal{V} = -e^2/r$, where e is the electric charge of the proton, $-e$ is the charge of the electron and r is their distance. For $r = 0.53 \times 10^{-10}\,m$, $|\mathcal{V}| = 27.2$ eV. The photon in this case does not appear in either the initial or the final state; when this happens, the particle is called a *virtual* one (here a virtual photon)

pictorial way of the ongoing physical process. Besides this advantage, Feynman diagrams facilitate a full quantum mechanical calculation of physical quantities through a series of rather complicated rules which are clearly beyond the level of this book. Thus, here, we shall restrict ourselves to using only the descriptive character of the Feynman diagrams.

In Fig. 2.2 we employed a Feynman diagram to present the electromagnetic interaction between two particles carrying electric charge; this interaction is mediated by the emission and subsequent absorption of a photon.[7]

In Fig. 2.3 we present through Feynman diagrams processes involving emission or absorption of photons by an electron or processes where an electron/positron pair appears or disappears.

In Fig. 2.4 an important and characteristic process involving creation and transformation of particles through weak interaction is shown. Indeed, the main function of the weak interaction is to transform a quark or a lepton to another quark or to another lepton respectively by the emission of one of its two charged ic-particles which collectively are called *vector bosons* (W^+, W^-). (The vector bosons include also the electrically neutral ic-particle Z^o). Almost immediately, the emitted ic-particle decays into a pair of quark/antiquark or to a pair of lepton/antilepton. We remind the reader that all *m*-particles (or their antiparticles) are subject to the weak interaction as well as the photons and all the vector bosons. Notice in Table 2.2 how massive the latter are, a fact which explains the extreme short range of the weak

[7]Throughout this book the so-called Gauss-CGS (G-CGS) system is preferred for E/M quantities, because it is physically more transparent, since in it the velocity of light appears explicitly in Maxwell's equations. On the other hand, the widely used system called SI employs more convenient and familiar units. For the correspondence of quantities appearing in the SI with those appearing in G-CGS see Appendix B; some of them are as follows (first the SI and second the G-CGS): $\varepsilon_0 \leftrightarrow 1/4\pi$; $\mu_0 \leftrightarrow 4\pi/c^2$; $\boldsymbol{E} \leftrightarrow \boldsymbol{E}$; $\boldsymbol{D} \leftrightarrow \boldsymbol{D}/4\pi$; $\boldsymbol{B} \leftrightarrow \boldsymbol{B}/c$; $\boldsymbol{H} \leftrightarrow c\boldsymbol{H}/4\pi$. The Coulomb potential in SI is given by $e^2/4\pi\varepsilon_0 r$. Widely used formulae will be given also in SI.

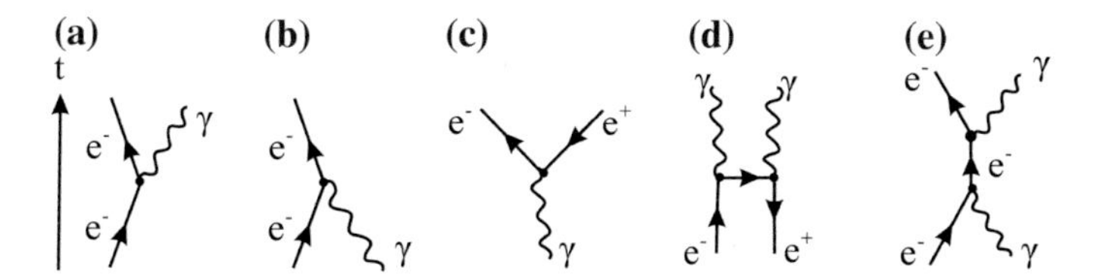

Fig. 2.3 Processes in which a free photon (*wavy solid line*) is participating: **a** Emission of a photon by a decelerating electron, as in X-rays production. **b** Absorption of a photon by an electron, as in the photoelectric phenomenon. **c** Creation of an electron/positron pair by the annihilation of an incoming high energy photon. **d** Annihilation of an electron/positron pair with the emission of a pair of two photons, as in the diagnostic method PET. **e** Absorption of a photon and emission of another photon from an electron, as in Raman scattering. Note that antiparticles, such as the positron e^+ in the present case, are denoted in the diagrams with an *arrow* of opposite direction to the arrow of time

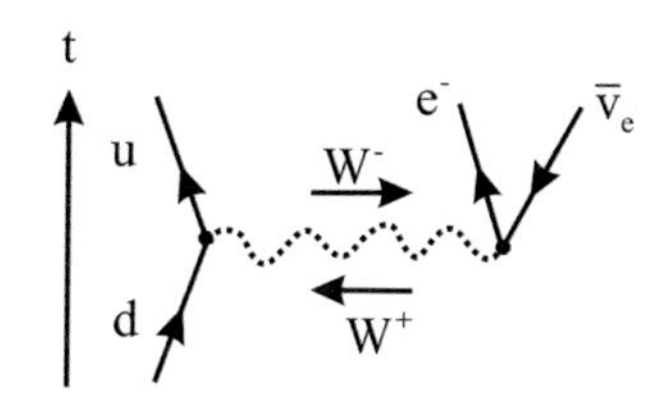

Fig. 2.4 The elementary process leading to the so-called β-decay, where a d quark is transformed to a u quark and a pair of electron/e-antineutrino is created. This process is responsible for the decay of a neutron to a proton accompanied by the emission of a pair of electron/e-antineutrino. The radioactivity of the byproducts in nuclear reactors is due to one of the neutrons in a daughter nucleus undergoing this process. The weak interaction (*wavy dotted line*) is responsible for this process, as the emission and subsequent absorption of a $W^{\pm}$ vector boson implies. Notice again that antiparticles, as the e-antineutrino in the present case, are denoted in the diagrams with an arrow of opposite direction to the arrow of time. Diagrams, as those of Figs. 2.2, 2.3 and 2.4 etc. are known as Feynman diagrams; they allow, through some explicit but rather complicated rules quantitative calculation of physical quantities such as the life-time of a free neutron

interaction through the formula $r_0 = \hbar/mc$ giving the range r_0 of the interaction. In Fig. 2.4 a d quark is transformed into a u quark by the emission of a W^- vector boson (or the absorption of an already existing W^+ vector boson); the emitted W^- particle decays in turn into a pair of electron/e-antineutrino. The whole process, known as beta-decay, changes a neutron (u, d, d) to a proton (u, u, d) with the emission of an electron and an e-antineutrino. This so-called beta-decay is taking place in the byproducts of a nuclear reactor and is responsible for their radioactivity. In most cases the beta-decay is followed by the emission of a very energetic photon, an event known as gamma radioactivity. The reason for the gamma-emission is that just after the beta-decay which took place in a radioactive nucleus, the latter is usually in an excited state.

In Fig. 2.5 the Feynman diagram for the process of a muon decay to an electron, an e-antineutrino, and a μ-neutrino taking place through the weak interaction is

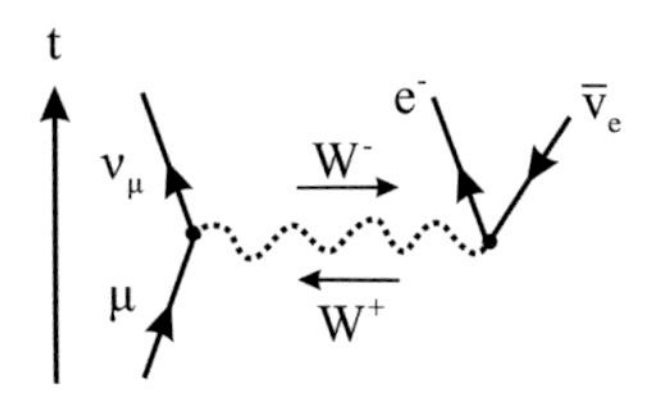

Fig. 2.5 A muon (μ) is transformed to μ-neutrino (ν_μ) by emitting a W^- vector boson. Subsequently, the W^- decays to an electron/e-antineutrino pair. This is how the muon decays. The same process can take place by the creation of an electron/e-antineutrino pair plus a W^+ vector boson which in turn by being absorbed by a muon leads to its annihilation and the creation of a μ-neutrino. Energy conservation requires that the rest energy of the muon must be equal to the sum of the rest energies of e^-, ν_μ, $\bar{\nu}_e$ plus their kinetic energies in the system where the muon was at rest

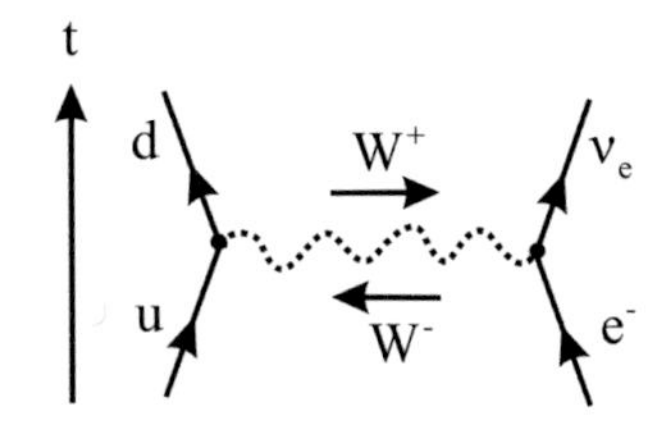

Fig. 2.6 A u quark is transformed into a d quark by emitting a W^+ vector boson, which in turn meets a preexisting electron and the pair W^+/electron disappears giving rise to an e-neutrino (ν_e). The same process can be realized by the emission of a W^- by the e^- and its transformation to an e-neutrino; the emitted W^- is absorbed by the u quark which is transformed to a d quark

shown. In Fig. 2.6 the process of a u quark and an electron being transformed to a d quark and an e-neutrino through the exchange of a charged vector boson is shown. This process takes place in some heavy nuclei changing a proton to neutron with the absorption of an inner orbiting electron and the emission of an e-neutrino.

As we already mentioned in the beginning of Sect. 2.3, in all processes of interaction associated with a vertex of a Feynman diagram, where an interaction-carrying particle is emitted, or absorbed, or decays, etc., certain quantities, such as the momentum or the angular momentum, the electric charge, the color charge, the baryon number, the lepton number (for each family separately) must be conserved. This means, e.g., that for each vertex, the total baryon number *in* must be equal to the total baryon number *out*. Notice that the energy may not need to be conserved at each elementary vertex (see Sect. 6.1). However, it must be conserved overall, meaning that the total energy before any reaction starts must be the same as the total energy after this reaction is concluded.

We are coming finally to the strong interactions. The corresponding ic-particles are the so-called gluons. There are eight types of gluons of different combinations of c-charge/c-anticharge. Six of them carry a net c-charge, since they carry different types c-charge/c-anticharge as follows: $R\bar{G}$, $R\bar{B}$, $G\bar{R}$, $G\bar{B}$, $B\bar{R}$, $B\bar{G}$. Two of them carry the same c-charge/c-anticharge and as such are colorless: $(R\bar{R} - G\bar{G})/\sqrt{2}$, $(R\bar{R} + G\bar{G} - 2B\bar{B})/\sqrt{6}$.

The gluons bind together three quarks to form the so-called baryons, such as the proton (consisting of two u quarks and one d quark, all of different color-charge as to make the proton colorless) or the neutron (consisting of two d quarks and one u quark, all of different color-charge as to make the neutron colorless). Free protons, in contrast to free neutrons, are stable (no decay of free proton has been observed), although there are theoretical models proposing a finite but extremely long lifetime. The mean lifetime of free neutron is 889 s, even though neutrons are stable within non-radioactive nuclei. There are many other short-lived baryons, all of them colorless (see Table 8.1).

A quark and an antiquark through the exchange of gluons may also form short-lived combinations called *mesons* (see Table 8.2). Mesons are also colorless. The lowest energy mesons consisting of a quark and an antiquark of the first family are called pions. There are three types of pions: $\pi^+ \equiv u\,\bar{d}$, $\pi^- \equiv d\,\bar{u}$, and $\pi^o \equiv (u\,\bar{u} - d\,\bar{d})/\sqrt{2}$.

2.5 Concluding Comments

We should mention here that the accepted theory of elementary particles, called the *Standard Model*, requires the existence of another ic-particle, known as the *Higgs particle*. Virtual Higgs particles interacting with the various elementary particles give rise to the mass of all these particles. According to the Standard Model, the spin of the Higgs particle must be zero and its electric charge must also be zero. Its rest energy was expected to be in the range between 114,000 and 158,000 MeV. Because of its very large rest mass the creation of Higgs required very high energies. It seems almost certain that the Higgs particle has been finally detected at the Large Hadron Collider (LHC) of CERN in 2012; its rest energy turned out to be 125,000 MeV. This provided another strong confirmation of the Standard Model.

In spite of the fact that the Standard Model accounts for all data concerning reactions among elementary and/or composite particle, there are deeper questions which remain unanswered: Why are there three and only three families of elementary *m*-particles? What determines the values of the rest masses of elementary particles? Why do the dimensionless strengths of the interactions have their observed values? Is there a deeper level at a much smaller scale where such questions may find satisfactory answers? There is no lack of ideas and proposed theoretical schemes going beyond the Standard Model designed to provide a more unified world-view even at the expense of introducing more abstract concepts and less familiar mathematics. One such scheme goes under the name of supersymmetry [1] and predicts the existence of more elementary particles: one additional fermion for each boson of the Standard Model and one additional boson for each fermion. It predicts also an energy of unification of the forces in the sense that the coupling constants of all interactions have the same value at this energy. Another scheme known as string theory [1] abandons the idea that the ultimate elementary building units of the World are point-like particles; instead it introduces extended objects such as strings and/or membranes at a scale of 10^{-35} m, i.e., 17 orders of magnitude smaller than our present day experimental capabilities. This scheme has several attractive features as to be actively studied over the last 30 years or so, in spite of the lack of experimental evidence supporting it [1].

Returning now to the established world-view of the hierarchical structure of matter, we can summarize the situation as follows: By combining three quarks (u, u, d) or (u, d, d) through exchange of gluons, we obtain two composite m-particles, the proton and the neutron, as a first step in the long journey towards the enormous variety of composite stable formations of the World. Following this first step, we continue with protons and neutrons attracting each other through residual strong interactions to form the various nuclei. The free protons and all the other composite nuclei, having positive electric charge attract the negatively charged electrons and trap them to form electrically neutral atoms. As we shall see, atoms attract each other through residual electrical interactions to form an enormous number of different molecules. This process continues with the wonderful emergence of living systems and seems to end up with the various types of astronomical objects, where all their matter is held together by the always attractive gravitational force. The crucial question is why all these attractive interactions do not continue to squeeze the various systems without limit, the same way that gravity squeezes the big dead stars to black holes? What changes the attraction to repulsion upon the particles "*being squeezed into one another*" to quote Feynman? What opposes the attractive interactions so that physical systems end up in stable equilibrium states? We thus returned to where we started, namely to the third question posed in the summary of this chapter:

What counterbalances the overall attraction of the interactions and establishes equilibrium?

Feynman implicitly suggested the answer to the above question by mentioning that atoms are "*little particles that move around in perpetual motion*". Indeed, perpetual motion of the "*little particles*" implies a pressure which tends to blow the system apart. If this pressure becomes equal to the squeezing pressure of the attractive forces, the system will come to equilibrium. By following this line of thought, it is only natural to ask ourselves questions such as these: How does this perpetual motion come into being? What physical mechanism produces it? The answers are found in the most fundamental, yet difficult to accept and digest, empirical discovery, that of *the particle-wave duality*, to be briefly presented in the next chapter.

2.6 Summary of Important Concepts, Relations, and Data

Some of the basic concepts introduced in this chapter are: Elementary m-paricles (see Table 2.1 especially the particles of the first family) and ic-particles (see Table 2.2), protons and neutrons, other baryons, mesons, antiparticles, electric charge, color charges, Feynman diagrams. The range r of an interaction is connected to the mass m of the corresponding ic-particle by the relation

$$\boxed{r = \hbar/mc.} \tag{2.1}$$

The exceptional case of strong interactions will be discussed in a later chapter.

2.7 Multiple-Choice Questions/Statements

1. The complete list of quarks consists of:
   (a) The up and the down quark (b) the charm and the strange quark
   (c) The top and the bottom quark (d) All the above
2. The baryon number of each quark is:
   (a) 1 (b) 0 (c) 1/3 (d) 2/3
3. The rest energy of the electron is (in MeV):
   (a) 10^{-30} (b) 0.511 (c) 1/1823 (d) 1836
4. The spin of each neutrino is:
   (a) 1/2 (b) 0 (c) 1 (d) 3/2
5. The rest energy of the muon is approximately (in MeV):
   (a) 0.511 (b) 1838 (c) 106 (d) 206
6. The electric charge of up quark is (in units of the proton charge)
   (a) 1 (b) -1 (c) 2/3 (d) -1/3
7. The difference between the rest energies of neutron and proton is (in MeV):
   (a) 0.112 (b) 10 (c) 1 (d) 1.3
8. The mean lifetime (in seconds) of a neutron within the carbon-12 nucleus is:
   (a) 10^{-23} (b) infinite (c) 889 (d) 10^{-8}
9. The dimensionless strength of the gravitational interactions is:
   (a) 5.9×10^{-39} (b) 1 (c) 1/137 (d) 10^{-18}
10. The dimensionless strength of the electromagnetic interactions is:
    (a) 1 (b) 10^{-5} (c) 5.9×10^{-39} (d) 1/137
11. The dimensionless strength of the strong interactions is:
    (a) 0.01 (b) 1 (c) 10^{5} (d) 1/137
12. The dimensionless strength of the weak interactions is:
    (a) 10^{-8} (b) 0.01 (c) 10^{-5} (d) 1/137

The correct answers to the multiple choice questions are given at the end of the book (see Appendix H).

2.8 Solved Problems

1. *Draw the Feynman diagrams for the decays of* π^0 and π^+ *and estimate the mean lifetime of these pions. What is the energy and the momentum of the decay products?*

Solution The zero pion consists of a quark (u or d) and its antiquark (($\bar{u}$ or $\bar{d}$) respectively. Such pairs are annihilated in a way similar to that shown in Fig. 2.3d, i.e. by producing two photons. By working in the system where the zero pion is at rest, it becomes clear that decay to just one photon would violate the conservation of momentum. We conclude that we need at least two photons as shown in the Feynman diagram a below. To conserve momentum the two photons must have opposite momenta and, hence, energies equal to 1/2 the rest energy of zero pion, i.e. about 67 MeV each. In general the typical mean lifetime for a process depends on the degree of complication of the corresponding Feynman diagram and the type of interaction involved: As a general rule the larger the strength of the interaction, the shorter the lifetime is. In the particular case shown in Fig. a below, which involves two vertices of electromagnetic interaction, the mean lifetime is expected to be of the order of 10^{-16} s. Actually it is 8.4×10^{-17}s.

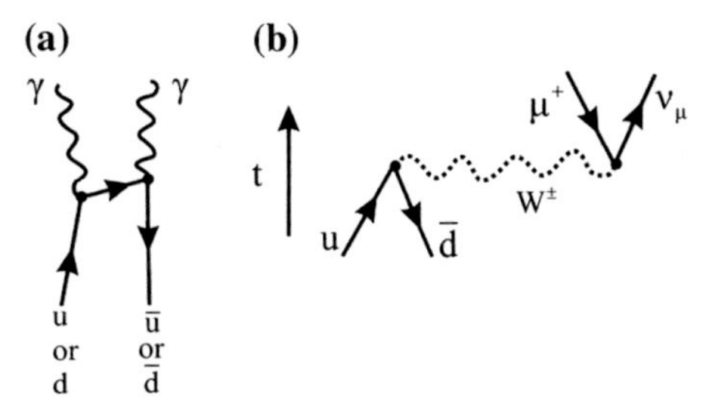

For the π^+ which consists of a quark u and an antiquark $\bar{d}$ there is no possibility for photons (which are chargeless) to be involved, neither, of course, for gluons (since the latter, as we shall see in Chap. 7, cannot transform a quark to a different quark). Hence, only vector bosons of the weak interactions can participate in the decay of π^+. Because of charge conservation, the π^+ can decay by emitting a W^+ vector boson, which in turn disappears by creating a pair of leptons, of zero total lepton number and +1 charge (see Figs. 7.4 and 7.6). Such pairs are the positron and the e-neutrino or the antimuon and the μ-neutrino. For reasons which are beyond the level of this book, it turns out that the process $\pi^+ \to \mu^+ + \nu_\mu$ is much more probable. The same process can occur by the creation of a antimuon/μ-neutrino pair associated by the emission of a W^-, which in turn annihilates the incoming π^+ by being absorbed by it. Thus the corresponding Feynman diagram is as shown in the diagram b above. The mean lifetime of $\pi^\pm$ is expected to be at least six orders of magnitude longer than that of the zero pion; the reason is that for both lifetimes two vertices are involved and each vertex contribute to the corresponding lifetime inversely proportional to the ratio of the dimensionless strengths of the interactions responsible for their decay. The mean lifetime of π^+ is actually 2.6×10^{-8} s.

2. *The range r of the weak interaction is about* 2.2×10^{-18} m. *What is the rest energy of the vector bosons?*

Solution From the formula $r = \hbar/mc = \hbar c/mc^2$ it follows that $mc^2 = \hbar c/r$. We shall work in the atomic system of units where $\hbar = e = m_e = 1$ (see Table I.2 in Appendix I).

In this system the velocity of light in vacuum is equal to about 137, i.e. the inverse of the fine structure constant. The unit of length is the Bohr radius 0.529 A (1 A = 10^{-10} m), and the unit of energy is $\hbar^2/m_e a_B^2 \approx 27.2$ eV. In the present case the range r in atomic units is about $2.2 \times 10^{-18}/0.529 \times 10^{-10} \approx 4.16 \times 10^{-8}$. Hence in atomic units the rest energy of the vector boson is $mc^2 = \hbar c/r = (1 \times 137/4.16) \times 10^8 = 32.93 \times 10^8$ atomic units of energy = 89,600 MeV. The actual values for the rest energy of the vector bosons are 91,000 and 80,000 MeV as shown in Table 2.2.

2.9 Unsolved Problems

1. Create a table giving the symbol, the rest energy, the mass, the spin, and the electric charge of all leptons and all quarks.
2. Create a table giving the names, the dimensionless strengths, the range, the corresponding ic-particles (their rest energies, their spin, their electric charge and their emitters and absorbers) of all four basic interactions.

References

1. S.W. Hawking, L. Mlodinow, *The Grand Design* (Bantam books, New York, 2010)
2. M. Veltman, *Facts and Mysteries in Elementary Particle Physics* (World Scientific, New Jersey, 2003)

Chapter 3
The Wave-Particle Duality

So reasonable the incomprehensible.

O. Elytis

Abstract The wave-particle duality asserts that every entity, in spite of being of particle character in its very fabric, nevertheless moves as a wave. This dual nature of the World, on which Quantum Mechanics is based, provides the required kinetic energy to counterbalance the attractive nature of the interactions and to stabilize both the microscopic and the *macroscopic* structures of matter. Proof of this is the presence of Planck's constant in formulae referring to properties of all these structures. The core of Quantum Mechanics can be condensed to the following three fundamental principles possessing amazing quantitative explanatory power: Those of Heisenberg, Pauli, and Schrödinger. *The content of this chapter is 'sine qua non' for what follows.*

3.1 Concepts and Formulae

A classical particle is characterized by the following two properties: (a) Its composition, which is either indivisible, that is elementary or consisting of bound indivisible entities. (b) Its motions which follows a clearly determined trajectory. On the other hand a classical wave is characterized by the following two properties: (i) It is of continuous nature and, hence, infinitely divisible and (ii) it moves by being spread in space and thus can overlap with other waves and may exhibit the phenomenon of interference.[1] Now, according to the wave-particle duality, all entities (considered classically either as particles or as waves) possess the discrete character of property (a), while at the same time they do not follow a trajectory but

[1]A vivid picture of interference is provided by throwing two stones in a quiet pond and observe the resulting two circular waves cancelling each other at some points and being reinforced at some other.

E.N. Economou, *From Quarks to the Universe*,
DOI 10.1007/978-3-319-20654-7_3

rather move as dictated by property (ii). In other words, they are partly particles (because of property (a)) and partly waves (because of property (ii)). This mixed character seems to contradict in a fundamental way our everyday perceptions. However, in spite of some serious initial resistance, it was imposed on us and it is currently widely accepted as true not only because of an enormous number of experimental results but also by its extraordinary quantitative explanatory power. Recently there is a revival of interest in the foundations of Quantum Mechanics as a result of new theoretical insights as well as technological challenges such as quantum cryptography and the possibility of quantum computers.

To implement the basic ideas of the wave-particle duality, we assert that a particle of energy ε and momentum $\mathbf{p}$ does not follow a trajectory, as classical physics claims, but propagates as a wave of angular frequency

$$\boxed{\omega = \frac{\varepsilon}{\hbar}} \tag{3.1}$$

and of wavevector[2]

$$\boxed{\mathbf{k} = \frac{\mathbf{p}}{\hbar}} \tag{3.2}$$

where $\hbar$ is one of the most fundamental universal physical constants called (reduced) Planck's constant. Thus, this "particle" is not a classical one, but what could be called a waveparticle. Moreover, a wave of angular frequency ω and wavevector $\mathbf{k}$ consists of discrete, indivisible entities of energy $\varepsilon = \hbar\omega$ and momentum $\mathbf{p} = \hbar\,\mathbf{k}$. It is only fair to call such a "wave" particlewave. Quantum Mechanics synthesizes the apparently contradictory concepts of *particle* and *wave* to a consistent, yet at odds with our perceptions, theory which can be distilled into three fundamental principles: *uncertainty*, *exclusion* (for fermions), and *spectral discreteness* (when the waveparticle is confined).

The first two of these principles account quantitatively for the required perpetual microscopic motion (when confined) necessary to counterbalance the squeezing pressure of the attractive forces and establish thus the various equilibrium structures of matter. The third principle makes composite microscopic structures to behave like elementary ones up to a limit. This exhibits an admirable balance between the need for stability and the need for change.

As was stated before, the wave-particle duality, in spite of the initial strong resistance to its counterintuitive features, has been established by numerous hard facts as one of the more basic laws of Nature. This duality asserts that the actual

[2]The magnitude k of the wavevector is related to the wavelength λ by the relation, $k = 2\pi/\lambda$, and its direction gives the direction of propagation of the wave.

entities of the World are endowed with some of the particle properties, such as discreteness and, hence, indivisibility when elementary, *as well as* some of the properties characteristic of waves, such as non-locality, which is a necessary condition for the phenomenon of interference. Thus, a more appropriate name—to be used frequently in this book—for these entities is waveparticles or, equivalently, particlewaves. The essence of the wave-particle duality is captured in Feynman's description of the *gedanken* two slit experiment [1]. This idealized version of a real experiment shows beyond any doubt that the familiar macroscopic laws of motion are completely inapplicable to the microscopic motions of microscopic particles. Moreover, the experiments clearly show that waveparticles exhibit the characteristic wave phenomenon of interference, which, thus, provides a direct experimental proof to the idea of the wave-particle duality. It is worth pointing out that the implications of the latter, and, hence, of Quantum Mechanics, are not restricted only to the microscopic world, but they extend to the *properties* of all structures of the World from the smallest to the largest. This follows directly from the atomic idea which attributes the properties of any part of the World, macroscopic or not, to the properties and the microscopic motions of the microscopic constituents of matter. We would like to stress this point emphatically.

3.2 The Properties of the Structures of the World at Every Scale Are of Quantum Nature. If They Were Not, We Would Not Exist

The validity of this statement becomes apparent by noticing[3] that the observed values of macroscopic properties, such as density, compressibility, specific heat, electrical resistance of solids and liquids as well as the mass of the stars, and their final fate (whether they eventually end up as white dwarfs, or neutron stars, or black holes) involve the universal constant $\hbar$, the "trade mark" of particle-wave duality and of Quantum Mechanics (QM). If $\hbar$ were zero, as in the classical worldview, the present structures of the World would collapse to the classical analogue of black holes. Of course, there are some *isolated* cases where Classical Mechanics (CM) produces results in very good agreement with the quantum mechanical ones. Examples of such cases, which are more than adequately accounted for by CM, are: The specific heat of insulators at elevated temperatures and the *macroscopic motions*, such as sea waves or planets orbiting the Sun. This does not imply that QM fails in these cases. Simply, in these cases QM reduces to CM. We must also clarify that there are some important aspects of QM, which are extremely difficult to say the least- to establish and maintain when the microscopic particles involved exceed a one digit number. Among them are the coherence, a prerequisite for wave interference, and the peculiar correlations present in the so-called entangled systems; these peculiar non-local correlations

[3]See Chaps. 12–15.

associated with entangled systems were first discussed by Einstein-Podolsky-Rosen in relation with a gedanken experiment [2] which was later realized by Aspect et al. [3] and are examined in several modern textbooks on Quantum Mechanics [4–7]. However, this by no means implies that the macroscopic world is classical. We repeat once more that a classical world, i.e. a world with $\hbar = 0$, would be devoid of any structure we know of.

We have already mentioned that QM synthesizes the contradictory concepts of wave and particle but at a cost which is twofold:

(a) It employs an inherently probabilistic description of the World either through the introduction of an abstract concept, the so-called wavefunction $\psi(\mathbf{r}, t)$ (such that $|\psi(\mathbf{r}, t)|^2$ gives the probability density of finding the waveparticle at the position $\mathbf{r}$ at time t), or by introducing infinitely many trajectories (instead of a single one as in classical physics) connecting the initial point A to the final point B; then the probability of going from A to B is calculated by taking the square of the absolute value of the sum of the so-called probability amplitudes associated with each one of the infinitely many trajectories.
(b) The required mathematical formalism goes clearly beyond the first two years of University calculus. What justifies this twofold cost is the extraordinary explanatory power of QM which extends from the structure of protons and neutrons to the Whole Universe. Fortunately, one can retain a large fraction of this explanatory power without most of the cost. This can be achieved, as we shall show in this book, by: (a) Focusing on three basic principles of QM, namely those of Heisenberg, Pauli, and Schrödinger, which can be deduced directly from the wave-particle duality and (b) By making frequent use of the method of dimensional analysis.

Before we leave this introductory section, we shall mention the basic equations of the non-relativistic formulation of Quantum Mechanics based on $\psi(\mathbf{r}, t)$. Given the potential energy $\mathcal{V}(\mathbf{r}, t)$, $\psi(\mathbf{r}, t)$ is determined by the so-called time-dependent Schrödinger equation (expressing how the immediate future, $\partial\psi/\partial t$, is determined by the present)

$$i\hbar\frac{\partial\psi(\mathbf{r},t)}{\partial t} = -\frac{\hbar^2}{2m}\left(\frac{\partial^2}{\partial x^2} + \frac{\partial^2}{\partial y^2} + \frac{\partial^2}{\partial z^2}\right)\psi(\mathbf{r},t) + \mathcal{V}(\mathbf{r},t)\psi(\mathbf{r},t) \tag{3.3}$$

In the case where the potential energy is time-independent, a set of solutions of (3.3) is obtained by setting $\psi(\mathbf{r}, t) = \psi(\mathbf{r})\exp(-iEt/\hbar)$ with E being the total energy of the particle of mass m and $\psi(\mathbf{r})$ satisfying the so-called time-independent Schrödinger equation

$$-\frac{\hbar^2}{2m}\left(\frac{\partial^2}{\partial x^2} + \frac{\partial^2}{\partial y^2} + \frac{\partial^2}{\partial z^2}\right)\psi(\mathbf{r}) + \mathcal{V}(\mathbf{r})\psi(\mathbf{r}) = E\psi(\mathbf{r}) \tag{3.4}$$

Having $\psi(\mathbf{r},t)$ or $\psi(\mathbf{r})$, the average value of any physical quantity $A(\mathbf{r})$ which depends only on the position vector $\mathbf{r}$ is given by the relation

$$<A> = \int d\mathbf{r}A(\mathbf{r})|\psi(\mathbf{r},t)|^2 \tag{3.5}$$

Having $\psi(\mathbf{r},t)$ or $\psi(\mathbf{r})$, the average values of the momentum $\mathbf{p}$ or the square $\mathbf{p}^2$ of the momentum are given by the relations

$$<\mathbf{p}> = \int d\mathbf{r}\, \psi^*(\mathbf{r})\, [-i\hbar\nabla\psi(\mathbf{r})], \quad <\mathbf{p}^2> = \int d\mathbf{r}\, \psi^*(\mathbf{r})\, \left[-\hbar^2\Delta\psi(\mathbf{r})\right] \tag{3.6}$$

Combining (3.5) and (3.6) with the non-relativistic expression for the total energy ε, $\varepsilon = (p^2/2m) + \mathcal{V}$, of a particle of mass m we find

$$<\varepsilon> = \int d\mathbf{r}\, \psi^*(\mathbf{r}) \left\{ -\frac{\hbar^2}{2m}\Delta\psi(\mathbf{r}) + \mathcal{V}(\mathbf{r})\psi(\mathbf{r}) \right\} \tag{3.7}$$

In writing (3.5)–(3.7) we assumed that ψ is normalized, i.e. $\int d\mathbf{r}\, |\psi(\mathbf{r})|^2 = 1$. Equations (3.3)–(3.7) were presented here for completeness; we shall try to avoid making use of them in the rest of this book.

3.3 Heisenberg's Uncertainty Principle[4]

Heisenberg's uncertainty principle states that the product of the uncertainty [5]Δx in the position of a particle (along a given direction) times the uncertainty Δp_x of the momentum component along the same direction is never smaller than $\hbar/2$:

$$\boxed{\Delta x \cdot \Delta p_x \geq \frac{\hbar}{2}} \tag{3.8}$$

The physical meaning of (3.8) is the following: A waveparticle confined in a straight segment of length l_x cannot be at rest. It must move back and forth perpetually so that its momentum squared p_x^2 to be on the average larger[6] than a

[4]Heisenberg's principle stems from the well-known wave relation stating that the confinement of a wavepacket within a spatial extent Δx requires the superposition of plane waves belonging to a range of k's of extent $\Delta k_x \approx 1/\Delta x$; combining this relation with the wave-particle (3.2) we obtain essentially (3.8).

[5]Δx is the so-called standard deviation defined by the relation: $\Delta x^2 \equiv <(x - <x>)^2> = <x^2> - <x>^2$. A similar definition applies to Δp_x with x replaced by p_x. The symbol <f>, for any random quantity f, denotes its average value.

[6]However, the average value of p_x can be zero, if the contributions of positive and negative values cancel each other.

quantity of the order of $\hbar^2/l_x^2$. More precisely, confinement of a waveparticle in a straight segment of length l_x necessarily forces the waveparticle to go back and forth perpetually as to satisfy the inequality[7] (3.8), i.e. $\Delta p_x^2 \geq (\hbar^2/4\Delta x^2)$.

For the usual three dimensional (3-D) case, where a particle may be confined within a sphere of radius R and volume $V = (4\pi/3)R^3$, we obtain from (3.8) (and the choice $<\mathbf{r}> = <\mathbf{p}> = 0$):

$$<\mathbf{p}^2> = <p_x^2 + p_y^2 + p_z^2> = 3<p_x^2> \geq \frac{3}{4}\frac{\hbar^2}{\Delta x^2} \tag{3.9}$$

where[8]

$$3\Delta x^2 = <x^2 + y^2 + z^2> = <r^2> = \frac{3}{5}R^2 \approx 0.23\, V^{2/3} \tag{3.10}$$

Combining (3.9) and (3.10) with the expression for the non-relativistic[9] kinetic energy, we obtain for the average value of the latter

$$\boxed{\varepsilon_K = \frac{<\mathbf{p}^2>}{2m} \geq 4.87\frac{\hbar^2}{mV^{2/3}} = 1.875\frac{\hbar^2}{m\,R^2}} \tag{3.11}$$

Notice that the numerical factors in (3.11) were obtained for a uniform probability density within the volume V; a different probability density will change in general these numerical factors. For this reason from now on we shall usually omit the numerical constants in our relations and, as a result, we shall have

---

[7]At this point it is not useless to clarify the relation between the confinement length $l_x \equiv 2a$ along the direction x and the standard deviation Δx. For a waveparticle confined in a *fixed* region of space we can always choose the coordinate system in such a way that $<x> = 0$; since this region is fixed, i.e. not moving, $<p_x>$ is also necessarily equal to zero. By the general definition of the standard deviation given in the footnote 5 and taking into account that $<x> = 0$ and $<p_x> = 0$, we obtain the relations $\Delta x^2 \equiv <x^2>$ and $\Delta p_x^2 = <p_x^2>$. Hence, it follows from (3.8) that $<p_x^2> \geq \hbar^2/(4<x^2>)$, where by definition (and the choice $<x> = 0), <x^2> \equiv \int_{l_x} dx|\psi(x)|^2 x^2$; for a uniform probability density where $|\psi(x,t)|^2 = 1/l_x$, we have $<x^2> = \Delta x^2 = l_x^2/12$. In general Δx is proportional to and smaller than the length of confinement. The proportionality factor depends on the probability density. It is important to keep in mind that the probability density in Quantum Mechanics (actually the wave function ψ for the lowest energy state to be precise) is everywhere as smooth as possible within the restrictions imposed by the potential. This feature of QM is implicitly assumed here and in the rest of this book as a supplement of the uncertainty principle. Otherwise, as it was pointed out by Lieb [11], the uncertainty principle per se cannot guarantee the stabilization of the structures of the world.

[8]Equation (3.10) was obtained by using the symmetry of the sphere and by assuming constant probability density for any value of r inside the sphere.

[9]The correct relativistic relation between kinetic energy and momentum is: $\varepsilon_K = (m_o^2c^4 + c^2p^2)^{1/2} - m_oc^2$. This relation in the non-relativistic limit, $m_oc^2 \gg cp$, becomes $\varepsilon_K \approx \mathbf{p}^2/2m_o$, while in the extreme relativistic limit, $m_oc^2 \ll cp$, becomes $\varepsilon_K = cp$; m_o is the rest mass of the particle.

proportionalities instead of equalities. This simplification helps also the reader to pay attention to the dependence on the physical parameters and the universal constants instead of memorizing numerical factors. In the problem section the reader may be asked to restore some of the numerical factors and to compare his results with exact formulae. Within this simplifying notation we have $<r^2> \; \propto R^2$ and

$$<\mathbf{p}^2> \; \propto \frac{\hbar^2}{R^2} \propto \frac{\hbar^2}{V^{2/3}} \tag{3.12}$$

Equation (3.11) is very important, because it accounts for the non-zero kinetic energy necessary for the stabilization of the various structures of the World. It asserts that a waveparticle confined within a volume V has a kinetic energy which cannot be smaller than a minimum non-zero value. This value is inversely proportional to the 2/3 power of the volume V, inversely proportional to the mass of the waveparticle, and proportional to the square of Planck's constant $\hbar$, the "trade mark" of QM. Thus, the smaller the volume and the lighter the mass, the larger the minimum value of the average kinetic energy is. If $\hbar$ were zero, as in classical physics, this minimum value would be zero and the World would be left unprotected against the collapse, which the attractive forces, left alone, would eventually bring on. Notice that the minimum kinetic energy becomes zero, if $V \to \infty$, i.e., if the waveparticle is spatially unconfined.

In summary, (3.11) for the minimum average *non-relativistic* kinetic energy $\varepsilon_K = p^2/2m$ of a particle of mass m confined within a finite volume V of radius R can be rewritten (by omitting the numerical factor) as

$$\varepsilon_K \propto \frac{\hbar^2}{mV^{2/3}} \propto \frac{\hbar^2}{mR^2} \tag{3.13}$$

It is important, as we shall see in future chapters, to calculate how (3.11) or (3.13) is modified in the extreme relativistic limit, $\varepsilon_K = cp$ (see Footnote 9). Taking into account that $<|\mathbf{p}|> \; \propto \sqrt{<\mathbf{p}^2>}$ and (3.12), we obtain that the minimum kinetic energy of a particle confined within a volume V of radius R in the extreme relativistic limit is:

$$\boxed{\varepsilon_K \propto \frac{c\hbar}{V^{1/3}} \propto \frac{c\hbar}{R},} \tag{3.14}$$

For a uniform probability distribution we have that $<|\mathbf{p}|> \; = 0.968\sqrt{<\mathbf{p}^2>}$ and $\varepsilon_K = 1.875\hbar c/R$. Notice in (3.14) that the minimum kinetic energy is now inversely proportional to the one third power of the volume within which the waveparticle is confined, or inversely proportional to its radius R. This change in the exponent of V or R is crucial in explaining the transition from white dwarfs to neutron stars and the collapse of neutron stars to black holes.

3.4 Pauli's Exclusion Principle

Identical waveparticles sharing the *same* volume V as they go on with their perpetual motion are *indistinguishable*. This feature is due to their wave-like motion and, consequently, the absence of a trajectory; the latter, if it were present, would allow us to tell which particle is which. The indistinguishability means that all physical properties must remain unchanged, if we interchange the "names" of any two of the identical particles. The quantity, $|\Psi(1,2,\ldots)|^2$, being a probability density as the absolute value of the square of the wavefunction Ψ, is such a physical property; $\Psi(1,2,\ldots)$ determines the state of a system consisting of several identical waveparticles, while the indices 1, 2, etc. denote the corresponding positions and projections of the spin of waveparticles numbered 1, numbered 2, etc. Since $|\Psi(1,2,\ldots)|^2 = |\Psi(2,1,\ldots)|^2$, it follows that

$$\Psi(1,2,\ldots) = \pm\Psi(2,1,\ldots) \tag{3.15}$$

It turns out that the upper sign is valid for *all* waveparticles (elementary or not) having integer spin and the lower one for *all* waveparticles (elementary or not) with spin equal to integer plus 1/2. The particles in the first category are called *bosons* and the ones in the second *fermions*. Thus all ic-waveparticles (such as photons, gluons, etc., see Table 2.2) are bosons and all elementary m-waveparticles (such as electrons, neutrinos, quarks, see Table 2.1) are fermions; protons and neutrons, consisting of three quarks are necessarily fermions. Consider now two identical *fermions* sharing the same volume and being in the *same* spatial/spin single-particle state ψ; then this pair is described as follows: $\Psi(1,2) = \psi(1)\psi(2)$. Now, if we interchange their names, their joint wavefunction will remain the same, since the two particles are in the same single-particle state ψ; however, (3.15) tells us that the joint wavefunction must change sign. The only way to satisfy both requirements is to set the joint wavefunction equal to zero, which means physically that the probability of finding two identical fermions in the same spatial/spin single-particle state is zero. Thus, the famous Pauli's exclusion principle was deduced:

> Out of many identical *fermions*, no more than one can occupy the same *spatial/spin* single-particle state. For spin 1/2 identical fermions (e.g., electrons, or protons, or neutrons, or quarks) no more than two can occupy the same single-particle *spatial* state, one with spin up and the other with spin down. Less accurately, we can state that no more than two identical spin 1/2 fermions can be in the same region of space.

3.5 Quantum Kinetic Energy in View of Heisenberg and Pauli Principles

What will happen when N spin 1/2 identical fermions are forced to share the same space of volume V? A reasonable approach to answer this question is (a) to obtain the single-particle states (which in the absence of interactions are characterized by their momentum $\mathbf{p} = \hbar\mathbf{k}$ and their spin orientation), (b) to arrange these states in order of increasing energy (which in the absence of interactions and in the non-relativistic limit is $\varepsilon_K = \mathbf{p}^2/2m = \hbar^2\mathbf{k}^2/2m$), and (c) to fill up the $N/2$ lowest energy of these states, each with two fermions (of opposite spin) starting from the one with the lowest energy and ending up at the one with the highest occupied energy $E_F = \mathbf{p}_F^2/2m = \hbar^2\mathbf{k}_F^2/2m$ called the *Fermi energy*, as shown in Fig. 3.1.

The Fermi momentum $\mathbf{p}_F = \hbar\mathbf{k}_F$ is obtained by taking into account that the sum of all $\boldsymbol{k}$'s between $k = 0$ and $k = k_F$ (times 2, because of spin) must obviously be equal to the number N of fermions:

$$N = 2\sum_{k=0}^{k=k_F} \tag{3.16}$$

Having obtained the Fermi momentum $p_F = \hbar k_F$ we can calculate the minimum total kinetic energy E_K by summing over all occupied states:

$$E_K = 2\sum_{k=0}^{k=k_F} \varepsilon_K = 2\sum_{k=0}^{k=k_F} \frac{\hbar^2 k^2}{2m} \tag{3.17}$$

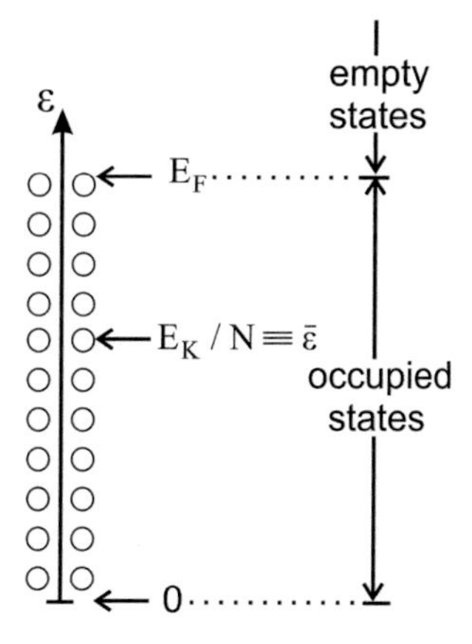

Fig. 3.1 The $N/2$ lowest energy single-particle spatial states are occupied by N spin ½ fermions as to obtain the system's ground state which is realized at $T = 0$ K

The summation over all occupied states appearing in (3.16) and (3.17) can be performed by the following general formula worth memorizing

$$\sum_{\mathbf{k}} f(\mathbf{k}) = \frac{V}{(2\pi)^3} \int d^3k f(\mathbf{k}) \tag{3.18}$$

Equation (3.18), valid for a three dimensional system and for any function $f(\mathbf{k})$, can be generalized to systems of any dimension (see multiple-choice questions 3.8.9 and 3.8.10). It is useful to introduce the average energy per particle defined as $\bar{\varepsilon} \equiv E_K/N$; obviously the value of $\bar{\varepsilon}$ is somewhere between 0 and E_F (see Fig. 3.1). The quantity $\bar{\varepsilon}$ can be interpreted as the common value of the kinetic energy of each fermion. Such a common value can be achieved in reality without violating Pauli's exclusion principle, if we divide the available space of volume V into equal $N/2$ subspaces each of volume $V/(N/2) = 2V/N$, and place one pair of identical fermions (one partner of the pair with spin up and the other with spin down) in each subspace. Then, the average kinetic energy per fermion, assuming its non-relativistic expression, will be obtained by replacing in (3.13) the total volume V by the reduced volume $2V/N$ in which each fermion is now confined. We have then for the average kinetic energy per particle

$$\bar{\varepsilon} \propto \frac{\hbar^2}{m(2V/N)^{2/3}} \propto \frac{\hbar^2 N^{2/3}}{mV^{2/3}} \propto \frac{\hbar^2 N^{2/3}}{m R^2} \tag{3.19}$$

The average total kinetic energy of the N fermions, E_K, is then equal to $N\,\bar{\varepsilon}$:

$$\boxed{E_K \propto N\frac{\hbar^2 N^{2/3}}{mV^{2/3}} = \frac{\hbar^2 N^{5/3}}{mV^{2/3}} \propto \frac{\hbar^2 N^{5/3}}{m R^2}} \tag{3.20}$$

The proportionality factor in (3.20) can be obtained by the following observation: If we add one more fermion identical to the N fermions already present in V, the *minimum* energy this additional fermion could have is necessarily E_F (because Pauli's principle forbids any lower value). It follows then that the increase in the total minimum kinetic energy $E_K(N+1) - E_K(N)$ must be equal to E_F, which implies (by dividing $E_K(N+1) - E_K(N)$ by one and keeping in mind that N is a very large number) that

$$\boxed{E_F = (\partial E_K/\partial N)_V} \tag{3.21}$$

Equation (3.21) is valid for systems of any dimensionality, even interacting ones (in which case E_K must be replaced by the total energy E_t). The only case in which (3.21) may require special care is whenever a spectral gap is present at $\varepsilon = E_F$, as in semiconductors and insulators; in this case, if N is odd the Fermi energy will coincide with the highest occupied state, while, if N is even, it will coincide with the lowest unoccupied state. Then (3.21) will take the mean value by placing the Fermi energy at the middle of the gap.

By combining (3.20) and (3.21) we obtain

$$E_F = \frac{5}{3}(E_K/N) \quad \text{or} \quad E_K = \frac{3}{5} N E_F \tag{3.22a}$$

The reader is encouraged to calculate E_F by combining (3.16) and (3.18). Then by replacing it in (3.22a) he/she must show that in the *non-relativistic limit* the result is:

$$\boxed{E_K = 2.871 \frac{\hbar^2 N^{5/3}}{m V^{2/3}} = 1.105 \frac{\hbar^2 N^{5/3}}{m R^2}} \tag{3.22b}$$

which coincides exactly with the outcome of (3.17). We must draw attention to (3.19) which (under the said conditions of identical fermions sharing the same space) boosts the minimum average kinetic energy per fermion (as compared with the Heisenberg-based (3.13)) by a factor proportional to $N^{2/3}$! This can be an enormous boost if N is very large, as in the case of "free" electrons in metals, where $N^{2/3}$ is of the order of 10^{15} per mole! This enormous boost is absolutely necessary in order to stabilize e.g., condensed matter at the observed values of density, compressibility, etc., as we shall see in a future chapter.

The reader is also asked to show (by using an approach similar to the one above) that the minimum total kinetic energy of spin ½ fermions sharing a spherical space of volume V and radius R is given in the *extreme relativistic limit* by:

$$\boxed{E_K = \frac{3}{4} N E_F = 2.32 \frac{\hbar c N^{4/3}}{V^{1/3}} = 1.44 \frac{\hbar c N^{4/3}}{R}} \tag{3.23}$$

3.6 Schrödinger's Principle of Spectral Discreteness

It is well known that a classical wave confined within a finite region of space acquires a discrete frequency spectrum, as opposed to the continuous spectrum it possesses if its extent is infinite (compare Fig. 3.3b with Fig. 3.3a). A very familiar example is a guitar string, which can produce only the fundamental frequency and its harmonics i.e., only the discrete frequencies $\omega_n = (\pi \upsilon/\ell)n$, resulting from the general relation $\omega = k\upsilon = (2\pi/\lambda)\upsilon$ and the requirement of an integer number of half wavelengths to fit exactly within the length ℓ of the string, $n(\lambda/2) = \ell$; n is a positive integer and υ is the velocity of the wave propagating along the string. Another example is the discrete spectrum of E/M waves confined within a cavity.

Because of the relation $\varepsilon = \hbar\omega$, we expect that a waveparticle, confined within a finite region of space, will acquire a discrete energy spectrum, in contrast to the continuous energy spectrum it would have according to the classical worldview.

In this book we will associate this wave-attributable discreteness with the name of Schrödinger, the discoverer of the equation which correctly describes the motion of non-relativistic waveparticles. We can restate Schrödinger's discreteness principle as follows:

> A waveparticle, confined within a finite region of space, can only have a series of discrete allowed energies $\varepsilon_o,\ \varepsilon_1,\ \varepsilon_2,\ldots$ These can be arranged in order of increasing magnitude with no energy allowed in between two successive discrete values.

(See Figs. 3.2 and 3.3d′). In information language, the energy of a confined waveparticle is "digital" not "analog". We already know that the lowest single-particle energy ε_o is higher than the chosen zero of energy by an amount equal to or larger than the minimum kinetic energy given by (3.13). It is also true that the energy difference $\delta\varepsilon \equiv \varepsilon_1 - \varepsilon_o$ between the first excited and the ground state energy of a confined waveparticle, in the non-relativistic case, is of the same order of magnitude:

$$\delta\varepsilon = c_1 \frac{\hbar^2}{m\,V^{2/3}} = c_2 \frac{\hbar^2}{m\,R^2} \tag{3.24}$$

where the numerical factors c_1 and c_2 are of order one; their exact values depend on the type of the potential responsible for the confinement (see Fig. 3.2). Notice that the volume V appearing in (3.24) may be the result of an external confining potential or the result of internal interactions as in the case of waveparticles making up composite systems.

Equation (3.24) is of similar crucial importance as the combination of Heisenberg's and Pauli's principles leading to (3.22b), which guarantees the stability and, hence, the existence of the structures of the World. Equation (3.24)

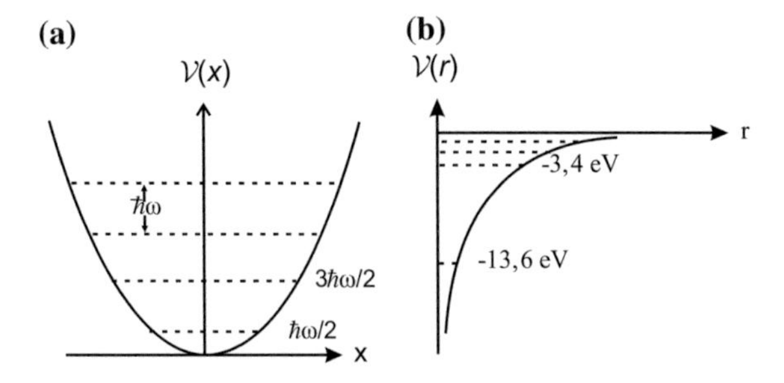

Fig. 3.2 **a** Schematic plot of the potential energy $\mathcal{V}(x) = \frac{1}{2}\kappa x^2$ and the corresponding discrete energy levels of one-dimensional harmonic oscillator. **b** The same for the Coulomb potential energy $\mathcal{V}(r) = -e^2/r$ and the corresponding discrete energy levels

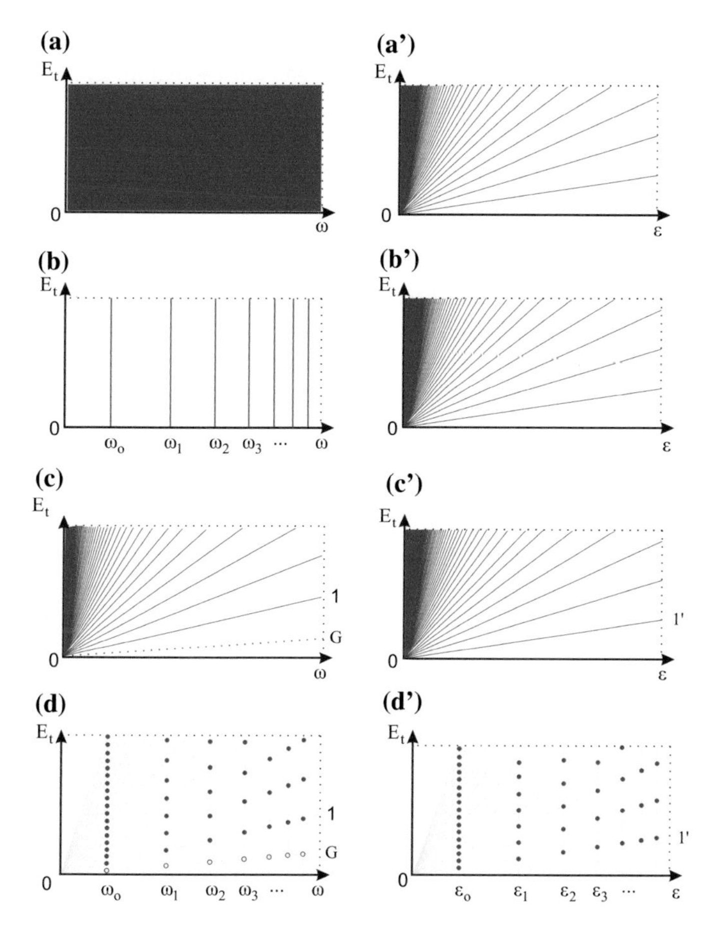

Fig. 3.3 **a–d** Allowed values in the plane, total energy E_t versus frequency ω: **a** unconfined classical wave; all points in the plane E_t versus ω are allowed. **b** Confined classical wave; only the points on the discrete lines perpendicular to the frequency axis are allowed (frequency discretization). **c** Unconfined particlewave; only the points on the lines $E_t = n\hbar\omega$ are allowed where $n = 1, 2, 3,\ldots$ (total energy quantization). **d** Confined particlewaves; only the points marked by dots are allowed (frequency discretization and total energy quantization). **a′–d′** Allowed values in the plane, total energy, E_t versus particle energy ε. **a′** Unconfined n ($n = 1, 2, 3,\ldots$) non-interacting classical particles, each of the same ε. **b′** As in **a′** but confined. **c′** Unconfined n ($n = 1, 2, 3,\ldots$) non-interacting waveparticles, each of energy ε. In cases **a′–c′**, only the points on the lines $E_t = n\varepsilon$ are allowed. **d′** As in **c′**, but confined; discretization along the E_t axis (as in cases **a′**, **b′**), due to particle aspect of waveparticles and quantization along the ε-axis (as a result of the wave aspect of waveparticles). Strictly speaking, only cases **c**, **d**, **c′**, **d′** correspond to reality, which is of quantum nature. Notice that, as ω or ε tend to zero, the classical and the quantum behavior tend to coincide

implies that any temporary perturbation of size smaller than $\delta\varepsilon$ will leave a composite waveparticle unchanged in its ground state, corresponding to the minimum energy ε_o, if it was in this state initially. This means that composite waveparticles, or even systems, behave as elementary ones up to a limit. In other words, composite waveparticles, although they possess the possibility of being changed–even breaking–they require for this to happen the application of an external perturbation *exceeding* a non-zero energy value, which, in some cases, can be much larger than any available perturbation. It is exactly this property which allows us to attribute a predictable chemical and physical behavior to, e.g., any atom of the periodic table, in spite of it being a composite system under the influence of quite different environments. In contrast, in a classical world, where $\hbar$ would be zero, the behavior of an atom would never be the same, because the ever present interactions with its environment-even the most minute ones-would force the atoms to be in a situation of continuous change. Equation (3.24) shows that the smaller a composite waveparticle is, the more stable it is. For a nucleus, where $m = m_p \approx 1837m_e$ and $R \approx 1$ fm, $\delta\varepsilon \approx 40$ MeV. This explains why non-radioactive nuclei cannot change under ordinary conditions and why the alchemists had no chance to succeed in their efforts to change (by employing chemical energy of the order of a few eV) other common substances to gold.

There are some differences regarding the picture of the ground state energy and the next excited energy level for waveparticles, such as electrons, and particle-waves, such as the photons, the quanta of the EM field. This difference stems from the fact that the number of electrons is conserved, while the number of photons is not. Thus, we can talk about the number of electrons in a system and its total energy as independent variables. E.g., we can consider just one electron being in its ground state (of energy ε_o in case of Fig. 3.3d′, or 0 in case of Fig. 3.3c′). *Any excitation of this one-electron system moves necessarily along the straight line* 01′ *of Fig.* 3.3c′, *or by jumping to the next point in the sequence of points lying on the straight line* 01′ *of Fig.* 3.3d′. In contrast to this picture, the number of photons is not an independent variable; its value is determined from the value of the total energy (and ω, if the latter is given). In particular, the ground state energy, E_G (i.e., the lowest possible energy) of the EM field involves zero number of photons, but it is not zero; it consists of the sum of all points on the line 0G (see Fig. 3.3c or d). This is due to the fact that, according to Quantum Mechanics, the electric field **E** and the magnetic field **B** cannot be simultaneously exactly zero, for the same reason that x and p_x cannot be zero simultaneously. In other words the sum of $\mathbf{E}^2 + \mathbf{B}^2$, which is proportional to the energy density of the electromagnetic field, cannot be exactly zero; its lowest value is equal to the sum of all $\frac{1}{2}N_i\hbar\omega_i$, where N_i is the number per unit volume of all the waves of frequency ω_i; for dimensional reasons $N_i \propto \omega_i^2/c^2$(see Chap. 6). In other words, the quantum "vacuum", as far as the EM field is concerned, is not a completely empty space, but contains the unavoidable minimum fluctuations of the EM field vectors around their average value, $\langle\mathbf{E}\rangle$,$\langle\mathbf{B}\rangle$; the latter, in contrast to $\langle \mathbf{E}^2 + \mathbf{B}^2 \rangle$, is zero, because of cancellation of positive and negative values. *The lowest excited energy level of the EM field for a given allowed* ω *is to move vertically from a point in the line 0G to the point on the line 01* (*see Fig.* 3.3c or d). The energy of this state is $E_G + \hbar\omega$ and it involves just one photon of frequency ω. As is clear from Fig. 3.3c or d, the higher the frequency, the higher the minimum excitation energy $\delta\varepsilon = \hbar\omega$ is. This means that in the quantum world, as opposed to the classical one, the higher the frequency of a particlewave such as a photon, the higher the external energy needed to excite it. The most common way to excite a system from its ground state is to bring it into contact with a heat bath of absolute (Kelvin) temperature T; then, the available energy that the heat bath can possibly transfer to a waveparticle of the system is within a range from 0 up to a maximum value which is of order $k_B T$, where k_B is the so-called Boltzmann's constant.[10]

Thus, a heat bath of temperature T in equilibrium with a photon system cannot appreciably excite photons of frequency higher than $\omega_M = c_1 k_B T/\hbar$, where c_1 is a constant of order one. This QM limit of the highest excitable frequency of a photonic system finds immediate application in calculating the total EM energy, I, emitted by a black body per unit time and per unit area. The quantity I can be obtained by integrating over ω the quantity $I_\omega d\omega$, corresponding to the contribution to I from the frequency range $[\omega,\ \omega + d\omega]$. The classical result for $I_\omega d\omega$ (for dimensional reasons) is proportional to $(k_B T/c^2)\,\omega^2 d\omega$ (the reader is encouraged to prove it). If this quantity is integrated from zero to infinity, as in classical physics, one gets infinity, an unphysical result. If the upper limit of integration is restricted to the QM one, $\omega_M = c_1 k_B T/\hbar$, the result will be proportional to $\frac{1}{3}(k_B T/c^2)\,\omega_M^3 = (c_1^3/3)(k_B T)^4/c^2\,\hbar^3$ in agreement with the experimental data (see also (6.3)).

We shall conclude this section by returning to the picture, proposed by R. Feynman, of how a waveparticle propagates from a point A to a point B. If it were a classical particle, it would follow a single unique trajectory from A to B. In contrast, the waveparticle will follow *all possible trajectories* joining the two points, each trajectory weighted by what is called a probability amplitude. The absolute value squared of the sum of all these probability amplitudes gives the probability for the waveparticle to propagate from A to B. Usually, to a good approximation, the trajectories which give an appreciable contribution to the sum are those which are within a tube of cross-section proportional to the wavelength squared around the classical trajectory. This alternative formulation of quantum mechanics, called the path integral formulation [8], is equivalent to the formulation through ψ obeying Schrödinger's equation. Conceptually, it is probably easier to reconcile our perceptions of particle

[10]For the numerical values of all universal constants see Table I.1 in Appendix I.

motion with Feynman's approach rather than with the ψ formulation, since the former still makes use of the trajectory concept, although not a single one. Computationally, it is usually simpler to work with ψ, especially for bound systems, as compared with the path integral formulation, although there are cases where the latter is more convenient.

3.7 Summary of Important Concepts and Formulae

The uncertainty principle implies that a waveparticle confined in a finite region of space acquires necessarily a non-zero kinetic energy which becomes progressively larger as the confinement length decreases. This feature is of critical importance because it provides the mechanism for equilibrium to be established in composite systems. Indeed, as the attractive interactions squeeze the system, the kinetic energy increases until the point where its pressure becomes equal to the squeezing pressure of the interactions and equilibrium is established.

The uncertainty principle states that the product of the standard deviation of a Cartesian component of the position vector, e.g., Δx, times the standard deviation of the corresponding component of the momentum, e.g., Δp_x, is always larger than $\hbar/2$:

$$\boxed{\Delta x \cdot \Delta p_x \geq \hbar/2} \tag{3.8}$$

We remind the reader that the standard deviations satisfy by definition the following relations: $\Delta x^2 \equiv <x^2> - <x>^2$ and $\Delta p_x^2 \equiv <p_x^2> - <p_x>^2$ (see Footnote 5). For a waveparticle confined in a fixed region of space we can always chose the coordinate system in such a way that $<x> = 0$ and $<p_x> = 0$; thus $\Delta x^2 = <x^2>$ and $\Delta p_x^2 = <p_x^2>$. It follows that $<p_x^2> \geq \left[\hbar^2/(4<x^2>)\right]$; thus the minimum value of $<p^2>$ is

$$<p^2> \propto \frac{\hbar^2}{R^2} \propto \frac{\hbar^2}{V^{2/3}} \tag{3.9}$$

From (3.9) it follows that the minimum average *non-relativistic* kinetic energy $\varepsilon_K = p^2/2m$ of a waveparticle of mass m confined within a finite volume V is necessarily non-zero and proportional to

$$\varepsilon_K \propto \frac{\hbar^2}{mV^{2/3}} \propto \frac{\hbar^2}{mR^2} \tag{3.13}$$

Also from the relation $\varepsilon_K = c\,|\boldsymbol{p}|$, valid in the extreme relativistic limit, and the relation $<|\mathbf{p}|> \propto \sqrt{<p^2>}$ we have that the minimum average *extreme relativistic* kinetic energy $\varepsilon_K = c\,|\mathbf{p}|$ of a particle confined within a finite volume V is necessarily non-zero and proportional to

$$\varepsilon_K \propto \frac{\hbar c}{V^{1/3}} \propto \frac{\hbar c}{R} \tag{3.14}$$

The exclusion principle is applicable only to identical fermions sharing the same volume V. Let us have N of them (each of spin ½ to be concrete). If we divide the volume V to $N/2$ equal subspaces each of volume $2\,V/N$ and place two fermions in each subspace (one with spin up and the other with spin down) we shall satisfy the exclusion principle, but now, because of the reduction in volume, the minimum average kinetic energy of each fermion will increase by a factor of $N^{2/3}$ (in the case of (3.13)) and by a factor of $N^{1/3}$ (in the case of (3.14)). Thus the minimum total kinetic energy $E_K = N\varepsilon_K$ of N identical spin ½ fermions sharing the same volume V is for the *non-relativistic* limit

$$\boxed{E_K \propto \frac{\hbar^2 N^{5/3}}{mV^{2/3}} \propto \frac{\hbar^2 N^{5/3}}{mR^2}} \tag{3.22}$$

while for the *extreme relativistic* limit is

$$\boxed{E_K \propto \frac{c\hbar N^{4/3}}{V^{1/3}} \propto \frac{c\hbar N^{4/3}}{R}} \tag{3.23}$$

The energy difference $\delta\varepsilon$ between the first excited and the ground state of a single particle (either elementary or composite) confined in a finite region of space of volume V and radius R is given in the non-relativistic limit by the formula

$$\delta\varepsilon = c_1 \frac{\hbar^2}{m\,V^{2/3}} = c_2 \frac{\hbar^2}{mR^2} \tag{3.24}$$

while in the extreme relativistic limit by the formula

$$\delta\varepsilon = c_1 \frac{\hbar c}{V^{1/3}} = c_2 \frac{\hbar c}{R} \tag{3.25}$$

3.8 Multiple-Choice Questions/Statements

1. An isolated composite system left undisturbed would eventually end up in a final state which is determined by one of the following mechanisms:

(a) *The fundamental forces among the particles of the system are attractive for relative distances* $d>a$ *and become repulsive for* $d<a$. *Thus the system finds an equilibrium state when* $d = a$. *A characteristic example is the case of a diatomic molecule.*

(b) *The final state is not uniquely determined, because the initial conditions as well as various random processes play a crucial role. A characteristic example is the case of a planet going around the Sun.*

(c) *Two antagonistic factors determine always the equilibrium state of a system: One is its potential energy (or more generally the presence of the interactions) which on the average are of attractive nature and would force the system to collapse, if unopposed, and the other is the total kinetic energy with respect to the center of mass which tends to blow the system apart. A characteristic example is the case of the hydrogen atom.*

(d) *Every system will eventually collapse because of the continuous emission of some kind of radiation or more generally because of the continuous decrease of its mechanical energy. A characteristic example is the case of an artificial satellite of Earth.*

For each of the above four answers/statements argue against or in favor.

2. The minimum kinetic energy of a system of N identical spin ½ particles (of zero total momentum and angular momentum) confined within a finite 3-D space of volume $V = \frac{4\pi}{3}R^3$ is given by one of the following formulas (if $\varepsilon_K = p^2/2m$):

(a) $E_K \geq 2.87N\dfrac{\hbar^2 N^{1/3}}{mV^{2/3}} = 1.105\dfrac{\hbar^2}{m}\dfrac{N^{4/3}}{R^2}$,

(b) $E_K \geq 2.87N\dfrac{\hbar^2 N^{2/3}}{mV^{2/3}} = 1.105\dfrac{\hbar^2}{m}\dfrac{N^{5/3}}{R^2}$,

(c) $E_K \geq 2.87N\dfrac{\hbar^2}{mV^{2/3}} = 1.105\dfrac{\hbar^2}{m}\dfrac{N}{R^2}$

(d) $E_K \geq 2.87N\dfrac{\hbar^2 N^{2/3}}{mV^{1/3}} = 1.105\dfrac{\hbar^2}{m}\dfrac{N^{5/3}}{R}$,

3. The minimum kinetic energy of a system of N identical spin ½ particles (of zero total momentum and angular momentum) confined within a finite 3-D space of volume $V = \frac{4\pi}{3}R^3$ is given by one of the following formulas (if $\varepsilon_K = c\,p$):

(a) $E_K \geq 2.87N\dfrac{\hbar c N^{2/3}}{V^{1/3}} = 1.105\dfrac{\hbar c N^{5/3}}{R}$,

(b) $E_K \geq 2.87N\dfrac{\hbar c N^{2/3}}{V^{2/3}} = 1.105\dfrac{\hbar c N^{5/3}}{R^2}$

(c) $E_K \geq 2.32N\dfrac{\hbar c N^{1/3}}{V^{1/3}} = 1.44\,\hbar c\dfrac{N^{4/3}}{R}$,

(d) $E_K \geq 2.32N\dfrac{\hbar c}{V^{1/3}} = 1.44\,\hbar c\dfrac{N}{R}$,

4. In a 1-D harmonic oscillator characterized by κ and m, Δx is the standard deviation from the classical equilibrium position for the ground state. The energy difference between the first excited and the ground state is given by one of the following formulas:

(a) $\frac{1}{2}\kappa(\Delta x)^2$ (b) $\frac{1}{2}(\hbar c/\Delta x)$ (c) $\frac{1}{2}(\hbar^2/m\Delta x^2)$ (d) $\frac{1}{2}(\hbar\sqrt{\kappa/m})$

(See solved problem 4 below).

5. In the ground state of a hydrogen atom, the electron is confined (with a probability of about 76%) within a spherical volume of radius $r = 2\hbar^2 / m_e e^2$ around the proton. The energy difference between the first excited and the ground state is given by one of the following formulas:

(a) $\delta\varepsilon = (c_1 / 10.39)(\hbar^2 / m_e r)$ (b) $\delta\varepsilon = (c_1 / 10.39)(\hbar^2 / m_e r^2)$

(c) $\delta\varepsilon = (c_1 / 10.39)(\hbar^2 / m_e r^3)$ (d) $\delta\varepsilon = (c_1 / 10.39)(\hbar / m_e r^2)$

6. The relation between the average kinetic energy per particle $\overline{\varepsilon} \equiv E_K / N$ of N identical spin ½ particles of zero total momentum and angular momentum confined within a finite 3-D space of volume $V = \frac{4\pi}{3} R^3$ and the corresponding Fermi energy is given by one of the following formulas (assume that $\varepsilon_K = p^2 / 2m$):

(a) $\overline{\varepsilon} = (3/5)E_F$ (b) $\overline{\varepsilon} = (1/2)E_F$ (c) $\overline{\varepsilon} = (3/4)E_F$ (d) $\overline{\varepsilon} = (2/5)E_F$

7. The relation between the average kinetic energy per particle $\overline{\varepsilon} \equiv E_K / N$ of N identical spin ½ particles of zero total momentum and angular momentum confined within a finite 3-D space of volume $V = \frac{4\pi}{3} R^3$ and the corresponding Fermi energy is given by one of the following formulas (assume that $\varepsilon_K = c\,p$):

(a) $\overline{\varepsilon} = (3/5)E_F$ (b) $\overline{\varepsilon} = (1/2)E_F$ (c) $\overline{\varepsilon} = (3/4)E_F$ (d) $\overline{\varepsilon} = (2/5)E_F$

8. The relation between the average kinetic energy per particle $\overline{\varepsilon} \equiv E_K / N$ of N identical spin ½ particles of zero total momentum and angular momentum confined within a finite 2-D (**two-dimensional**) space of area $A = \pi R^2$ and the corresponding Fermi energy is given by one of the following formulas (assume that $\varepsilon_K = p^2 / 2m$):

(a) $\overline{\varepsilon} = (3/5)E_F$ (b) $\overline{\varepsilon} = (1/2)E_F$ (c) $\overline{\varepsilon} = (3/4)E_F$ (d) $\overline{\varepsilon} = (2/5)E_F$

9. The Fermi energy of N identical spin ½ particles of zero total momentum and angular momentum confined within a finite 2-D (**two-dimensional**) space of area $A = \pi R^2$ is given by one of the following formulae (assume that $\varepsilon_K = p^2 / 2m$):

(a) $E_F = 2N\hbar^2 / mR^2$ (b) $E_F = N\hbar^2 / mR^2$

(c) $E_F = \pi N\hbar^2 / mR^2$ (d) $E_F = 3N\hbar^2 / mR^2$

(For the 2-D case we have $\sum_k \ldots = \frac{A}{(2\pi)^2} \int d^2k \ldots$)

10. The relation between the average kinetic energy per particle $\overline{\varepsilon} \equiv E_K / N$ of N identical spin ½ particles of zero total momentum confined within a finite 1-D (**one-dimensional**) space of length $L = 2a$ and the corresponding Fermi energy is given by one of the following formulas (assume that $\varepsilon_K = p^2 / 2m$):

(a) $\overline{\varepsilon} = (1/5)E_F$ (b) $\overline{\varepsilon} = (1/3)E_F$ (c) $\overline{\varepsilon} = (1/2)E_F$ (d) $\overline{\varepsilon} = (2/5)E_F$

(For the 1-D case we have $\sum_k \ldots = \frac{L}{2\pi} \int dk \ldots$)

The correct answers to the multiple choice questions are given at the end of the book (see Appendix H).

3.9 Solved Problems

1. *By employing Heisenberg's uncertainty principle, obtain the minimum average total energy* $<\varepsilon>$ *of a one-dimensional harmonic oscillator in terms of the uncertainty* Δx *in its position. The oscillator is characterized by its 'spring' constant* κ *and its mass m. Minimize* $<\varepsilon>$ *with respect to* Δx *in order to determine* Δx *and* $<\varepsilon>$.
   Solution By definition $\Delta x^2 \equiv <x^2> - <x>^2$. For a harmonic oscillator x is going back and forth symmetrically around the classical equilibrium position which is taken to be zero. Hence $<x> = 0$ and $\Delta x^2 = <x^2>$. By the same arguments $<p^2> = \Delta p^2$. It follows that $<\varepsilon> = <p^2/2m> + <\kappa x^2/2> = (\Delta p^2/2m) + (\kappa\, \Delta x^2/2) \geq (\hbar^2/8\, m\, \Delta x^2) + (\kappa\, \Delta x^2/2)$. In the last relation Heisenberg's uncertainty relation $\Delta x \Delta p \geq \hbar/2$ was used. Next, we minimize $(\hbar^2/8\, m\, \Delta x^2) + (\kappa\, \Delta x^2/2)$ by setting its derivative with respect to Δx equal to zero, and we find $\Delta x^2 = \hbar/(2\sqrt{\kappa\, m})$, $(\hbar^2/8\, m\, \Delta x^2) = (\kappa\, \Delta x^2/2) = \hbar\, \omega/4$, and $<\varepsilon> \; \geq (\hbar\, \omega/2)$, where $\omega \equiv \sqrt{\kappa/m}$. The exact result is indeed $<\varepsilon> = \hbar\, \omega/2$ and the average kinetic energy is equal to the average potential energy.
2. *[11] *Consider a one-dimensional shallow* $(V \ll (\hbar^2/m\, a^2) \equiv \varepsilon)$ *potential energy well* $\mathcal{V}(x)$ *of depth* $-V$ *and extent* $2a$*. For* $|x| > a$, $\mathcal{V}(x) = 0$*. Can such a potential well bound a particle of mass m?*
   Solution A particle in a bound state means that it is not free to escape to infinity. Hence, its energy E must be less than the value of the potential energy at infinity (which is zero), since otherwise the particle will not be confined to a finite region of space. On the other hand, it is clear that the extent L of its confinement must be much larger than a, since, if $L \approx a$, its quantum kinetic energy, which is of order $\hbar^2/m\, L^2$, would produce a positive total energy: $\hbar^2/m\, a^2 - V \gg 0$. Given that $L \gg a$, the potential energy is of order $-V(a/L)$, i.e. $-V$ (when the particle is in the potential well) times the probability (which is of order a/L) of the particle to be inside the potential well. Minimizing the sum $\hbar^2/2mL^2 - c_P Va/L$ of the kinetic and the potential average energy with respect to the extent L of the confinement, we find: $L = (1/c_P)(\hbar^2/ma^2 V)a = (1/c_P)(\varepsilon/V)a$ and $E = -2c_P^2 V^2/4\varepsilon)$. The exact results (see Economou [9]) in the limit $V/\varepsilon \to 0$ are obtained for $c_P = 2$, i.e. $L/a = \varepsilon/2V$ and $E = -2\, V^2/\varepsilon$, which means that there is always at least one bound state no matter how shallow the one-dimensional potential well is; the binding energy for very shallow one-dimensional potential wells is equal to the ratio $V^2/(\hbar^2/2m\, a^2)$.

[11] Problems indicated by an asterisk have broader physical implications, but require more than a simple application of a formula.

3. **Consider a three-dimensional potential well* ($\mathcal{V}(r) = -V$, *for* $r<a$ *and* $\mathcal{V}(r) = 0$, for $r > a$). *Under which conditions can such a potential well sustain at least one bound state for a waveparticle of mass m?*

   Solution As we shall show, in three dimensions (in contrast to what happens in one-dimension) a very shallow potential well cannot sustain a bound state; the ratio $V/\left(\hbar^2/m\,a^2\right) \equiv V/\varepsilon$ has to exceed a critical value of order one for at least one bound state to appear. The main reason for this difference is that in 3-D the probability density $|\psi(\boldsymbol{r})|^2$ acquires an extra factor $1/r^2$ compared to 1-D. This factor can be understood by considering a spherically symmetric electromagnetic (EM) wave stemming from a spherical object; conservation of energy dictates that the energy flux $4\pi r^2 c(|\boldsymbol{E}(r)|^2/4\pi)$ through any concentric spherical surface must be a constant, which implies that $|\boldsymbol{E}(r)|^2/4\pi$ (which is the EM energy density) must be proportional to $1/r^2$. A similar argument stemming from the conservation of the probability demands the presence of a factor $1/r^2$ in $|\psi(\mathbf{r})|^2$. The presence of this factor in combination with the continuity of $\psi(\mathbf{r})$ and $\nabla\psi(\mathbf{r})$ shows that the assumption of a negligible value of the kinetic energy associated with the region $r<a$ (used implicitly in the 1-D case) is not valid in 3-D. On the contrary, the contribution of the region $r<a$ to the kinetic energy is equal on the average to about $(\hbar^2/2ma^2)$ times the probability that the waveparticle is inside the potential well; this probability is of order $a^3|\psi(a)|^2/L^3|\psi(L)|^2 \approx a/L$. In the last equation we took into account the $1/r^2$ factor in $|\psi(\boldsymbol{r})|^2$ and the reasonable expectation that $L \gg a$ for V just a little larger than the critical value. To the resulting kinetic energy $c_K(a/L)(\hbar^2/2ma^2)$ coming from the region $r<a$ we must add the kinetic energy $c'_K(\hbar^2/2mL^2)$ coming from the region $r > a$ and the potential energy $-c_P(a/L)V$ to end up with the total energy E. Minimizing E with respect to L we find $L/a = \{2c'_K\varepsilon/(2c_P V - c_K\varepsilon)\}$ and $E = -\{(2c_P V - c_K\varepsilon)^2/8c'_K\varepsilon\}$. The exact results are obtained for $c_P = c'_K = 1$ and $c_K = \pi^2/4$ and the critical value of V is $V_c/\left(\hbar^2/m\,a^2\right) = \pi^2/8$.

4. **Consider a neutral atom of large atomic number Z. Its total energy depends on an appropriate mean distance r between an electron and the nucleus, on an appropriate average distance a between two electrons, and on the "radius" R of the atom. Three terms contribute to the total energy: The quantum kinetic energy of the electrons; the Coulomb attractive potential energy* $-Z^2e^2/r$ *of the Z electrons with the charge Ze of the nucleus; and the repulsive mutual potential energy among the Z electrons which is equal to* $Z(Z-1)e^2/2a \approx Z^2e^2/2a$. *Obtain the total energy of the atom in terms of R and the ratios* $r/R = x$ *and* $a/R = y$.

   Solution It is more convenient to work in atomic units ($\hbar = e = m_e = 1$) and restore the ordinary units at the end. The total kinetic energy according to (3.22) is $E_K = 1.105\,Z^{5/3}/R^2$. The Coulomb interactions between electrons and the nucleus is $-Z^2/r = -Z^2/xR$ and the Coulomb repulsions among electrons is $Z^2/2a = Z^2/2yR$. Thus the total energy is $E = 1.105(Z^{5/3}/R^2) - b(Z^2/R)$, where $b \equiv \{(1/x) - (1/2y)\}$. Minimizing the total energy by setting its derivative with respect to R equal to zero (under x and y constant) we find for R: $R = (2.21/b)\,Z^{-1/3}$. Substituting this value of R in E we find that

$$E = -0.226\,b^2 Z^{7/3}$$

This same problem is solved in the book by L.D. Landau and E.M. Lifshitz *Quantum Mechanics*, 3rd ed., [10], pp 261–264 by applying an advanced method known as the Thomas-Fermi approximation, which in addition to the results obtained by our simpler method, it also determines the distribution of electrons around the atom. Our simple method obtains the same dependence on Z for both R and E as the sophisticated Thomas-Fermi approach. We shall meet again these (−1/3), (7/3) exponents in the case of white dwarfs and neutron stars. The experimental value of E is obtained by choosing b = 1.61 which corresponds to the reasonable values $r \approx a \approx 0.31\,R$. Restoring the ordinary units and employing the value b = 1.61 we find $R = 0.73\,Z^{-1/3}$A and $E \approx -16\,Z^{7/3}$ eV. Notice that this small value of R defines a spherical surface within which the majority of the electrons can be found; it is completely unrelated to the radius of the atom as determined by its higher occupied orbital.

3.10 Unsolved Problems

1. Prove (3.22a) by employing (3.20) and (3.21)
2. Prove (3.23).
3. Prove (3.11).
4. For a one-dimensional attractive potential of the form $V(x) \propto \pm|x|^{\beta}$, where β is a constant, show that the energy of the nth state is

$$E_n \propto n^a, \text{ where } a = 2\beta/(\beta+2) \tag{3.26}$$

For the potential well of infinite depth we have $\beta \to \infty$, and $a = 2$; for a harmonic oscillator $\beta = 2$, and $a = 1$; and for a Coulomb-like potential $\beta = -1$, and $a = -2$.
Prove (3.26).
(*Hint* $(dE/dn)\delta n \propto (dV/dx)\lambda$, and $\lambda \propto E^{-1/2}$).

5. Consider a non-relativistic particle of mass m confined within an ellipsoid with main semi axes a, b, and c. What is its minimum kinetic energy? How does this kinetic energy compare with what would result if the ellipsoid becomes a sphere

of equal volume? The probability density to find the particle within the ellipsoid or the sphere is assumed constant.

6. *Consider a two-dimensional shallow $(V \ll (\hbar^2/m a^2) \equiv \varepsilon)$ potential energy well of depth $-V$ and radius a. Is such a potential well capable of binding a particle of mass m?
   (*Hint* The kinetic energy is proportional to $\hbar^2/2mL^2$, where $L \gg a$ and the potential energy is proportional to $-(a^2/L^2)V\ln(L/L_0)$, where the logarithmic factor appears because the wave function, being a Bessel function contains such a term (see Economou [9]). The 2D case is borderline between the 1D and the 3D cases as far as the bound state in a shallow potential well is concerned. The problem of a bound state in 1D, or 2D, or 3D potential wells has many important physical applications.
7. *Low frequency sound waves in solids resemble electromagnetic (EM) waves in the vacuum in the sense that their frequency is proportional to their wavevector k. In analogy with the discussion in Sect. 3.7 for the temperature dependence of the average energy of EM waves, show that at low temperatures the average energy in a solid due to the thermal excitation of sound waves is proportional to the fourth power of temperature and, hence, the corresponding specific heat is proportional to the third power. What is the corresponding result according to classical physics?

Problems indicated by an asterisk have broader physical implications, but require more than a simple application of a relevant formula.

References

1. R. Feynman, R. Leighton, M. Sands, *The Feynman Lectures on Physics* (Addison-Wesley, Reading, 1964)
2. A. Einstein, B. Podolsky, N. Rosen, Phys. Rev. **47**, 777 (1935)
3. A. Aspect, P. Grangier, G. Roger, Phys. Rev. Lett. **49**, 91 (1982)
4. R. Griffiths, *Consistent Quantum Theory* (Gambridge University Press, Cambridge, 2002)
5. P. Ghose, *Testing Quantum Mechanics on New Ground* (Gambridge University Press, Cambridge, 1999)
6. A. Peres, *Quantum Theory: Concepts and Methods* (Kluwer Academic Publishers, Dordrecht, 1995)
7. H. Paul, *Introduction to Quantum Theory* (Gambridge University Press, Cambridge, 2008)
8. R. Shankar, *Principles of Quantum Mechanics*, 2nd edn. (Springer, NY, 1994)
9. E.N. Economou, *Green's Functions in Quantum Physics*, 3rd edn. (Springer, Berlin, 2006)
10. L.D. Landau, E.M. Lifshitz, *Quantum Mechanics*, 3rd edn (Pergamon Press, Oxford 1980) (Butterworth-Heinemann, Great Britain, 3rd rev. and enl. edn 2003)
11. E.H. Lieb, Rev Modern Phys **48**, 553 (1976)

Chapter 4
Equilibrium and Minimization of Total Energy

Abstract Combining the First and the Second Law of Thermodynamics and choosing $đ W = P\,dV$ we find that

$$dU \leq T\,dS - P\,dV + \mu\,dN \qquad (4.1)$$

where μ is the so-called chemical potential. The inequality in (4.1) is valid in the course towards equilibrium and equality when equilibrium is reached. Thus, under the conditions $T\,dS = P\,dV = dN = 0$, the energy U is decreasing and reaches its minimum value when equilibrium is achieved. We can change independent variables from S to T and/or from V to P and/or from N to μ by subtracting respectively from both sides of (4.1) the differentials of the product TS, and/or adding the differentials of PV, and/or subtracting the differentials of μN. Thus various differentials of thermodynamic potentials result, e.g. that of the Gibbs free energy $G \equiv U + PV - TS$ satisfying the relation $dG \leq -SdT + VdP + \mu dN$ which means that, under conditions of constant temperature and pressure and no exchange of matter, equilibrium corresponds to the minimum of G.

4.1 Concepts

Equation (4.1) implies that the stable equilibrium state of a physical system of negligible absolute temperature, under conditions of receiving no work and no m-particles from the environment, corresponds to the minimum of its total energy U.[1]

---

[1]More accurately, U is the so-called internal energy, which is defined as the average value of the total energy E_t of the system, under conditions in which the total momentum is equal to zero on the average. (For macroscopic systems, the total angular momentum must also be zero on the average). Notice that the value of U or E_t is fully determined only after we choose a reference state, the energy of which is by definition zero. Usually, the reference state is the one in which all particles of the system are at infinite distance from each other and each one is in its ground state. There are three types of contributions to U or E_t: The relativistic rest energy $E_o = \sum m_{oi}c^2$, the kinetic energy E_K, and the potential energy E_P, which may include interactions both with the environment and among the particles of the system itself. For convenience, it is not uncommon to ignore the rest energy, if it remains constant, by incorporating it in the reference state.

E.N. Economou, *From Quarks to the Universe*,
DOI 10.1007/978-3-319-20654-7_4

Under the more common conditions of a physical system being in contact with an environment of constant temperature T_o and constant pressure P_o, the system's stable equilibrium state corresponds to the minimum of a quantity A called *availability* and defined as follows [1]:

$$A \equiv U + P_o V - T_o S \tag{4.2}$$

where U is the total energy (see Footnote 1) of the system, V is its volume, and S is its entropy. These minimization principles stem from the 2nd law of thermodynamics, i.e., the law of entropy increase. Expanding $U(V, S)$ in powers of V and S up to second order and substituting the result in (4.2), we find that the minimization of A implies that $P = P_o$ and $T = T_o$ as well as some general thermodynamic inequalities.[2] Thus, at equilibrium, the availability A coincides with the so-called *Gibbs free energy*, $G \equiv U + PV - TS$ which will be discussed in Sect. 4.4.

In this short chapter the 1st and 2nd laws of thermodynamics are presented. Their central role for equilibrium structures is pointed out. Some other important thermodynamic relations are also introduced. (For more details see [2]).

4.2 Conservation of Energy and the First Law

The first law is essentially the law of conservation of energy. It states that the infinitesimal change dU of the internal energy U of a system is due to the following three exchanges with the environment: (a) receiving an infinitesimal quantity of heat $đQ$; (b) performing an infinitesimal amount of work $đW$ on the environment; and (c) receiving an infinitesimal amount of energy $đE_m$ associated with the inflow of an infinitesimally small number of m-particles. Then, conservation of energy implies that

$$\boxed{dU = đQ - đW + đE_m} \tag{4.3}$$

The infinitesimal quantities $đQ,\ đW,\ đE_m$ are not perfect differentials of certain functions $Q,\ W,\ E_m$ respectively; such functions do not exist. We cannot separate the internal energy of a system into these three components. Only during the process of *exchange* or *transformation* of energy such a separation makes sense for the energy being exchanged or transformed. Mathematically speaking, $đQ,\ đW,\ đE_m$ are so-called differential forms and not perfect differentials. It is exactly this distinction which is denoted by the symbol $đ$ instead of d in front of $Q,\ W,\ E_m$.

[2]These inequalities are: $T \geq 0,\quad C_V > 0,\quad C_P > C_V,\quad (\partial P/\partial V)_T < 0$ [3].

4.3 Entropy and the Second Law

The second law, one of the most basic laws of Nature, dealing with the consequences of connecting the macroscopic world to its microscopic constituents, can be stated in several equivalent ways. One of the most revealing is through the concept of entropy. The latter is usually defined in systems consisting of so many microscopic particles that a detailed microscopic description of their state becomes practically impossible and a gross average macroscopic description becomes imperative. The entropy of such a system is proportional to the logarithm of the total number, $\Gamma(U, V, N, \ldots)$, of microscopic states of the system corresponding to a given macroscopic state fully described by the macroscopic independent variables[3] $U, V, N, \ldots$:

$$\boxed{S \equiv k_B \ln \Gamma(U, V, N, \ldots),} \tag{4.4}$$

$k_B = 1.38 \times 10^{-23}\,\mathrm{m^2kg\,s^{-2}K^{-1}}$ is a universal physical constant called Boltzmann's constant.

The second law states that the entropy of a system, *which does not exchange heat and m-particles with the environment*, is always increasing until the system reaches the equilibrium state at which the entropy has its maximum value. Thus, for *such an isolated* system the maximum entropy implies equilibrium and vice versa. From the 2nd law, it follows that processes taking place within the system in its way towards equilibrium tend to increase its entropy. Hence, the inverse processes, if they could happen without external intervention, would decrease the entropy and violate the 2nd law, and as such they are impossible. We conclude that processes driving an isolated system towards equilibrium are *irreversible*; only in extreme limiting cases such processes can be considered as reversible, and only if they leave the entropy unchanged. Examples of irreversible processes are: Heat transfer among parts of the system at different temperatures, diffusion of particles among parts of the system having different chemical potentials, chemical reactions within the system, etc.

What happens when a system exchanges heat with the environment but no m-particles? In this case the entropy changes in two distinct ways: The internal one, as before through irreversible processes taking place within the system, and the

[3] A simple system, such as a perfect gas, when *in equilibrium,* can be described macroscopically by only three independent macroscopic variables, e.g. U, V, N; for a photon system in equilibrium N is not an independent variable and, hence, only two independent variables are sufficient for its macroscopic description (see Sect. 3.6). Other more complicated systems in equilibrium may require a larger number of independent macroscopic variables. Non-equilibrium macroscopic states of a system require more independent macroscopic variables than the ones required for the equilibrium state of the same system.

external one due to the exchange of heat. (Exchange of work per se does not affect the entropy).

$$\boxed{dS = đS_i + đS_e, \quad \text{where} \quad đS_i \geq 0, \quad dS_e = \frac{đQ}{T} \text{ assuming } đE_m = 0} \quad (4.5)$$

Since $đQ$ can be either positive or negative, the change of entropy of an open system exchanging heat with its environment can be positive, or negative, or zero. From (4.5) and the positivity of temperature at equilibrium, another equivalent version of the second law follows:

$$T\,dS \geq đQ \quad (4.6)$$

Combining (4.3) and (4.6), i.e., the 1st and the 2nd law, we obtain:

$$\boxed{dU \leq T\,dS - đW + đE_m} \quad (4.7)$$

Equation (4.7) shows that when $TdS = đW = đE_m = 0$, $dU \leq 0$; this means that, under the conditions $TdS = đW = đE_m = 0$, U keeps decreasing until the equilibrium state is reached, at which the energy becomes minimum. In particular, (4.8) implies that in the case where U *at equilibrium* is a function of only $S,\ V,\ N$, the internal energy U of a system on its way to equilibrium is still decreasing even if all of $S,\ V,\ N$ are kept constant.

4.4 Inequality (4.7) as a Source of Thermodynamic Relations

In this section we will make some remarks which may help the reader in manipulating thermodynamic inequalities or equalities and therefore reducing the need for memorization.

To take advantage of the relation (4.7) we need to have explicit expressions for $đW$ and $đE_m$. The most familiar formula for $đW$ is the expression PdV appropriate for the mechanical work done by gases and liquids. However, the reader may keep in mind that there are other types of work: E.g. the electrical work done by a system is $-\phi dq$, where ϕ is its electrostatic potential and dq is the change of the system's charge. Similarly, the infinitesimal quantity $đE_m$, in the simplest case, is written as $đE_m = \mu\, dN$, where μ is the so-called chemical potential to be discussed in the next section. Replacing these explicit expressions in (4.7) we find

$$dU \leq TdS - PdV + \mu dN \quad (4.8)$$

There are two cases where (4.8) reduces to equality: (a) If the physical processes taking place in the system are such that the system remains at equilibrium or are such that a negligible amount of internal change in entropy is produced. (b) If the function $U = U(S, V, N)$ has been obtained theoretically under the assumption of thermodynamic equilibrium and dU appearing in (4.8) is the differential of this function. Then $dU = (\partial U/\partial S)_{V,N} dS + (\partial U/\partial V)_{S,N} dV + (\partial U/\partial N)_{S,V} dN$ and by comparing with $dU = TdS - PdV + \mu dN$ we have:

$$(\partial U/\partial S)_{V,N} = T > 0, \quad (\partial U/\partial V)_{S,N} = -P < 0, \quad (\partial U/\partial N)_{S,V} = \mu \tag{4.9}$$

Useful relations are obtained by taking the derivatives of (4.9). E.g.

$$(\partial T/\partial V)_N = (\partial^2 U/\partial S \partial V)_N = (\partial^2 U/\partial V \partial S)_N = -(\partial P/\partial S)_N \tag{4.10}$$

In (4.8), which gives an upper bound on the quantity dU, the differentials of the independent variables S, V, N appear on its right-hand-side. For this reason we characterize S, V, N as the natural independent variables of U. However, from the experimental point of view another set of independent variables may be much easier to control, e.g. T, V, N instead of S, V, N. There is a trick which allows us to implement easily this change (in the present case going from S to T). We either add to or subtract from U the product TS whose differential is $d(TS) = TdS + SdT$. We have then $d(U \pm TS) \leq TdS \pm TdS \pm SdT - PdV + \mu dN$. It is clear that in order to get rid of TdS we must subtract TS from U. By defining the quantity F as $U - TS$ we have

$$dF \equiv d(U - TS) \leq -SdT - PdV + \mu dN \tag{4.11}$$

The quantity F having as natural independent variables the set T, V, N is called *Helmholtz free energy*. The relation (4.11) reduces to equality under the same conditions as (4.8) does. The above procedure of changing the natural independent variables by adding or subtracting one or more products of two appropriate thermodynamic quantities can be applied no matter how many or what the natural independent variables in U are. E.g. we can go from the set S, V, N associated with U to the set T, P, N(which is the easiest one to control experimentally) by adding to U the quantity $-TS + PV$ or we can go to the set T, P, N by adding to F the quantity PV. The quantity $G \equiv U - TS + PV \equiv F + PV$ is the so-called *Gibbs free energy* introduced in Sect. 4.1; its differential satisfies the relation

$$\boxed{dG \leq -SdT + VdP + đE_m} \tag{4.12}$$

Under conditions of constant temperature and pressure and no exchange of m-particles, it follows from (4.12) that $dG \leq 0$; this means that the Gibbs free energy, under the said conditions (which are the ones prevailing in experiments done under normal conditions of pressure and temperature), is always decreasing on the way towards equilibrium (only in extreme limiting cases it may remain unchanged) and it reaches its minimum value at equilibrium.

4.5 Maximum Work, Gibbs' Free Energy, and Chemical Potential

Let us consider now a system *not in equilibrium* and in contact with an environment of constant temperature T_o and constant pressure P_o. What is the maximum work, W_M, which can be extracted from such a system by exploiting its non-equilibrium state? It turns out [1] that this maximum work is obtained only in the limiting case where the total entropy of the system and the environment together remains unchanged during the whole process; this condition implies that W_M is given by the formula

$$W_M = A_{\text{initial}} - A_{\text{final}} \tag{4.13}$$

where A is the availability defined in (4.2). (See Unsolved Problem 3). If $P = P_o$ and $T = T_o$ then the maximum work becomes (taking into account (4.2) and $G \equiv U + PV - TS$)

$$W_M = G_{\text{initial}} - G_{\text{final}} \tag{4.14}$$

Equations (4.13) and (4.14) provide a justification for the names availability and free energy for A and G respectively.

As an elementary application of (4.7), let us minimize $U = E_K + E_P$ with respect to the volume V of the system, under the conditions $T_o = P_o = đ\, E_m = 0$. We have

$$(\partial U/\partial V) = (\partial E_K/\partial V) + (\partial E_P/\partial V) = 0 \tag{4.15}$$

However, under processes where the system stays at equilibrium and under the conditions $T_o = P_o = đ\, E_m = 0$, $(\partial U/\partial V) = -P$, $(\partial E_K/\partial V) = -P_K$, and $(\partial E_P/\partial V) = -P_P$. Thus, the minimization of the energy U with respect to the volume, under the conditions, $T_o = P_o = đ\, E_m = 0$ implies the obvious relation that, at equilibrium, $P_K = -P_P$, which means that the expanding pressure of the kinetic energy equals the absolute value of the squeezing pressure of the potential energy.

As we mentioned before for processes starting from an initial equilibrium state and ending at another equilibrium state, while keeping $đ\, S_i = 0$ during the whole duration of the process, the equality sign in (4.7), (4.8), (4.11), and (4.12) holds. For such processes, taking into account (4.12), we have:

$$\left(\frac{\partial G}{\partial T}\right)_{P,N} = -S < 0, \quad \left(\frac{\partial G}{\partial P}\right)_{T,N} = V > 0, \quad \left(\frac{\partial G}{\partial N}\right)_{T,P} = \mu \tag{4.16}$$

where the subscripts indicate which variables are kept constant during the differentiation. By differentiating once more (4.16) we can obtain further thermodynamic relations in analogy with (4.10).

The quantity μ, called the *chemical potential*, is a very important one, because at equilibrium it must have the same value throughout the system, similarly to the temperature. Moreover, in non-equilibrium states where μ may have different values at different parts of the system, particles flow from regions of high μ to regions of low μ until equalization of μ is achieved. Finally, it is easy to show that μ is simply related to Gibbs' free energy: $G = N\,\mu(T, P)$. For photons, as well as for any other type of particles whose number is not conserved, the chemical potential is zero; this follows from the fact that the number of such particles is not an independent variable and, hence, the term $đ\, E_m = \mu\, dN$ must not be present in (4.7) and (4.8).

4.6 Extensive and Intensive Thermodynamic Quantities

Most systems studied within the framework of thermodynamics are held together by short-range forces. This is true in particular for macroscopic systems in equilibrium where the forces responsible for their existence are of electromagnetic nature. To explain this take into account that such systems, in order to be in equilibrium, have to be essentially electrically neutral locally. The proximity of equal numbers of positive and negative charges cancel the long-range character of the EM forces and renders them short-range of the order of Angstroms. However, there are systems where the forces are long-range, and the usual thermodynamic relations cannot be applied at least globally. Such systems are those where gravitational forces become important or dominant; these will be examined in Part V of this book. Another exception where long-range forces become significant is the atomic nuclei to be treated in Chap. 9.

Returning to the usual thermodynamic systems where the forces holding them together are of short range, we observe that their various properties are either proportional to the number of particles N or independent of the number of particles. Examples of the former are the internal energy, the entropy, the volume, the Gibbs free energy, etc.; examples of the latter are the pressure, the temperature, the chemical potential, etc. This observation allows us to draw some conclusions regarding the dependence of, let's say, the Gibbs free energy on its independent variables $T,\ P,\ N$. Since the Gibbs free energy is an extensive quantity it must be proportional to N; the proportionality coefficient must be an intensive quantity, which is a function of the other two intensive variables T and P and some universal constants, {u.c.}, characterizing the constituent particles. Hence G ought to have the form $G = Nf(T,P;\ \{\text{u.c.}\})$. Taking into account the last of the three equations in (4.16) we prove immediately that $f(T,P;\ \{\text{u.c.}\}) = \mu(T,P;\ \{\text{u.c.}\})$ and $G = N\,\mu(T,P)$.

We shall return to the subject of how to exploit the intensive or extensive character of the thermodynamic quantities in the questions and problem sections and in later chapters where dimensional analysis will be involved as well. In any case, notice that the combinations $N^{5/3}/V^{2/3},\ N^{4/3}/V^{1/3}$ appearing in Chap. 3 for the total kinetic energy of systems consisting of identical fermions are indeed extensive quantities.

4.7 Summary of Important Relations

The conservation of energy as expressed by the 1st Law,

$$\boxed{dU = đQ - đW + đE_m} \tag{4.3}$$

in combination with the definition of entropy,

$$\boxed{S \equiv k_B \ln \Gamma(U,\ V, N, \ldots),} \tag{4.4}$$

and the 2nd Law in the form

$$T\,dS \geq đQ, \tag{4.6}$$

leads to the basic relation

$$\boxed{dU \leq T\,dS - đW + đE_m} \tag{4.7}$$

which, after $đW$ and $đE_m$ are replaced by explicit expressions, is the starting point for the direct derivation of a large number of thermodynamic relations.

In the simple and usual case of $đW = PdV$ and $đE_m = \mu dN$, (4.7) becomes

$$dU \leq T\,dS - PdV + \mu dN \tag{4.8}$$

Starting from (4.8) with its set S, V, N of natural independent variables associated with the internal energy U, we can end up with any set of three independent variables. More explicitly, by applying the general trick of adding or subtracting one or more products of appropriate two thermodynamic quantities we can end up with the sets T, V, N or T, P, N or T, V, μ associated with $F \equiv U - TS$, or $G \equiv F + PV \equiv U - TS + PV$, or $\Omega \equiv F - \mu N \equiv U - TS - \mu N$ respectively:

$$dF \leq -SdT - PdV + \mu dN \tag{4.17}$$

$$\boxed{dG \leq -SdT + VdP + \mu dN} \tag{4.18}$$

$$d\Omega \leq -SdT - PdV - Nd\mu \tag{4.19}$$

Equation (4.19) with its set of natural independent variables T, V, μ is introduced because certain second derivatives such as $\left(\partial^2\Omega/\partial \mathrm{T}^2\right)_{V,\mu} = -(\partial S/\partial T)_{V,\mu}$ are much easier to calculate, at least in the case of non-interacting fermions.

If no change of internal entropy takes place, relations (4.8), (4.17), (4.18), (4.19) become equalities. Notice that, in the case of equalities, by taking first and second derivatives, each of (4.8), (4.17), (4.18), (4.19) produces six more thermodynamic relations by analogy to (4.9) and (4.10).

We conclude by emphasizing a point which is of central importance for what follows in this book: We possess a recipe to predict (or to justify a posteriori) the equilibrium structures of matter and their properties. Here is the recipe:

For the system under study obtain the Gibbs free energy G (under conditions of constant P and T and $đE_m = 0$), or the energy U (under conditions $P \approx T \approx 0$ and $đE_m = 0$), as a function of various other free parameters (such as the volume, the positions of the atoms or ions, the electronic concentration $n_e(\boldsymbol{r})$, etc.); minimize G or U with respect to all these free parameters; the resulting state is one of stable equilibrium to be observed in nature.

4.8 Multiple-Choice Questions/Statements

1. Which one of the following schematic graphs of S vs U (under constant V and N) is consistent with physical reality?

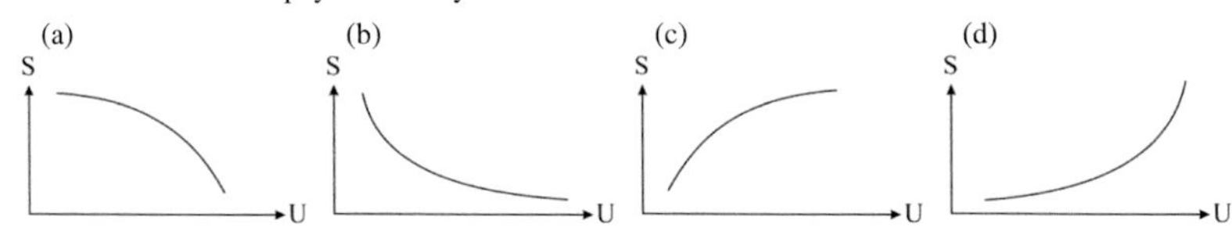

2. Which one of the following schematic graphs of G vs T (under constant P and N) is consistent with physical reality?

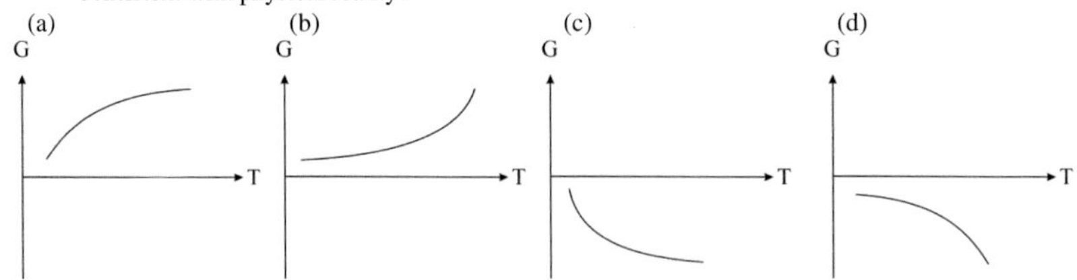

3. Which one of the following schematic graphs of G vs P (under constant T and N) is consistent with physical reality?

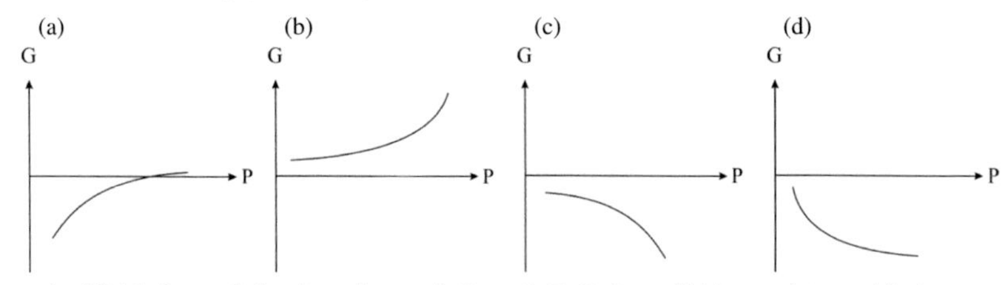

4. Which form of the dependence of U on S, V, N is explicitly consistent with the intensive/extensive character of thermodynamic quantities?
   (a) $U = Nf_1(S,V)$ (b) $U = Vf_2(N,S)$
   (c) $U = Nf_3(S/N, V/N)$ (d) $U = Nf_4(S/N, V)$
5. Which form of the dependence of F on T, V, N is explicitly consistent with the intensive/extensive character of thermodynamic quantities?
   (a) $F = Nf(T,V)$ (b) $F = (N/V)f(T,N)$
   (c) $F = Nf(T, V/N)$ (d) $F = f(N,V,T)$
6. Which form of the dependence of Ω on T, V, μ is explicitly consistent with the intensive/extensive character of thermodynamic quantities?
   (a) $\Omega = Vf(T,\mu)$ (b) $\Omega = f(V,T,\mu)$
   (c) $\Omega = Vf(T/V,\mu)$ (d) $\Omega = Vf(T,\mu/V)$
7. In the graphs below, the entropy vs internal energy for a thermaly isolated system with $đE_m = 0$ is shown together with a point P representing an initial non-equilibrium state. Which one is consistent with physical reality?

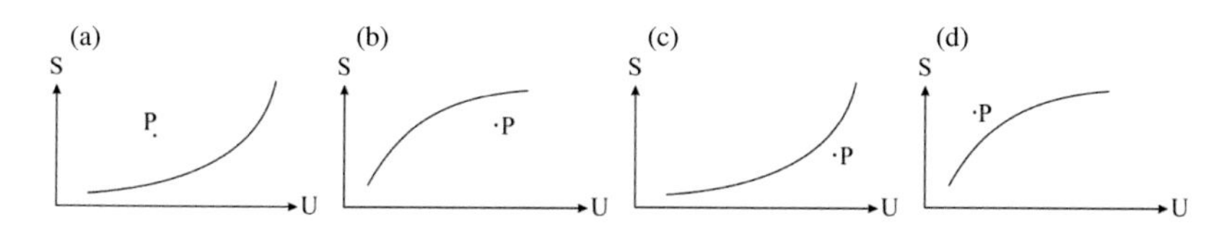

8. In the graphs of question 7 indicate the maximum work which can be obtained by exploiting the non-equilibrium initial state. ($W_{max} = U_i - U_f$)

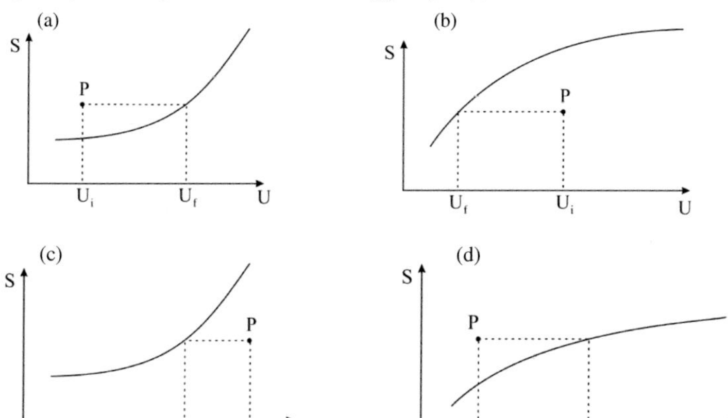

9. In the graphs below, the Gibbs free energy vs temperature (under constant pressure) is plotted for each of the three phases of matter (solid,S, liquid,L, gas,G). Taking into account the definition of entropy and the properties of *G*, indicate which graph is consistent with physical reality.

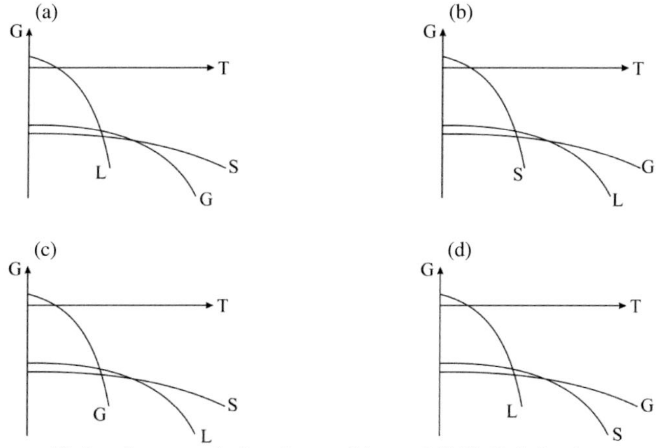

10. As equilibrium is approached under conditions of S, V, N being kept constant, the energy U is decreasing. This decrease to be consistent with energy conservation must be accompanied by one of the following processes:

(a) Work is done by the system
(b) Outflow of mass is taking place
(c) Outflow of heat occurs
(d) The system lowers its pressure

4.9 Solved Problems

1. *Using the intensive/extensive feature and (4.19) prove that* $\Omega = -PV$

Solution Assuming that the natural independent variables for Ω are only three, namely T, V, μ, and taking into account that out of these three only V is extensive, we necessarily have that the extensive quantity Ω must be of the form $\Omega = Vf(T, \mu)$. Taking into account this last relation and (4.19) we have that $f(T, \mu) = (\partial \Omega / \partial V)_{T,\mu} = -P$. QED.

2. *A homogeneous and isotropic material of volume V and of magnetic susceptibility* χ *(where by definition* $M = \chi H$ *with M being the magnetization and H the auxiliary magnetic field) is placed in a uniform externally controlled magnetic field H. It is given that* $\chi = A/T$, *where A is a positive constant and T is the temperature, and that* $đW = -VHdM$.
   Calculate (a) the entropy S(H, T) in terms of S(0, T) and in terms of the other relevant parameters, (b) the derivative $(\partial T/\partial H)_S$. *Is there any physical application of this quantity?*

Solution Equation (4.7), assuming no transfer of matter ($đE_m = 0$) and no production of internal entropy, becomes in the present case,

$$dU = TdS + VHdM \tag{4.20}$$

It is clear that both S and M are inconvenient independent variables because not only they are very difficult to control but they are not easily related to what is needed. The obvious convenient independent variables are the temperature T and the magnetic field H. The standard trick for changing from S, M to T, H is to subtract from U the sum of $TS + VHM$: $U - TS - VHM \equiv J$ and (assuming that V is constant)

$$dJ = -SdT - VMdH \tag{4.21}$$

from which it follows by taking second derivatives as in (4.10) that

$$(\partial S/\partial H)_T = V(\partial M/\partial T)_H = VH(\partial \chi/\partial T) = -VHA/T^2 \tag{4.22}$$

Integrating (4.22) with respect to H from $H = 0$ to its final value under $T = \text{const.}$ and taking into account that $A/T^2 = \chi/T$, we have

$$S(H,T) = S(0,T) - \frac{1}{2}VH^2(\chi/T) \tag{4.23}$$

The differential of (4.23) is

$$dS = [dS(0,T)/\mathrm{d}T]\mathrm{d}T + (VH^2\chi/T^2)\mathrm{d}T - (VH\chi/T)\mathrm{d}H \tag{4.24}$$

Setting $dS = 0$ in (4.5) we obtain the desired derivative

$$(dT/dH)_S = \frac{VH\chi}{C_V + (VH^2\chi/T)} \tag{4.25}$$

In (4.24) we made use of the relation $C_V = (\partial U/\partial T)_V = T(\partial S/\partial T)_V$.

Notice that the derivative in (4.25) is positive. Thus by reducing the magnetic field from its maximum value to zero under constant entropy we reduce the temperature. This demagnetization is a standard way to obtain very low temperatures of the order of a small fraction of a Kelvin (see [2], pp. 477–485).

3. *Two thermally isolated (* $đQ = 0$ and $đE_m = 0$*) identical bodies A and B are at different temperatures* $T_A > T_B$*; their specific heat C is temperature independent. The system of these two bodies is coming to thermodynamic equilibrium under the condition of extracting the maximum work through the help of a third body which undergoes cyclic operations. What is the final temperature? What percentage of the energy lost by body A became work?*

Solution Since maximum work was extracted from the system on its way to equilibrium, the total initial and the total final entropy must be equal. By integrating the relation $C = T(\partial S/\partial T)$ and taking into account that C is temperature independent we find:

$S_i = C[\ln(\mathrm{T}_A/\mathrm{T}_0) + \ln(\mathrm{T}_B/\mathrm{T}_0)] + 2S_0$ and $S_f = C[\ln(\mathrm{T}_f/\mathrm{T}_0) + \ln(\mathrm{T}_f/\mathrm{T}_0)] + 2S_0$where S_0 is the entropy of each body at the reference temperature T_0. By equating the two expressions for the total entropy we obtain

$$T_f = \sqrt{T_A T_B}$$

The energy lost by body A, $C(T_A - T_f)$, became partly work W and partly was transferred to body B, $C(T_f - T_B)$: Thus $W = C(T_A - T_f) - C(T_f - T_B) = C(T_A + T_B - 2T_f) = C(\sqrt{T_A} - \sqrt{T_B})^2$. The percentage which became work is

$$\eta = (\sqrt{T_A} - \sqrt{T_B})^2 / \sqrt{T_A}(\sqrt{T_A} - \sqrt{T_B}) = (\sqrt{T_A} - \sqrt{T_B}) / \sqrt{T_A}$$

4.10 Unsolved Problems

1. The mass exchange term $đ\, E_m$ in the general case of more than one kind of m-particles becomes $đ\, E_m = \sum_i \mu_i dN_i$. In this case prove that $G = \sum_i \mu_i N_i$
   Hint From the extensive/intensive discussion, we have the relation

   $$xG = G(T, P, xN_1, xN_2, \ldots)$$

   Take the derivative with respect to x and take into account that $dG\,)_{P,T} = d\,E_m$
2. The energy U of a system is decreasing as it is approaching its equilibrium state under the conditions $dS = dV = dN$. Argue why this decrease does not violate the law of conservation of energy.
3. Prove (4.13). *Hint*: Take into account that $\Delta U = -W + Q_{b\leftarrow e} - W_{b\rightarrow e} \Rightarrow -W = \Delta U - Q_{b\leftarrow e} + W_{b\rightarrow e} = \Delta U + Q_{e\leftarrow b} - W_{e\rightarrow b}$ where $Q_{e\leftarrow b} = T_0 \Delta S_0 \geq -T_0 \Delta S$, $W_{e\rightarrow b} = P_0 \Delta V_0 = -P_0 \Delta V$. Substituting these last relations in the expression for $-W$ and multiplying by -1 we have $W \leq -\Delta U + T_0 \Delta S - P_0 \Delta V = -\Delta(U + P_0 V - T_0 S) = A_{\text{initial}} - A_{\text{final}}$

4.11 The Three Phases of Matter (Solid (*s*), Liquid (*l*), Gas (*g*))

In the next page we plot the Gibbs free energy vs. temperature T for the three phases of matter under constant pressure P: (a) very low pressure, (b) intermediate pressure, (c) very high pressure. Justify the following relations and then the plots of next page:

$$G_s(P) < G_l(P) \ll G_g(P), \quad T = 0; \quad S_s(P,T) < S_l(P,T) < S_g(P,T) \tag{4.26}$$

$$(\partial G/\partial T)_P = -S < 0; \quad \left(\partial^2 G/\partial T^2\right)_P = -C_p/T < 0; \quad (\partial G/\partial P)_T = V > 0 \tag{4.27}$$

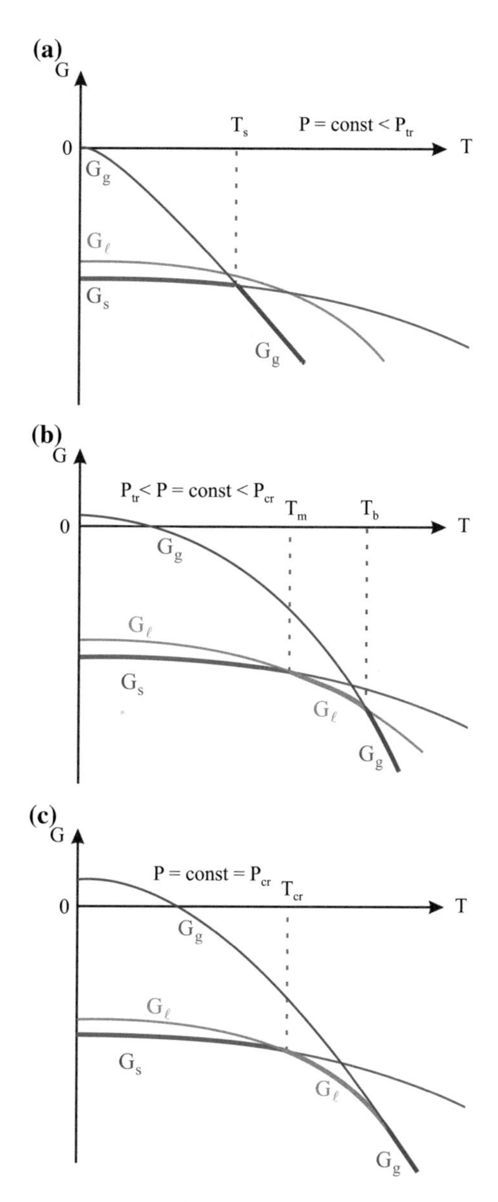

Case (a): The lower value of G in the range $0 \leq T < T_s$ is for the solid phase (which, therefore, is the equilibrium one). At $T = T_s$ there is a transition directly from the solid phase to the gas one and, therefore, for $T_s < T$ the lowest G phase (i.e. the equilibrium one) is the gas phase. As the pressure is raised, the G_g is increasing much faster than the other two, which they hardly change (why?). Thus at some pressure denoted by P_{tr} the three curves have a common crossing point called triple point at which all three phases coexist. Further increase of pressure leads to the case (b), where for $0 < T < T_m$ the solid phase is the equilibrium one, at $T = T_m$ solid and liquid phase coexist and for $T_m < T < T_b$ the liquid phase is the equilibrium one; at T_b there is a transition to the gas phase. At an elevated pressure denoted by P_{cr} there is no crossing anymore of the curves G_l, G_g as in case (b) but a smooth merging as in (c) indicating physically that beyond this point (called critical point) there is no distinction between the liquid phase and the very compressed gas phase.

Summarizing this description which is based exclusively on (4.26, 4.27) we have the following phase diagram in the plane T, P:

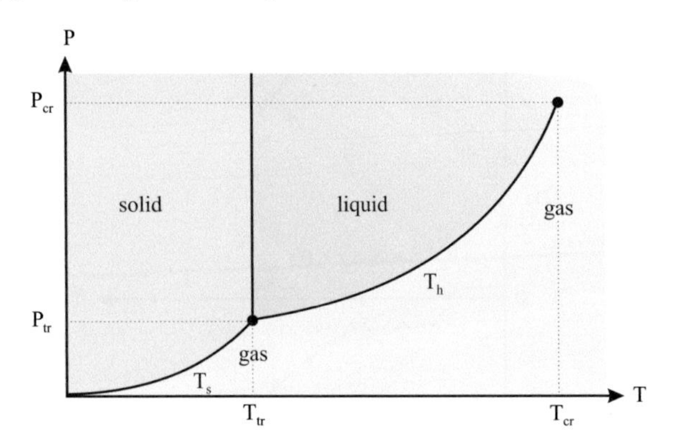

What is the physical reason for the T_m curve (on which the solid and the liquid phase coexist) to be almost vertical?

By considering two neighboring points on any curve on which two phases coexist (i.e. have the same G) we end up with the Clayperon-Clausius formula for the slope of any of these curves:

$$\frac{dP}{dT} = \frac{S_2 - S_1}{V_2 - V_1} = \frac{T\,\Delta S}{T\,\Delta V} = \frac{Q}{T\,\Delta V} \tag{4.28}$$

Given that ice is lighter than water, how we can freeze water of pressure 1 atm and temperature 0.1 °C? By increasing or decreasing the applied pressure?

References

1. L.D. Landau, E.M. Lifshitz, *Statistical Physics*, Part 1 (Pergamon Press, Oxford 1980)
2. E. Fermi, *Thermodynamics* (Dover Publications, NY, 1956)
3. M.W. Zemansky, R.H. Dittman, *Heat and Thermodynamics*, 7th edn. (McGraw-Hill, New York, 1997)

Chapter 5
Dimensional Analysis: A Short-Cut to Physics Relations

Abstract The method of dimensional analysis, which under certain conditions is capable of producing physics formulae in an amazingly quick way, is presented. The method is based on *identifying* and creating the proper combination of the relevant quantities and physical constants on which the quantity under examination may depend. The dimensions of the latter ought to be the same as those of this proper combination. In this chapter several examples of the application of dimensional analysis will be given to familiarize the reader with this powerful method. Throughout this book extensive use of dimensional analysis will be made in order to stimulate physical insight and allow derivations rather than memorizations of formulae.

5.1 Outline of the Method

Under certain circumstances dimensional analysis allows us to determine how a given quantity X depends on the "relevant" physical constants and independent variables. As the reader may have guessed, the crucial and far from trivial task is to identify all the "relevant" quantities on which the value of the given quantity may depend. Once this task has been completed, the next step is to find the more general combination of these "relevant" quantities having the same dimensions (i.e., the same powers of length, time, and mass) as the given quantity.

To be more explicit we start with the observation that the units of any physical quantity can be written in terms of the units of the fundamental quantities length, time, mass (in the system Gauss-CGS) and, additionally, electric current (in the system SI).[1]

$$[\text{units of any physical quantity}\, X] = [\text{time unit}]^a [\text{length unit}]^b [\text{mass unit}]^c \quad (5.1)$$

[1] One main difference between the two systems is the way Coulomb's law is written: in G-CGS $F = q_1 q_2 / r^2$, while in SI $F = q_1 q_2 / 4\pi\, \varepsilon_0 r^2$ where ε_0 is the permittivity of the vacuum.

E.N. Economou, *From Quarks to the Universe*,
DOI 10.1007/978-3-319-20654-7_5

Thus the 'dimensions' of every physical quantity are determined by the three numbers [*a, b, c*] (in G-CGS) or four numbers [*a, b, c, d*] (in SI), where *d* refers to the power of the electric current:

$$[X] = [\text{time}]^a [\text{length}]^b [\text{mass}]^c, \text{ (G-CGS)} \tag{5.1a}$$

$$[X] = [\text{time}]^a [\text{length}]^b [\text{mass}]^c [\text{current}]^d, \text{ (SI)} \tag{5.1b}$$

where the equality of bracketed quantities means equality of their dimensions. Assume now that there is a physical relation of the form $A = B$ or $A > B$ or $A < B$. Then, obviously, as far as the dimensions are concerned, $[A] = [B]$. Let us assume next that we have identified all possible quantities $A_1, \ldots, A_n$ on which the quantity of interest, X, may depend and let us further assume that all of them are of different dimensions. (If two or more of them, e.g. A_5 and A_7 are of the same dimensions, we form dimensionless ratios, e.g. A_5/A_7).

Then, if $n \leq 3$ (in the G-CGS system) or $n \leq 4$ (in the SI system) there is usually a *unique* combination of the form $c_1 A_1^{\mu_1} \ldots A_n^{\mu_n}$ which has the same dimensions as X. Thus, because of this uniqueness, the formula for X has been determined up to the quantity c_1:

$$X = c_1 A_1^{\mu_1} \ldots A_n^{\mu_n} \tag{5.2}$$

where c_1 is a function of all dimensionless quantities (if there is none, c_1 is a numerical factor) and all $n \leq 3$ or $n \leq 4$ quantities $A_1, \ldots A_n$ are of different dimensions.

If $n > 3$ (in the G-CGS system) we choose three quantities A_1, A_2, A_3 which define a system of units (in the sense that three different combinations of the form $A_1^{\mu_1} A_2^{\mu_2} A_3^{\mu_3}$ can be found which have the dimensions of length, time, and mass respectively). Then we have to perform the following steps:

(1) By a proper choice of ν_1, ν_2, ν_3 we form the combination $A_1^{\nu_1} A_2^{\nu_2} A_3^{\nu_3} \equiv X_o$, where X_o has the same dimensions as the quantity of interest X.
(2) We create also combinations $A_1^{\xi_{1n}} A_2^{\xi_{2n}} A_3^{\xi_{3n}} \equiv A_{no}$ which have the same dimensions as A_n, ($n = 4, 5, \ldots$) and we define the dimensionless quantities A_n/A_{no} ($n = 4, 5, \ldots$).
(3) Having determined the quantities X_o, A_{40}, A_{50},... in terms of the chosen quantities A_1, A_2, A_3, we are ready to express X in terms of $A_1, \ldots A_n$ as follows:

$$\boxed{X = X_0 f\left(\frac{A_4}{A_{40}}, \frac{A_5}{A_{50}}, \ldots\right)} \tag{5.3}$$

where the unknown function f cannot be determined from dimensional analysis; additional information or even a complete physical theory is needed to find out what f is. A similar approach is followed in the SI system with $n > 3$ replaced by $n > 4$.

Several examples of the method of dimensional analysis are presented in this chapter demonstrating its extraordinary power.

5.2 Relations Regarding Some Eigenfrequencies

Determine by dimensional analysis the natural frequency (called eigenfrequency) of oscillation of a pendulum (Fig. A.1a in Appendix A).

For the pendulum case the relevant quantities seem to be: The length l of the string, the mass m, the acceleration of gravity g (since it is gravity that provides the potential energy; in its absence the pendulum would be in equilibrium for any angle between the string and the perpendicular z-axis, since in this case no force would be present); a not so obvious fourth parameter is the dimensionless *maximum* angle θ between the string and the z-axis. The only set of values of μ_1, μ_2, μ_3 which will give to the combination $l^{\mu_1} g^{\mu_2} m^{\mu_3}$ dimensions of one over time (as that of frequency) is $\mu_1 = -\frac{1}{2}$, $\mu_2 = \frac{1}{2}$, $\mu_3 = 0$. Hence, $\omega = \sqrt{g/l} f(\theta)$, where $f(\theta)$ is a function which cannot be determined from dimensional analysis. In the limit of small angle θ $(\theta \to 0)$ this function tends to $f(0)$, i.e. to a constant. A detailed calculation shows that this unavoidable numerical factor $f(0)$ turns out to be equal to one.

Determine by dimensional analysis the natural frequency (called eigenfrequency) of oscillation of an LC circuit (Fig. A.1b in Appendix A).

For the LC circuit the obvious relevant quantities are the capacitance C and the self-inductance L, both with dimensions of length in the G-CGS system; in this system a third relevant quantity is needed in order to end up with the required dimension of frequency, i.e. of time to minus one. This is obviously the velocity of light c (since we are dealing with an E/M phenomenon; the latter is governed by Maxwell's equations which in the G-CGS system contain the velocity of light; see Appendix B). The quantity c/L has dimension of time to minus one, the same as frequency. However, given that the ratio C/L is dimensionless, the result of the dimensional analysis within the G-CGS is $\omega = (c/L)\,\mathrm{f}\,(C/L)$, which leaves undetermined an arbitrary function of the ratio C/L.

Let us try the other system, the SI, to see if it will do better than the G-CGS: In SI the dimension of the capacitance is permittivity ε times length and the dimension of self-inductance is permeability μ times length. It is clear that we need the dimensions of ε and μ which can be found from Coulomb's law in SI, $[F] = [q]^2[\varepsilon]^{-1}[l]^{-2}$ and the law of force per unit length in SI between two currents $[F][l]^{-1} = [\mu][I]^2[l]^{-1}$ respectively. From these two relations the reader may deduce the dimensions of ε and μ as well as two equivalent relations for ε and μ which are much easier to remember: The product of permittivity times permeability has dimensions of inverse velocity squared and the ratio permeability/permittivity has dimensions of resistance squared. In particular,

$$\boxed{\varepsilon_0\mu_0 = \frac{1}{c^2}, \quad \text{and} \quad \sqrt{\frac{\mu_0}{\varepsilon_0}} = Z_0 = 376.7\,\text{ohms}} \tag{5.4}$$

where c is the velocity of light in vacuum and Z_0 is the so-called impedance of the vacuum; the latter is indicated here by the subscript 0. Now, if we assume that the eigenfrequency depends only on L and C, we find by dimensional analysis that

$$\omega = c_1/\sqrt{LC},$$

where c_1 is a numerical factor which turns out to be one, $c_1 = 1$. Thus it follows that the function $\mathrm{f}\,(C/L)$ in the G-CGS result is equal to $\mathrm{f}\,(C/L) = \sqrt{L/C}$.

5.3 Some Relations in Fluid Dynamics

Derive Poiseuille's formula which gives the flow rate Π (volume of fluid passing per unit time) of an incompressible fluid flowing through a pipe of circular cross-section of radius r in the presence of friction.

In order to establish and maintain a steady flow in the presence of friction a pressure difference $\delta P(l)$ between the ends of the pipe is needed. The longer the length l of the pipe is, the larger this pressure difference must be. To be more specific, the required pressure difference $\delta P(l)$ must be proportional to the length l, since the friction forces act uniformly along the length of the pipe. Thus the gradient of pressure, $\delta P(l)/l$, is the relevant quantity which compensate the friction; the latter depends on a characteristic property of the fluid called viscosity η with dimensions of pressure $\times$ time. (Physically the friction forces arise because neighboring layers of fluid move with different velocities; in other words it is the existence of a *velocity gradient* which produces the friction. In the present case such a velocity gradient appears since at the center of each cross-section of the pipe the velocity of the fluid is maximum, while at the walls of the pipe the velocity is zero. It turns out that the friction force per unit area is proportional to viscosity $\times$ velocity gradient). Thus the quantity Π seems to depend on only three parameters: $\delta P(l)/l$, η, and r. Hence, following the general recipe (5.2) we have

$$[\Pi] \equiv [l]^3[t]^{-1} = [\delta P/l]^n[\eta]^m[r]^k \equiv [P]^n[l]^{-n}[P]^m[t]^m[l]^k$$

which implies $n + m = 0$, $k - n = 3$, and $m = -1$. So $n = 1$, $m = -1$, $k = 4$, and

$$\Pi = c_1\frac{\delta P(l)}{l}\frac{r^4}{\eta} \tag{5.5}$$

where a detailed calculation gives for the numerical constant c_1 the value $\pi/8$. Notice that the quantity of liquid passing through the pipe per unit time is proportional to the square of its cross-sectional area and not proportional to its first power as one naively might expect.

Derive the formula for the drag force acting on a solid object as it moves with a constant velocity υ within a fluid of viscosity η.

Obviously the drag force must depend on the velocity of the solid body (no velocity, no force), on its size and shape, and on the properties of the fluid such as viscosity and density. The continuation of the motion requires expenditure of energy because of the presence of friction. The latter is due to two distinct physical mechanisms:

The first one is due to the presence of a velocity gradient within the fluid, as in the case of the flow in a pipe. This velocity gradient comes because the fluid at the immediate vicinity of the body is carried on with it and, hence, it moves with the velocity of the body; on the other hand, the parts of the fluid far away from the body have zero velocity. As was discussed in the previous example, this velocity gradient multiplied by the viscosity η of the fluid and an area of the body produces a frictional force. This type of drag force must obviously depend on η, the velocity of the body υ and an area A; this area has to do with the body's size and shape. Hence, this type of drag force must be given by a relation of the form

$$F_1 = c_1 \eta^a \upsilon^b A^c \tag{5.6}$$

where the equality of dimensions in both sides of (5.6) leads to the following values of the exponents: $a = 1, \; b = 1, \; c = 1/2$; the dimensionless numerical factor c_1 depends on the shape of the body. For a sphere of radius r, c_1 is equal to 6π, if we chose $A^{1/2} = r$.

The second physical mechanism giving rise to friction is due to the kinetic energy continuously offered to the fluid by the moving body without the latter retrieving it (the velocity of the body is maintained by the work done on the body by an external force compensating this type of drag force). In this case the drag force depends on the density ρ of the fluid (since the kinetic energy of the fluid per unit volume is proportional to the density $\times$ velocity squared), on the cross-sectional area A of the body, and, of course, on its velocity υ. Hence

$$F_2 = c_2 \rho^k \upsilon^m A^n \tag{5.7}$$

where the equality of dimensions in both sides of (5.7) leads to the following values of the exponents: $k = 1, \; m = 2, \; n = 1$. The dimensionless numerical factor c_2, which is usually written as $\frac{1}{2}C_D$, depends on the dimensional ratio $\eta / \left(\rho \upsilon \sqrt{A}\right)$ and the shape of the body. In the limit where this ratio is approaching zero and for a sphere of radius r and area $A = \pi r^2, \; C_D$ is about equal to 0.48.

The ratio of the two types of drag forces (by omitting the coefficients $c_1, \; c_2$) is the so-called Reynolds number R_e:

$$R_e \equiv \frac{F_2}{F_1} = \frac{\rho \upsilon^2 A}{\eta \upsilon \sqrt{A}} = \frac{\upsilon \sqrt{A}}{\eta/\rho} = \frac{\upsilon \sqrt{A}}{\nu} \tag{5.8}$$

Notice that the Reynolds number is the inverse of the ratio $\eta/(\rho\upsilon\sqrt{A})$. The quantity $\nu \equiv \eta/\rho$ is called kinematic viscosity. For water the viscosity under normal conditions in units of $\mathrm{kg\,m^{-1}s^{-1}}$ is 0.001 and the kinematic viscosity in units of $\mathrm{m^2s^{-1}}$ is 10^{-6}. The corresponding numbers for air are 1.8×10^{-5} and 1.5×10^{-5}.

The viscosity of blood at 37 °C is $2.7 \times 10^{-3}\mathrm{kg\,m^{-1}s^{-1}}$.

Equation (5.8) shows that the second mechanism dominates for large bodies and large velocities ($R_e \geq 1000$), while the first mechanism dominates for small bodies and small velocities ($R_e < 0.2$). In the intermediate regime, $0.2 < R_e < 1000$, where both mechanisms contribute, we are usually writing the *total* drag force in the form of (5.7) with c_2 written as $\frac{1}{2}C_D$, but with C_D being now a function of the Reynolds number. This function starts from a value proportional to $1/R_e$ at the lower end of this regime (as to recover (5.6)) and is monotonically decreasing up to the point $R_e \approx 1000$ where the second mechanism fully dominates.

5.4 Thermodynamic Relations Revisited

The dimensional analysis combined with the extensive or intensive character of thermodynamic quantities allows us to pin down further their formulae. For example the Gibbs free energy which is a function of two intensive quantities (T and P) and one extensive (N) ought to have the form

$$G = N\varepsilon f\left(\frac{k_B T}{\varepsilon}, \frac{Pa^3}{\varepsilon}\right) \tag{5.9}$$

where f is an undetermined function of the two dimensionless intensive quantities shown in (5.9). Indeed, by making G proportional to $N\varepsilon$ (where ε is a characteristic energy per particle) we made sure that G is an extensive quantity having dimensions of energy (as it should), provided that the rest of the independent variables are intensive (as not to jeopardize the already insured extensive character of G) and dimensionless (in order not to jeopardize the already energy dimensions of G); a^3 is a characteristic volume per particle. We expect ε and a to be related with universal constants such as $\hbar$ and e (if E/M interactions are significant) and particle characteristics such as their mass m.

Let us present one more example of thermodynamic quantity, namely that of the entropy S, which is extensive and a function of N, U, V and has the same dimensions as Boltzmann's constant k_B. The extensive character combined with dimensional analysis gives:

$$S = N\,k_B f\left(\frac{U}{N\varepsilon}, \frac{V}{Na^3}\right) \tag{5.10}$$

For a perfect atomic or molecular gas, where there are no interactions, we expect the parameters ε and a to be related as follows: $\varepsilon a^2 \propto \hbar^2/m$ so that only the combination $\hbar^2/m$ (besides the natural independent variables and k_B) to enter in (5.9) and (5.10).

5.5 Waves in Extended, Discrete or Continuous, Media

Determine by dimensional analysis the velocity of propagation υ (phase velocity) of a longitudinal wave in the system of coupled pendulums (see Fig. A.2c in Appendix A).

By simple inspection of Fig. A.2c in Appendix A we conclude that υ must depend on the lengths l and a, on the mass m and the spring constant κ; it is expected to depend also on the acceleration of gravity g and possibly on the wavenumber k. The so-called phase velocity υ, is by definition equal to ω/k, where ω, the eigenfrequency, is the common frequency of oscillation of all the masses m. Before we proceed we observe that there are two additive contributions to the potential energy: One is due to gravity and the other to the springs. Since the frequency squared is proportional to the kinetic energy which in turn is proportional to the potential energy, we expect ω^2 to be also additive, $\omega^2 = \omega_g^2 + \omega_\kappa^2$, where ω_g^2 would be the frequency squared, if only gravity were present and the springs were absent; ω_κ^2 would be the frequency squared, in the absence of gravity. We have already shown above that $\omega_g^2 = g/l$. It remains to obtain ω_κ^2 which depends on four quantities: $\kappa,\ m,\ a,$ and k; we observe that κ/m has dimensions of frequency squared and the combination $a\,k$ is dimensionless. Hence the most general expression for ω_κ^2 is of the form: $\omega_\kappa^2 = (\kappa/m)f(a\,k)$. As far as dimensional analysis is concerned, f is an arbitrary positive definite function of ak. From physical considerations we expect that $f \to 0$ as $k \to 0$, since in this limit all masses move in phase and the springs are inactive. The maximum value of f is obtained when neighboring masses are completely out of phase; this occurs when $\lambda/2 = a$, or, equivalently, when $a\,k = \pi$. The function $\sin^2(a\,k/2)$ satisfies all these requirements for f. Thus, it is not unlikely for someone to guess that f may be proportional to $\sin^2(a\,k/2)$. It turns out that this is actually the case. So

$$\omega^2 = \omega_g^2 + c_1\frac{\kappa}{m}\sin^2\left(\frac{a\,k}{2}\right) \tag{5.11}$$

A detailed calculation shows that the numerical constant c_1 is equal to 4. In the absence of gravity and for $a\,k \ll 1$ (i.e. for wavelengths much larger than a) we have $\omega = \left(\sqrt{\kappa/m}\right) a\, k = \left(\sqrt{(\kappa\, a)/(m/a)}\right) k$; hence, in this case, the velocity of propagation, $\upsilon = \omega\, /\, k$, of the wave depends only on the properties of the system and not on the wavelength:

$$\upsilon = \sqrt{\frac{B}{\rho}}, \quad B = \kappa a, \quad \rho \equiv \frac{m}{a} \tag{5.12}$$

where B is the one-dimensional (1-D) analog of the so-called bulk modulus (which is the inverse of the compressibility) and ρ is the 1-D mass density. We mention here that the sound velocity in fluids is indeed given by the formula we just derived:

$$\upsilon = \sqrt{\frac{B}{\rho}} \tag{5.13}$$

Determine by dimensional analysis the velocity of propagation υ *of sea waves.*

This is another example where there are two additive contributions to the potential energy: One is the gravitational potential energy and the other is the potential energy due to surface tension; both reach their minimum equilibrium value when the surface becomes flat. As it was argued before, the additivity of the potential energies implies the additivity of the squares of the frequencies: $\omega^2 = \omega_g^2 + \omega_s^2$. The gravity part ω_g^2 depends for sure on the acceleration of gravity g, possibly on the wavenumber k, and the depth of the sea d; the density ρ does not enter, because its presence would introduce the dimension of mass only on the right side of the equation for ω_g^2. The combination $g\, k$ has dimensions of inverse time squared and the product $k\, d$ is dimensionless. Hence, the most general expression for ω_g^2 is of the form $\omega_g^2 = g\, k f_g(k\, d)$, where the function f_g cannot be determined from dimensional analysis. The surface tension part ω_s^2 depends, obviously, on the surface tension coefficient σ, possibly on the wavenumber k, on the depth of the sea d, and on the density ρ. The combination $\sigma^{\nu_1} k^{\nu_2} \rho^{\nu_3}$ with $\nu_1 = 1,\ \nu_2 = 3,\ \nu_3 = -1$ has dimensions of inverse time squared and the product $k\, d$ is dimensionless. Thus the most general expression for ω_s^2 has the form: $\omega_s^2 = (\sigma\, k^3/\rho) f_s(k\, d)$. The functions $f_g(k\, d)$ and $f_s(k\, d)$ must be reduced to a constant in the limit of very large $k\, d$, since, when the depth of the sea is much larger than the wavelength, the depth d plays no role at all; this constant turns out to be equal to one for both $\mathrm{f_g}$ and $\mathrm{f_s}$. Thus

$$\upsilon^2 \equiv \frac{\omega^2}{k^2} = \frac{g}{k} + \frac{\sigma\, k}{\rho}, \quad k\, d \equiv \frac{2\pi d}{\lambda} \gg 1 \tag{5.14}$$

In the opposite limit of $kd \ll 1$ i.e., very long wavelengths, it is expected that ω_s^2 will be negligible in comparison with ω_g^2 (since $\omega_s^2/\omega_g^2 \propto 1/\lambda^2$) and that the depth of the sea d will replace the $(\lambda/2\pi) = (1/k)$ in g/k as the only relevant length.. For

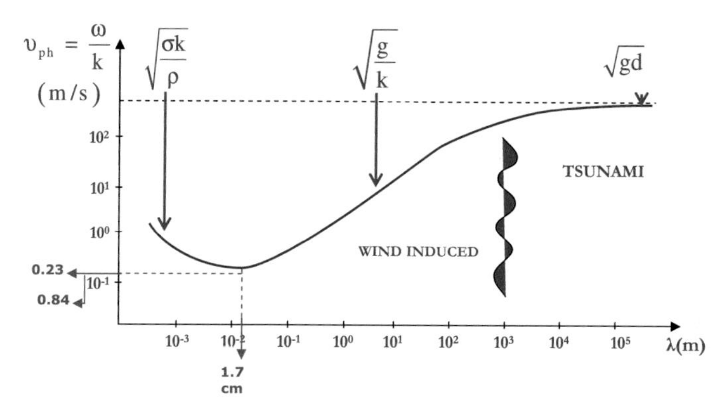

Fig. 5.1 Sea wave velocity versus wavelength. Very short wavelength waves are controlled by surface tension. Very long wavelength waves, as in tsunamis, propagate very fast; e.g. for $d = 2.5$ km, and $\lambda \gg d$, $\upsilon = 158$ m/s $= 569$ km/h. Minimum velocity $\upsilon_{min} = 0.23\,\mathrm{m/s} = 0.84\,\mathrm{km/h}$

this to happen $f_g(kd)$ must be proportional to kd in this limit as to eliminate the k dependence from υ; the proportionality constant turns out to be equal to one. So we have

$$\upsilon^2 = g\,d, \quad k\,d \equiv \frac{2\pi d}{\lambda} \ll 1 \tag{5.15}$$

Equation (5.15) gives the velocity of propagation of a tsunami, which usually has a wavelength much larger than the depth of the sea, d. A detailed calculation shows that $f_g(k\,d) = f_s(k\,d) = \tanh(k\,d)$; this function reproduces the two limits obtained by simple physical considerations. In Fig. 5.1 we plot the velocity υ of see waves versus their wavelength λ, taking into account that the surface tension coefficient for water at 293 K is equal to 0.073 J/m^2. Such a plot explains why the wind, as it starts blowing, will set up first sea waves of wavelength of the order of centimeters.

5.6 Summary of Important Formulae

The total drag force acting on a solid body of cross-sectional area A moving with velocity υ within a fluid of density ρ and viscosity η is given by the formula

$$F_D = \frac{1}{2} C_D \rho A \upsilon^2 \tag{5.16}$$

where C_D is a dimensionless quantity which in general is a function of the so-called Reynolds number $R_e \equiv \upsilon\sqrt{A}/\nu$; $\nu \equiv \eta/\rho$. For small values of the Reynolds number $(R_e \leq 1)$, $C_D = 2c_1/R_e$, as to recover (5.6), while for $R_e \geq 1000$, C_D does not depend on R_e as to recover (5.7). (An exception appears for R_e around 10^5 to 10^6, the so-called drag crisis).

The eigenfrequency of a wave propagating in the system of coupled pendulums (see Fig. A.2c in Appendix A) depends on the wavevector $k \equiv 2\pi/\lambda$ as follows:

$$\omega^2 = \omega_g^2 + 4\frac{\kappa}{m}\sin^2\left(\frac{a\,k}{2}\right) \tag{5.17}$$

where $\omega_g^2 \equiv g/l$. Equation (5.17) for $a\,k \ll 1$ and in the absence of gravity reduces to $(\omega/k) = \sqrt{\kappa\,a/\rho}$, $\rho \equiv m/a$ which immediately implies that the velocity $\upsilon \equiv \omega/k$ is $\sqrt{\kappa\,a/\rho}$. This is generalized to give the velocity of sound in fluids:

$$c_{sound} = \sqrt{\frac{B}{\rho}} \tag{5.18}$$

where B is the bulk modulus defined as $B \equiv -V(\partial P/\partial V)_S$.

Equation (5.17) is important because in the absence of gravity it describes the wavy ionic motions in a solid. Moreover, as we shall see in Sect. 12.8 of Chap. 12, it can also describe (by applying certain analogies) the wavy electronic motion in a solid. Thus a single classical equation, namely that of coupled pendulums, can provide the foundation for the quantum behavior of electrons and ions in the solid state of matter.

5.7 Multiple-Choice Questions/Statements

1. A high-frequency current running along a conducting wire is concentrated mostly near the surface of the wire up to a depth δ (skin-depth). Given that the permeability of the wire is μ and that its conductivity σ is very high $(\mathrm{Re}\{\sigma\} \gg \omega\mathrm{Re}\{\varepsilon\})$ the skin-depth is (For a detailed solution see solved problem 3 in Sect. 6.11, p.87)
(a) $\delta \propto c/\omega$ (b) $\delta \propto 1/\omega\sqrt{\varepsilon\mu}$ (c) $\delta \propto 1/\sqrt{\omega\sigma\mu}$ (d) $\delta \propto \lambda$

2. The total drag force on a car of weight B, cross-section A, running with 100 km/h is
(a) $F = c_1\eta\upsilon\sqrt{A}$ (b) $F = c_2\rho\upsilon^2 A$ (c) $F = c_1 B\upsilon/c_{sound}$ (d) $F = c_2\rho\,A\upsilon c_{sound}$

3. The radius of the large artery from the heart to the lungs is about 2 mm for a normal person. With age this radius may be reduced by 10%. In order for the heart to deliver the same volume of blood per unit time to the lungs the pressure difference must increase by about
(a) 10% (b) 24% (c) 52% (d) 82%

4. The radius of the large artery from the heart to the lungs is about 2 mm for a normal person and its length is about 10 cm. The viscosity of the blood is 0.0027 in SI units. With a pressure difference of about 400 Pa between the heart and the lungs the volume (in cm^3) of blood delivered to the lungs per second is approximately
(a) 0.093 (b) 0.93 (c) 9.3 (d) 93

5. Consider a 'spherical' bacterium of radius $r = 1\,\mu m$ moving in a water solution with speed of about 10μm/s. The drag force in pN is about
(a) 25 (b) 2.5 (c) 0.25 (d) 0.025

6. The propagation velocity of an ordinary sea wave of wavelength λ such that $1\,m \leq \lambda \ll d$, where d is the depth of the sea is given by
(a) $\upsilon = \sqrt{g\,d}$ (b) $\upsilon = \sqrt{g\,\lambda / 2\pi}$ (c) $\upsilon = \sqrt{g\,k}$ (d) $\upsilon = c_{sound}$

7. The propagation speed of a tsunami such that $d \ll \lambda$, where λ is its wavelength and d is the depth of the sea is given by
(a) $\upsilon = \sqrt{g\,d}$ (b) $\upsilon = \sqrt{g\,\lambda}$ (c) $\upsilon = \sqrt{g / k}$ (d) $\upsilon = c_{sound}$

8. Consider two parallel perfect metallic plates at a distance d apart, where d is much smaller than the linear extent of the plates. Although the plates do not carry any electric charge there is an attractive force between them due to quantum fluctuations of the EM field. This force per unit area (i.e. the pressure) is given by one of the following relations:
(a) $P \propto \hbar c / d^4$ (b) $P \propto \hbar^2 c^2 / e^2 d^4$
(c) $P \propto G M^2 / A^2$ (d) $P \propto e^2 / d^4$

5.8 Solved Problems

1. *Obtain the thermodynamic quantities for a perfect monoatomic gas.*
Solution We shall start with the entropy which is directly related with a microscopic quantity, namely the number of microscopic states. We select randomly one atom out of the N present in the gas. Classically its state is determined by specifying both its position $\mathbf{r}$ within the volume V and its momentum $\mathbf{p}$; hence the number of possible states this single atom may have is proportional to the product of volume in real space × volume in momentum space, i.e. proportional to $V \times p^3$. The magnitude of the maximum momentum p is related with the maximum energy u a single atom may possess; u is essentially proportional to the average energy per particle U/N. Since $u = p^2/2m \propto U/N$, we conclude that the number $\Gamma(1)$ of possible states a single atom may have classically is proportional to $V \times (m\,U/N)^{3/2}$. Since each of the possible states of a single atom can be combined with any state of a second atom and so on we can conclude that $\Gamma(\mathrm{N}) = \Gamma(1)^N$. Combining this last result with the definition of entropy, (4.3), we have that $S = Nk_B \ln\{\mathrm{const.}\, V(mU/N)^{3/2}\}$. However, according to (5.10) the variable V must necessarily enter in the (dimensionless) combination $V/N\,a^3$ and the

variable $(mU/N)^{3/2}$ in the (dimensionless) combination $(mU/N\,m\,\varepsilon)^{3/2}$. The resulting extra factors $N^{-1}(m\,\varepsilon\,a^2)^{-3/2}$ have dimensions of (energy × time)$^{-3}$ and hence is proportional to $\hbar^{-3}$, the only relevant universal constant having dimensions of energy × time. Making these obligatory corrections to the classical result we obtain

$$S = Nk_B\left\{\ln\left(\frac{V}{N}\right) + \frac{3}{2}\ln\left(\frac{U}{N}\right) + \frac{3}{2}\ln\left(c_1\frac{m}{\hbar^2}\right)\right\} \tag{5.19}$$

where c_1 is a constant numerical factor which turns out to be equal to $e^{5/3}/3\pi$, and e is Euler's number, e = 2.7182....By taking the derivative of S with respect to U we obtain

$$\frac{1}{T} = \left(\frac{\partial S}{\partial U}\right)_{N,V} = \frac{\frac{3}{2}N\,k_B}{U} \Rightarrow U = \frac{3}{2}N\,k_B T \tag{5.20}$$

By taking the derivative of S with respect to V we obtain the pressure

$$\frac{P}{T} = \left(\frac{\partial S}{\partial V}\right)_{N,U} = \frac{N\,k_B}{V} \quad \Rightarrow \; P\,V = N\,k_B T \tag{5.21}$$

The Helhmoltz free energy F can then be determined from $F = U - TS$ and the Gibbs free energy from $G = F + PV$. (The reader may easily obtain the specific heats as well).

$$G = N\mu = N\,k_B T\left\{\ln P - \frac{5}{2}\ln T - \frac{3}{2}\ln\left(m\,k_B^{5/3}/2\pi\hbar^2\right)\right\} \tag{5.22}$$

The above dimensional correction factor $N^{-1}(m\,\varepsilon\,a^2)^{-3/2}$ to the classical result is due to Quantum Mechanics, as the appearance of $\hbar$ attests. Indeed, instead of defining a single particle state by one point in real space and one point in momentum space, as classical physics dictates, we ought to use quantum mechanics, namely the uncertainty relation, which imposes the restriction that a single state is determined by the product of a non-zero region Δr^3 in real space and a non-zero region Δp^3 in momentum space such that $(\Delta r \times \Delta p)^3$ to be of the order of $\hbar^3$. Hence the number of states $\Gamma(1)$ of a single atom, because of Heisenberg's principle, becomes $\Gamma(1) = \left\{V \times \left(\frac{mU}{N}\right)^{3/2}\right\}/(\Delta r \times \Delta p)^3 \propto \left\{V \times \left(\frac{mU}{N}\right)^{3/2}\right\}/\hbar^3$. Moreover, again because of Pauli's principle, since the atoms are identical and they share the same volume, they are indistinguishable; as a result, any of the $N!$ ways of renaming the N atoms does not change $\Gamma(N)$, which means that the classical result for $\Gamma(N)$ overcounts the number of microstates of the system by a factor of $N!$ Thus the actual number of microstates in accordance with Pauli's principle is $\Gamma(N) = \Gamma(1)^N/N! \approx \{e\Gamma(1)/N\}^N$; in the last relation we took into account that $N! \approx (N/\mathrm{e})^N$ where e is Euler's number, e = 2.7182…. Combining these quantum corrections we obtain again (5.19).

2. *Estimate the life-time of a classical model of a hydrogen atom. Assume that, if there were no radiation, the electron will follow a circular orbit of radius a_B.*

Solution An electron behaving as a classical particle in a hydrogen atom will eventually fall to the proton as a result of energy loss due to its E/M radiated power. The time interval t_ℓ for the electron to fall on the proton must depend on the radius a_B of its initial circular orbit, the mass m_e (since the acceleration a depends on m_e), the charge $-e$ (both the acceleration a as well as the radiation I depend on e^2) and, finally, on the velocity of light c (the radiation I depends on c). Out of the *three* quantities a_B, m_e, and e we can make a combination with dimensions of time (in the G-CGS), $t_o = a_B^{3/2} m_e^{1/2} \big/ e$ and another one with dimensions of velocity, $\upsilon_o = a_\mathbf{B}/t_o = e \big/ a_B^{1/2} m_e^{1/2}$. Hence, the most general expression for the time t_ℓ has the form, $t_\ell = t_o f\left(\frac{\upsilon_o}{c}\right)$, where $f(\upsilon_o/c)$ is an arbitrary function of the dimensionless variable $\upsilon_o/c = e/a_B^{1/2} m_e^{1/2} c$.
It is reasonable to assume that the life-time t_ℓ is inversely proportional to the radiated power, which, according to (6.6) in the next chapter, is inversely proportional to the third power of the velocity of light. This implies that $f(\upsilon_o/c) = b\,.\,c^3/\upsilon_o^3$, where b is a numerical factor. Hence

$$t_\ell = b a_B^3 m_e^2 c^3 / e^4 \tag{5.23}$$

The period τ of the classical electronic motion in the absence of E/M radiation can be determined from Newton's equation of motion, which can be brought to the form

$$m_e \omega^2 r^2 = \frac{e^2}{r} \Rightarrow \omega^2 = \frac{e^2}{m_e r^3}$$

or, for $t = 0$

$$\omega^2 = \frac{e^2}{m_e a_B^3} \Rightarrow \tau \equiv \frac{2\pi}{\omega} = 2\pi \left(\frac{m_e a_B^3}{e^2} \right)^{1/2} \tag{5.24}$$

To obtain an explicit result for the time t_ℓ we can start from the equation of energy loss due to the radiated E/M power, I, by the electron:

$$\frac{d}{dt}\left(-\frac{e^2}{2r} \right) = -I$$

where, according to (6.6) in the next chapter, $I = \frac{2}{3}e^2a^2/c^3$. The equation above is valid assuming that the fall to the center is a very "slow" process relative to the period $\tau, t_\ell/\tau \gg 1$, so that in every revolution to have the relation $\varepsilon_{total} \approx \varepsilon_P/2$. Under the same inequality, $t_\ell/\tau \gg 1$, the acceleration a is essentially the centripetal one:

$$a \approx \frac{e^2}{m_e r^2}$$

Using the last three equations we end up with the following equation for r:

$$\frac{dr}{dt} = -\frac{4}{3}\frac{e^4}{m_e^2 c^3 r^2}$$

the solution of which is

$$r^3 = a_B^3 - \frac{4e^4}{m_e^2 c^3} t \tag{5.25}$$

By setting $r = 0$ we find that t_ℓ is given by (5.19) with the numerical factor b being equal to $1/4$. We must check whether $t_\ell/\tau \gg 1$. Working in atomic units ($e = m_e = a_B = 1, c \approx 137$) we find that indeed

$$t_\ell/\tau = \frac{c^3}{4 \times 2\pi} \approx 10^5$$

and $t_\ell = 137^3/4 = 6.42838 \times 10^5 \text{a.u.} = 1.555 \times 10^{-11}\text{s}$

3. *Estimate the lifting force on the wings of a plane. Assume that the wings are rectangular flat rigid metallic sheets.*

Solution The lift force F_ℓ on the wings of a plane must depend on the area S of the wings, and on the velocity υ of the plane. Moreover, it must depend on the presence of air (in the absence or air there is no lift force). The relevant properties of air are its density ρ and its viscosity η. For large velocities we expect that the explicit role of viscosity to be negligible as it was argued in Sect. 5.3. Finally, the lift force must depend on the angle of attack ϕ, i.e. the angle of the wings relative to the vector velocity of the plane. For symmetry reasons the lifting power is zero, if $\phi = 0$. For small angles ϕ we can keep only the term proportional to ϕ. Thus

$$F_\ell = c_1 \rho S \upsilon^2 \phi \tag{5.26}$$

where c_1 is a numerical factor. In reality the shape of the wings, which can be adjusted by the position of the flaps, determine an effective angle ϕ.

5.9 Unsolved Problems

1. Obtain the speed of sound in air.
   Hints. For air the term $\frac{3}{2}\ln(U/N)$ in (5.19) of problem 1 above must be replaced by $\frac{5}{2}\ln(U/N)$, because air consists mainly of diatomic molecules $(N_2,\ O_2)$. Diatomic molecules, besides their three translational degrees of freedom, have in addition two rotational degrees of freedom. In (5.19) substitute U/N by $\frac{3}{2}PV/N$ and then calculate the differential of S in order to obtain $B_S = -V(\partial P/\partial V)_{S,N}$
2. For a car moving with 108 km/h estimate how much power is needed to overcome the air resistance. (1 hp = 746 W).
3. Raindrops have a diameter usually between 0.5 to 2.5 mm (it can even reach 5 mm in thunderstorms). Estimate their terminal speed.

Part II
Interactions

Chapter 6
Photons: Messengers and Connectors

Abstract The equilibrium properties of a photon gas are derived. Emission of photons by black bodies, by accelerating charges and by vibrating electrical dipoles are also obtained. The scattering cross-section and the corresponding mean free path of photons as they propagate through a medium are calculated. Important quantities for the response of condensed matter to photons, such as permittivity and conductivity, are defined and determined. To obtain these results we make extensive use of the method of dimensional analysis.

6.1 Introductory Remarks

In this chapter we shall examine the electromagnetic (E/M) interactions separately from the other three interactions, which will be briefly presented in the next chapter. This special treatment of E/M interactions is justified because they are the ones to provide the necessary attraction for sustaining structures of matter extending over fifteen orders of magnitude from the level of atoms to that of an asteroid. Besides this unique role in the natural and biological world, E/M interactions are involved directly or indirectly in the core of almost every aspect of the technological man-made world. This chapter is longer and more detailed than most others for another practical reason too: In teaching this course, I found that most students encounter conceptual difficulties with Electromagnetism of continuous media; it is hoped that the fresh basic view offered in this chapter together with the contents of Appendix B may be of help.

After these short comments let us return to Physics. Interactions in general and E/M interactions in particular, can be studied at three levels:

The first and most accurate is the fully quantum mechanical one, in which both the wave and the particle character of reality is explicitly taken into account, as e.g. in the study of emission of a photon during the deexcitation of an atom. The usual way to treat interactions at this advanced level is through the employment of Feynman diagrams along with the calculational rules that accompany them. This

E.N. Economou, *From Quarks to the Universe*,
DOI 10.1007/978-3-319-20654-7_6

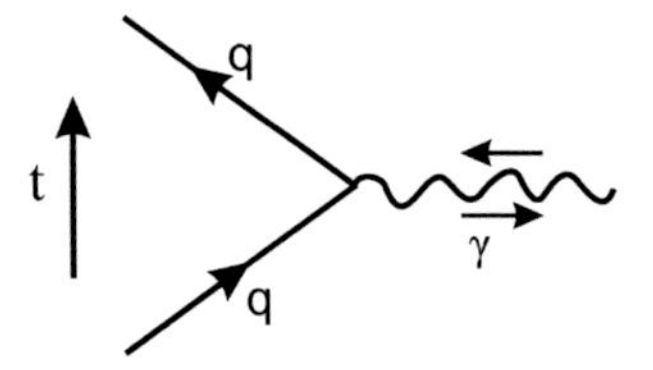

Fig. 6.1 The elementary Feynman diagram for the E/M interaction: An electrically charged particle q at some point in time emits or absorbs a photon γ. Depending on the direction of the time arrow the same diagram represents other elementary processes as well

approach, because of its advanced character, will not be used in this book. Here, we restrict ourselves to present only the diagrams in order to provide a vivid picture of the physical processes taking place without usually attempting to obtain any quantitative information from them.

In Fig. 6.1 the elementary Feynman diagram for the E/M interaction is presented: A charged particle q (straight line) emits or absorbs a photon γ (wavy solid line) remaining the same particle with the same electrical charge. The charged particle can be a lepton (e.g. an electron) or a quark or a charged boson (e.g. a W^+). If the time arrow is horizontal from left to right the same diagram represents a charged particle/antiparticle annihilation accompanied by the creation of a photon (the straight line with the arrow running opposite to the time arrow is by convention the antiparticle). Similarly, if the time arrow in Fig. 6.1 runs from top to bottom, the diagram in Fig. 6.1 would represent the emission or absorption of a photon by a charged antiparticle. Finally, if the time arrow is horizontal from right to left, the diagram describes the annihilation of a photon accompanied by the creation of a charged particle/antiparticle pair.

The E/M interactions, as well as all four interactions, conserve at each elementary junction, called *vertex*, the following quantities:

- Momentum
- Angular momentum (not simultaneously with momentum)
- Electric charge
- Color charge
- Baryon number
- Lepton number (separately for each family[1])

The E/M interaction also conserves other physical quantities such as parity (which means that a system and its image copy have identical E/M properties), time reversal (which means that the transformation $t \to -t$ maps the system to an identical system from the point of view of E/M interactions) and charge conjugation (which means that a particle and its antiparticle have the same properties except for the charge which is opposite).

[1]Note that the neutrinos transform from one family to another as time goes on (e.g. a ν_e neutrino may change to a ν_μ). However, this change does not happen on a Feynman diagram vertex.

Some comments are in order regarding conservation of energy, which is obeyed by all interactions in the sense that the energy of a closed system before it enters in any kind of physical reaction is equal to the energy of the system after the reaction is completed. However, conservation of energy at every elementary vertex may be treated in two different but essentially equivalent ways: Either you conserve energy at each elementary vertex but then you must abandon the energy/ momentum relation and treat the set $\{\mathbf{p}, E\}$ as four independent (conserved) variables, or you continue to use the relation $E = \mathrm{f}(\mathbf{p})$ but then you do not conserve energy at each elementary vertex, only overall.

The second level of dealing with interactions, especially E/M interactions, is less accurate than the first fully quantum-mechanical one, since it takes into account only their wave nature and ignores the particle nature. This approximation is absolutely satisfactory in cases where the number of photons involved is huge, as in the case of the E/M waves carrying information to our mobile phone or to our TV, etc. Obviously, in this case there is no practical distinction between the actual discrete nature of the incoming energy and the assumed continuous character of it.

The third level of treatment is even less accurate than the second, because it replaces the time dependent character of a wave by a static potential, as in the case of the hydrogen atom, where the complicated E/M interaction between the proton and the electron is replaced by a Coulomb potential of the form, $-e^2/r$. Nevertheless, even this drastic approximation produces results in acceptable agreement with experimental data regarding e.g. the energy levels of the hydrogen atom. If we wish to increase the accuracy of these results the wave character of the interaction must be taken into account, and, if this is not enough, the wave-particle character of the interaction is also taken into account to obtain accuracy better than one part in a million.

6.2 Photons in Equilibrium

6.2.1 Determine the Thermodynamic Quantities of a Photon Gas in Equilibrium by Employing Dimensional Analysis

As was discussed in Chap. 4 (Footnote 3), only two independent thermodynamic variables determine the equilibrium properties of a photon gas: One is the volume V and the other is the temperature T. Moreover, due to the extensive character of the thermodynamic potentials (such as the energy, the free energy, etc), all these quantities must be proportional to the volume. E.g., $U = V\,\varepsilon$, where by definition ε is the energy density with dimensions energy over length to the third power. The energy density ε must depend on the combination $k_B T$ (since physically k_B and T go together as a characteristic unit of thermal energy), on the velocity of light c (since we are dealing with an E/M quantity), and on $\hbar$ (since photons are the quanta of the

E/M field). Taking into account that $c\hbar/k_BT$ has dimensions of length, we see that the combination $(k_BT)/(c\hbar/k_BT)^3$ has the required dimensions for ε and hence we immediately conclude that $\varepsilon = c_1\,(k_BT)/(c\hbar/k_BT)^3 = c_1\,(k_BT)^4/c^3\,\hbar^3$. If this very quick derivation does not fully convince you, then try the standard procedure of writing ε in the form $\varepsilon = c_1\,(k_BT)^a c^b \hbar^d$ so that $[\varepsilon] = [m][l]^{-1}[t]^{-2} = [m]^a[l]^{2a}[t]^{-2a}[l]^b[t]^{-b}[m]^d[l]^{2d}[t]^{-d}$, which implies that $a+d=1,\ 2a+b+2d=-1,\ 2a+b+d=2$, and therefore $a=4,\ b=-3,\ d=-3$ in agreement with our previous quick result. A detailed theoretical calculation shows that the numerical constant c_1 is of the order of one, i.e. equal to $\pi^2/15$. Thus, the result for the energy of the photon gas in equilibrium is the following:

$$\boxed{U = \frac{\pi^2}{15}\,\frac{V(k_BT)^4}{c^3\,\hbar^3}} \tag{6.1}$$

Taking into account (6.1) and that $T = (\partial U/\partial S)_V = (\partial U/\partial T)_V(\partial T/\partial S)_V$, we obtain for the entropy S that $S = \frac{4}{3}U/T \propto T^3$; then Helmholtz's free energy is $F = U - TS = -\frac{1}{3}U$ and the pressure P is $P = -(\partial F/\partial V)_T = \frac{1}{3}\varepsilon = \frac{1}{3}(U/V)$. The specific heat is $C_V = (\partial U/\partial T)_V = 4U/T \propto T^3$. The Gibbs free energy is $G \equiv F + PV = -\frac{1}{3}U + \frac{1}{3}U = 0$, in agreement with the fact that the chemical potential, $\mu \propto G$, is zero for the photon gas, due to the non-conservation of the number of photons.

6.2.2 *Determine the Total E/M Energy, I, Emitted by a Black Body of Temperature T Per Unit Time and Per Unit Area by Employing Dimensional Analysis*

A black body is by definition one which absorbs 100 % of the E/M radiation falling on every elementary area δA of its surface. We shall calculate the energy per unit time emitted by δA by the following argument: Consider a black body of temperature T surrounded and being in equilibrium with a photon gas of the same temperature T, (since in equilibrium). The energy per unit time emitted by δA must be equal (as a result of being in equilibrium) to the energy per unit time absorbed by δA; the latter by definition of the black body is the same as the one falling per unit time on δA. However, the energy per unit time and per unit area falling on it is just the flux of energy I which obeys the general relation

$$\boxed{\text{Flux of any quantity} = \text{density of this quantity} \times \text{velocity of the flow}} \tag{6.2}$$

Equation (6.2) in the present case gives

$$\boxed{I = \frac{1}{4}\varepsilon\, c = \frac{\pi^2}{60}\frac{(k_B T)^4}{\hbar^3 c^2}} \tag{6.3}$$

The extra factor (1/4) = (1/2) × (1/2) appears because the flux falling on δA is the ingoing one and not the outgoing (this accounts for one of the two ½ factors) and because only the normal to δA component is absorbed (this takes care of the other ½ factor).

Of course, the same result will be reached by applying dimensional analysis without taking advantage of (6.1): The quantity I depends on the following three quantities: The combination $k_B T$ (physically k_B and T go together) the velocity of light c (as an E/M phenomenon), and Planck's constant $\hbar$ (as an explicitly quantum phenomenon). The combination $(k_B T)^{\mu_1} \hbar^{\mu_2} c^{\mu_3}$ which has the same dimensions as I, i.e., energy over time and over length squared is obtained only if $\mu_1 = 4$, $\mu_2 = -3$, and $\mu_3 = -2$. These values become obvious by taking into account that $\hbar / k_B T$ has dimension of time and $\hbar\, c / k_B T$ has dimension of length. Hence

$$I = a' \frac{k_B T}{(\hbar / k_B T)(\hbar\, c / k_B T)^2} = a' \frac{(k_B T)^4}{\hbar^3 c^2} = \sigma\, T^4 \tag{6.4}$$

where $\sigma = (\pi^2/60)\, k_B^4 / \hbar^3 c^2 \approx 5.67 \times 10^{-8}\,\mathrm{Js^{-1} m^{-2} K^{-4}}$ is the so-called Stefan-Boltzmann constant. The numerical factor in (6.4) cannot be obtained by dimensional analysis; an explicit integration of (6.6) below over all frequencies is needed as well as taking the outward normal component of the flux. This procedure gives the same result as in (6.4), $a' = \pi^2/60$.

6.2.3 *How Is the Emitted I_ω Black Body E/M Energy Per Unit Time, Per Unit Frequency, and Per Unit Area Distributed Among the Various Frequencies?*

We repeat here that a black body is one which absorbs 100 % of the E/M radiation falling on it; the emission properties of such a body are characterized by only one parameter, its absolute temperature. We stress again that in Nature the absolute temperature appears always as the product $k_B T$ with dimensions of energy, where k_B is the Boltzmann constant. Thus, one quantity on which I_ω depends is the product $k_B T$; another is, obviously, the frequency ω; finally, the velocity of light c is expected to appear in I_ω, since c is the main quantity characterizing the E/M radiation. From the point of view of classical physics these three quantities are the only ones determining I_ω. The dimensions of I_ω are energy per length squared.

To obtain the same dimensions from the combination $(k_B T)^{\mu_1} \omega^{\mu_2} c^{\mu_3}$ we must choose $\mu_1 = 1$, $\mu_2 = 2$, $\mu_3 = -2$. Hence, the classical result for I_ω is

$$\boxed{I_\omega = a \frac{k_B T\, \omega^2}{c^2}} \tag{6.5}$$

where a is a numerical factor which turns out to be equal to $1/(4\,\pi^2)$. It is clear that the classical result (6.5) cannot be right for high frequencies ($\hbar\omega \geq k_B T$), since its integral over all frequencies would produce an infinite result for the total radiation emitted by a black body. There are two solutions around this problem: One is the introduction of the cut-off we discussed in Chap. 3, $\hbar\omega_c = c_1 k_B T$; integration of (6.5) over ω from zero to the cut-off $c_1 k_B T$ gives for I the same result as (6.4) apart from a numerical factor. The other solution is to extend the integration over all frequencies but with a I_ω incorporating the correct quantum behavior at all frequencies, which means that a fourth quantity, that of $\hbar$, will enter in the expression for I_ω. Thus the correct formula for I_ω is the following:

$$I_\omega = a \frac{k_B T\, \omega^2}{c^2} f\left(\frac{\hbar\,\omega}{k_B T}\right) \tag{6.6}$$

where f is a function which cannot be obtained by dimensional analysis. We know though that (6.6) must reduce to (6.5), when $\hbar\,\omega/k_B T \ll 1$, so that $f \to 1$, for $\hbar\,\omega/k_B T \to 0$; moreover, in the opposite limit, $\hbar\,\omega/k_B T \gg 1$, f must approach zero. It turns out through a detailed theoretical calculation that

$$f\left(\frac{\hbar\,\omega}{k_B T}\right) = \frac{\hbar\,\omega}{k_B T} \frac{1}{e^{\hbar\,\omega/k_B T} - 1}. \tag{6.6a}$$

6.3 Emission of Photons by Accelerating Charges

6.3.1 *Radiation by a Moving Particle of Electric Charge* q

The electromagnetic (E/M) energy per unit time J emitted by such a particle must depend on q (since the electric charge is the source and the acceptor of E/M energy), on c (since the velocity of light is characterizing the E/M radiation), and on the acceleration, a, of the particle (since, from all kinematic quantities only a is directly connected to the force acting on the particle and providing the necessary work to compensate for the emitted energy). There is only one combination of these three quantities having dimensions of energy over time: $q^2 a^2/c^3$. Hence

$$J = \eta \frac{q^2 a^2}{c^3} \tag{6.7}$$

(In SI this formula becomes $J = \eta\,(q^2/4\pi\,\varepsilon_0)a^2/c^3$.

The numerical factor turns out to be 2/3, if a^2 is the time average of the acceleration squared assuming a sinusoidal time dependence.

6.3.2 *Radiation by a Neutral System with an Oscillating Electric Dipole Moment* $\boldsymbol{p} = \sum q_i \boldsymbol{r}_i$

From (6.7) and taking into account the definition of the dipole moment as well as that $\boldsymbol{a} = -\omega^2 \boldsymbol{r}$ (if the oscillation is sinusoidal) we have that $-\omega^2 \boldsymbol{p} = \sum q_i \boldsymbol{a}_i$. Thus we expect that

$$J = \eta \frac{\boldsymbol{p}^2 \omega^4}{c^3} \tag{6.8}$$

(In SI this formula becomes $J = \eta\,(\boldsymbol{p}^2/4\pi\,\varepsilon_0)\omega^4/c^3$.)

Note that we would arrive to the same formula by taking into account that J in this case depends on $\boldsymbol{p}$, ω, and c.

Use (6.8) *to obtain the life-time of an excited state.*

The product of J times the life-time τ of an excited state gives the energy emitted during the time τ; but this energy is just the energy $\hbar\omega$ of the emitted photon. Assuming that the deexcitation occurs by the emission of a photon through a dipole transition we obtain

$$\tau = \frac{\hbar\omega}{J} = \eta^{-1} \frac{\hbar\, c^3}{\boldsymbol{p}^2 \omega^3} \tag{6.9}$$

(In SI the quantity $\boldsymbol{p}^2$ must be replaced by $\boldsymbol{p}^2/4\pi\,\varepsilon_0$). Applying (6.9) to the transition $2p \rightarrow 1s$ of the hydrogen atom, where $\hbar\omega$ is equal to 10.2 eV (see Chap. 10, in particular (10.1)) and taking $\boldsymbol{p}^2 \approx e^2 a_B^2$ and $\eta \approx 2/3$ we find $\tau \approx 1.77 \times 10^{-9}$s versus $\tau = 1.59 \times 10^{-9}$s obtained by the detailed quantum mechanical calculation.

6.4 Scattering of Photons by Charged Particles, Atoms, Molecules

Any scattering event is characterized quantitatively by the so-called total scattering cross-section, σ_s, which can be thought of as the area of the incoming beam intercepted by each scatterer. More precisely, σ_s is defined as the ratio Γ/F, where Γ is the number of scattered incoming particles per unit time and F is the flux of these incoming particles. The flux, as was shown in (6.2), is in general equal to the concentration of the incoming particles times their common velocity. In particular for photons, we have that the number of scattered photons per unit time is the emitted energy per unit time J over the energy of each photon, $\Gamma = J/\hbar\omega$; similarly, the concentration of photons is equal to their energy density, $\boldsymbol{E}_o^2/4\pi$, (in SI the average energy density of the incoming EM field is $\varepsilon_0 \boldsymbol{E}_o^2$) divided by the energy $\hbar\omega$ of each photon. Hence, $F =$ concentration of photons $\times \mathrm{c} = \mathrm{c}\,\mathbf{E}_o^2/4\pi\hbar\omega$. Thus

$$\sigma_s = \frac{J\,4\pi\hbar\omega}{\hbar\omega\, c\boldsymbol{E}_o^2} = \frac{4\pi J}{c\boldsymbol{E}_o^2} \tag{6.10}$$

(In SI (6.10) is $\sigma_s = J/c\,\varepsilon_0 \boldsymbol{E}_o^2$).

If the scatterer is a *charged particle*, the emitted energy J in (6.10) is given by $J = \frac{2}{3} q^2 a^2/c^3$, where the acceleration a is connected to the electric field: $a = q\boldsymbol{E}_o/m$. In particular, if the scatterer is an electron, we have $q = e$ and $m = m_e$; thus

$$\sigma_s = \frac{8\pi}{3}\left(\frac{e^2}{m_e c^2}\right)^2 \equiv \frac{8\pi}{3} r_e^2 = 6.65 \times 10^{-29}\,\mathrm{m}^2 \tag{6.11}$$

where $e^2/m_e c^2 (= e^2/4\pi\,\varepsilon_0 m_e c^2$ in SI$) \equiv r_e = 2.82 \times 10^{-15}\,\mathrm{m}$ is the classical "radius" of the electron (resulting by equating the rest energy $m_e c^2$ to the fictitious electrostatic energy e^2/r_e). A more general version of (6.11) can be deduced by dimensional analysis taking into account that the scattering cross-section σ_s must depend on q (for $q = 0$ there is no scattering), on m (for $m = \infty$ there is no motion of the particle/scatterer and, therefore, no scattered radiation), and on c; moreover, the cross-section may also depend on two dimensionless parameters proportional to the frequency of the incoming photon: a classical one equal to $x_c = \omega\, t_c$ where $t_c \equiv r_e/c$ and a quantum one $x_q = \hbar\,\omega/m\,c^2$. Since $x_c = \alpha\, x_q$, where α is the fine structure constant, $\alpha \equiv e^2/\hbar\, c$ ($\equiv e^2/4\pi\,\varepsilon_0 \hbar\, c$ in SI), x_c is negligible in comparison with x_q. Thus, dimensional analysis predicts for σ_s the following expression

$$\sigma_s = c_1 \left(\frac{q^2}{m\,c^2}\right)^2 f\left(\frac{\hbar\omega}{m\,c^2}\right) \tag{6.12}$$

where the function f of the dimensionless ratio $\hbar\omega/m\,c^2$ cannot be determined by dimensional analysis, but has been obtained by an explicit quantum electrodynamic calculation which led to the so-called Klein-Nishina formula. Actually, if c_1 is taken to be equal to $8\pi/3, f$ is a monotonically decreasing function of $\hbar\omega/mc^2$ which is one for $\hbar\omega = 0$ and zero for $\hbar\omega/mc^2 \to \infty$. Notice that the square of the mass of the scatterer is in the denominator of (6.12) which means that the electron is by far the most efficient scatterer among the singly charged particles.

The scattering cross-section of a photon by a neutral atom or a molecule of zero average dipole moment is due to the *induced* dipole moment $\boldsymbol{p}$ by the external field $\boldsymbol{E}_o$

$$\boldsymbol{p} \equiv a_p \boldsymbol{E}_o \tag{6.13}$$

where a_p is by definition the so-called polarizability of the atom or the molecule. Substituting (6.8) and (6.13) in (6.10) we obtain

$$\sigma_s = \frac{8\pi\omega^4}{3\ c^4} a_p^2 \tag{6.14}$$

It follows from (6.13) that the polarizability a_p has dimensions of length cubed (or length cubed $\times$ permittivity in SI)

$$a_p \simeq c_1 r_o^3 f_R(\omega) \tag{6.15}$$

where c_1 is a numerical factor in the range roughly between 4 and 10, r_o is the radius of the atom (for the molecules N_2 and O_2, $a_p = 11.75 a_B^3$ and $10.67 a_B^3$ respectively) and $f_R(\omega)$ is a typical dimensionless resonance function of the type appearing in any forced harmonic oscillation as the one under consideration

$$f_R(\omega) \approx \frac{\omega_o^2}{\omega_o^2 - \omega^2 - i\omega/\tau} \tag{6.16}$$

which takes into account the strongly increased response of the atom (or molecule) when the frequency ω of the incoming E/M field coincides with one of the eigenfrequencies $\omega_o = \left|\varepsilon_\mu - \varepsilon_v\right|/\hbar$ of the atom (or molecule). For $\omega \ll \omega_o$, $f_R(\omega) \approx 1$ while for $\omega = \omega_o$

$$|f_R(\omega)| = \omega_o \tau \tag{6.17}$$

where τ is the lifetime given by (6.9). Thus, for $\omega \ll \omega_o$ and for the hydrogen atom for which $c_1 r_o^3 \approx 5.4 a_B^3$ we have by introducing the wavelengths $\lambda = 2\pi c/\omega$ and $\lambda_o = 2\pi c/\omega_o$ that

$$\sigma_s \approx 3.8 \times 10^5 \frac{a_B^6}{\lambda^4}, \quad \lambda \gg \lambda_o \approx 120\,\text{nm} \tag{6.18}$$

For the atmospheric air the numerical coefficient in (6.18) will be 1.72×10^6. Notice in (6.18) the characteristic λ^{-4} dependence valid for long wavelengths (Rayleigh scattering).

At the resonance, using (6.17) in combination with (6.9) and setting $\boldsymbol{p}^2 = c_2 \hbar \omega_o r_o^3$ (since $\hbar\omega_o \approx e^2/r_o$), we obtain $|f_R(\omega)| = \omega_o \tau = c^3/\eta\, c_2 r_o^3 \omega_o^3$. Substituting this expression in (6.15) and then in (6.14) we get

$$\sigma_s = \frac{8\pi}{3}\left(\frac{c_1}{2\pi\,\eta\,c_2}\right)^2 \lambda_o^2, \qquad \lambda = \lambda_o \tag{6.19}$$

For the hydrogen atom, where $c_2 \approx 2.66$, $c_1 \approx 5.4$, we obtain $\sigma_s \approx 1.96\lambda_o^2 \approx 2.8 \times 10^{-14}\mathrm{m}^2$.

Equation (6.18) implies that the off-resonance scattering cross-section of an *optical* photon, $\lambda \approx 600\,\mathrm{nm}$, by a neutral N_2 molecule is about equal to $2.88 \times 10^{-31}\,\mathrm{m}^2$, i.e., about eleven orders of magnitude smaller than its geometrical cross-section πr_o^2. On the other hand, at the first resonance of the hydrogen atom, $\hbar\omega_o = 10.2\,\mathrm{eV}$, the scattering cross-section is, according to (6.19), six orders of magnitude larger than the geometrical cross-section!

6.5 Scattering of Photons by Macroscopic Particles

Consider a spherical macroscopic particle of diameter $d = 2a$, permittivity ε_a (real) and permeability $\mu = 1$. The formula which gives the scattering of an E/M wave propagating in the vacuum and impending on this particle is expected to be proportional to some power of $\varepsilon_a - 1$ (since $\varepsilon_a = 1$ means that the particle behaves as the vacuum and, hence, produces no scattering) and to d^6/λ^4 (in analogy with (6.18), as long as the wavelength is much larger than the size of the particle, $\lambda \gg d$). The actual formula is

$$\sigma_s = 4\left(\frac{\varepsilon_a - 1}{\varepsilon_a + 2}\right)^2 \frac{d^6}{\lambda^4}, \qquad \lambda \gg d \tag{6.20}$$

If the wavelength becomes comparable to or smaller than the size of the particle, the scattering cross-section is comparable to the geometrical cross-section $\pi d^2/4$ with peaks and deeps as in Fig. 6.2. Notice that the peaks appear roughly when integer number of half wavelengths fit within the size of the particle. This is due to geometrical resonances known as Mie resonances.

6.6 Total Scattering Cross-Section and Mean Free Path

Important transport properties, such as electrical and thermal conductivities in solids, as well as propagation of E/M waves in matter are characterized by the so-called mean free path ℓ. For an electromagnetic wave propagating in the x direction within a medium, the *mean free path* is defined by the relation $\left|E^2(x)\right| = \left|E^2(0)\right| \exp(-x/\ell)$, i.e. through the exponential decay of the energy density of the beam as it propagates through the medium. If the medium is gaseous, the mean free

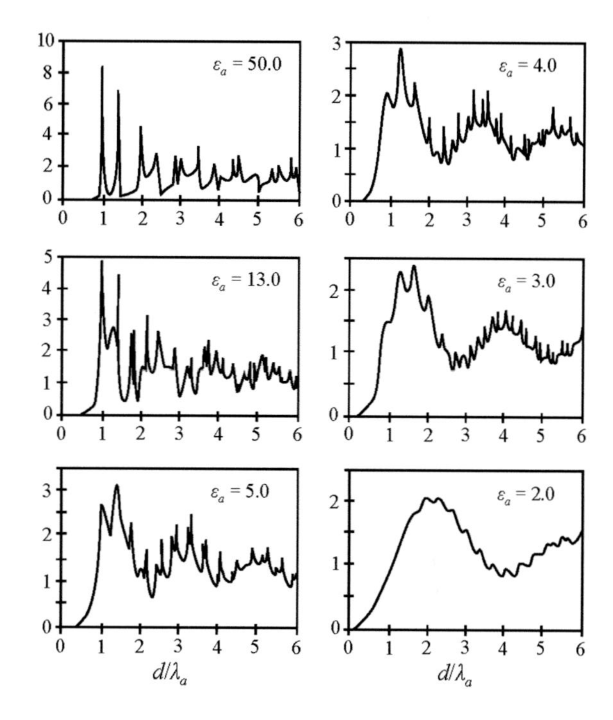

Fig. 6.2 Dimensionless scattering cross-section, $\sigma_s/\pi a^2$, versus the ratio d/λ_a, for a spherical particle of radius a, where $\lambda_a = \lambda/\sqrt{\varepsilon_a} = \lambda/n$ is the wavelength within the particle and n is its refractive index (for water at visible frequencies, $n \approx 1.33$). Notice the sharp peaks especially for high values of ε_a

path of photons is connected to the total scattering cross-section σ_s by each scatterer as follows:

$$\boxed{\frac{1}{\ell} \approx n_s \sigma_s} \tag{6.21}$$

where n_s is the concentration of scatterers. Equation(6.21) is valid for all types of waves or particles, but for weak scattering, i.e., when the mean free path is much larger than the wavelength of the propagating wave. The justification of (6.21) is based on the following argument: An incoming beam of waves or particles of total frontal area S after propagating for a length ℓ in the scattering medium has covered a volume $S\ \ell$ and has met $N_s = S\,\ell\, n_s$ individual scatterers. Since each scatterer intercepts from the propagating beam an area σ_s, the N_s scatterers will intercept a total area of $N_s\sigma_s = S\,\ell\, n_s\sigma_s$. By definition the mean free path is the minimum length required to just intercept the whole frontal area S of the incoming beam. Hence, $S\,\ell\, n_s\sigma_s = S$, which is the same as (6.21). This derivation of (6.21) assumes that each scatterer scatters only once, i.e. it omits multiple-scattering; for a dense system this omission is not generally justified.

For Earth's atmosphere at sea level the concentration of molecules can be found from the law of perfect gases, $n_s = P/k_B T = 2.4 \times 10^{25}\mathrm{m}^{-3}$, for P = 1 atm and T = 300 K. Using our previous estimate for σ_s, we obtain for the mean free path of visible photons in air $\ell \approx 140\,\mathrm{km}$. This estimate is an dealized upper limit; actually the mean free path of visible light in the lower atmosphere can be much smaller, because other terms similar to $n_s \sigma_s$ must be added to the right-hand side of (6.21); these terms are due to the concentration of dust particles and water droplets and their corresponding scattering cross-section (see (6.20)).

6.7 Quantities Characterizing the E/M Behavior of Solids and Liquids

The electrical properties of materials are expressed through various quantities such as the permittivity, ε, the conductivity, σ, (which is the inverse of the resistivity, ρ), and the electric susceptibility, χ_e. In most cases, these quantities depend on the electronic motion induced by an electric field. In various materials there are in general two groups of electrons: Those which contribute to the electric current even under the action of a static field (i.e. a time independent field) and they are usually named free electrons, and those which contribute to the current only under the action of a time dependent field and they are referred to as bound electrons. Materials, such as metals, possessing appreciable concentration of electrons of the first group are called conductors, while materials not possessing (or possessing negligible concentration) of electrons of the first group are usually characterized either as semiconductors or insulators. In several books the conductivity is related only to the first group of electrons through the equation $j_f = \sigma_f E$, where j_f is the current density associated with the free electrons and E is the electric field, and the susceptibility is related only to the second group of electrons through the equation $P_b = \chi_{e,b} E$ valid in G-CGS ($P_b = \chi_{e,b} \varepsilon_0 E$ in SI) where P_b is the polarization associated with the bound electrons and defined as the sum of individual dipole moments (induced by the field acting on the bound electrons only) divided by the total volume. Notice that the time derivative of P_b gives the contribution j_b of the bound electrons to the current density. (To prove this last statement use the definitions of P_b and j_b).

In this book we shall not follow this distinction but we shall treat all electrons in a unified way, since all of them may contribute to the total conductivity $\sigma \equiv \sigma_f + \sigma_b$ and all of them may contribute to the time derivative of the total polarization: $\partial P / \partial t \equiv \partial P_f / \partial t + \partial P_b / \partial t$ which is equal to the total current density $j = \sigma E$. From now on, we shall assume (without loss of generality) that the time dependence is of the form $\exp(-i\omega t)$, where ω is the angular frequency of the electric field. It follows from the above that the total σ is directly related to the total χ_e:

$$j = \sigma E = \frac{\partial P}{\partial t} = -i\omega P = -i\omega \chi_e E \quad \Rightarrow \quad \sigma = -i\omega \chi_e \quad \Rightarrow \quad \chi_e = \frac{i\sigma}{\omega}.$$

The permittivity is defined as follows:

$$\boxed{\varepsilon \equiv 1 + 4\pi\,\chi_e = 1 + \frac{4\pi\, i\,\sigma}{\omega}\,\text{G-CGS}}, \tag{6.22a}$$

$$\boxed{\varepsilon \equiv \frac{\varepsilon_{SI}}{\varepsilon_0} \equiv 1 + \chi_{e,SI} = 1 + \frac{i\,\sigma}{\varepsilon_0\omega},\ \text{SI}} \tag{6.22b}$$

Thus, in the framework of treating all the electrons in a unified way, all three electrical quantities $\varepsilon,\ \chi_e,\ \sigma$ are connected by simple relations as shown above. Therefore, it is sufficient to calculate just one of them to obtain the other two. However, because both free and bound electrons contribute, these electrical quantities become complex possessing in general both a real and an imaginary part. Note that the numerical values of the dimensionless quantities $\varepsilon,\ 4\pi\sigma/\omega$ (in G-CGS) and $\varepsilon_{SI}/\varepsilon_0,\ \sigma/\varepsilon_0\omega$ (in SI) are respectively equal to each other. This implies the following relation $\chi_{e,\mathrm{SI}} = 4\pi\,\chi_{e,G-CGS}$ between the dimensionless susceptibilities in the two systems.

At this point it is not useless to add two remarks:

(i) The permittivity of solids and liquids is a very important quantity which serves as a direct bridge between the microscopic motions and the macroscopic properties. E.g. the zeroes of ε give the collective oscillations in a solid or liquid, such as sound waves or plasma waves.

(ii) There are magnetic quantities such as the permeability μ and the magnetic susceptibility χ_m which are analogous to the permittivity and the electric susceptibility (see Appendix B). We shall not deal with the magnetic quantities here, because for most materials χ_m is much smaller than one and μ is very close to one (in G-CGS). However, there are both natural and artificial materials which exhibit large values of μ either positive or negative with important technological applications.

6.8 Calculation of the Conductivity σ and the Permittivity ε

Let us consider a bound electron in a solid and let us apply the classical Newton's law for its motion

$$m_e(dv/dt) = -e\,E - \kappa\,x + F_{\mathrm{f}} \tag{6.23}$$

where $v = dx/dt$ is the average velocity of the electron induced by the electric field, x is its average displacement from the average equilibrium position, $-e\,E$ is the force due to the electric field, $-\kappa\,x$ is the restoring force on the bound electron which is proportional to its displacement, and F_{f} is a friction force which according to (5.6) is proportional to υ, $F_{\mathrm{f}} = -\alpha\,\upsilon$ with the proportionality constant having dimensions of mass over time, $\alpha = m_e/\tau$, and τ is the so-called relaxation time which together with m_e determine the proportionality constant of the friction force. Taking into account the $\exp(-i\,\omega\,t)$ time dependence we have that $d\upsilon/dt = -i\,\omega\,\upsilon$, $x = i\,\upsilon/\omega$. Substituting in (6.23) we obtain for the velocity

$$\upsilon = \frac{i\,\omega\,eE}{m_e\{\omega_0^2 - \omega^2 - i\,(\omega/\tau)\}} \tag{6.24}$$

where $\omega_0^2 \equiv \kappa/m_e$. In a solid there are several groups of electrons: The free electrons, if any, will have $\omega_0 = 0$ (there is no restoring force on free electrons). The bound electrons can be classified in different groups according to the value of their restoring force coefficient. Summing all of them together, we have for the current density

$$\begin{aligned} j \equiv \sum_i -e\,n_i\upsilon_i &= \sum_i \frac{-i\,\omega\,e^2 n_i E}{m_e\left\{\omega_{0,i}^2 - \omega^2 - i\,(\omega/\tau_i)\right\}} \\ &= \frac{i\,\omega\,e^2 n_f E}{m_e\{\omega^2 + i\,(\omega/\tau_f)\}} + \sum_j \frac{-i\,\omega\,e^2 n_j E}{m_e\left\{\omega_{0,j}^2 - \omega^2 - i\,(\omega/\tau_j)\right\}} \end{aligned}$$

In the last equation we separated the contribution of free electrons (of concentration n_f) from the contribution of the various groups of bound electrons (last terms). Dividing the current j by E we have the final expression for the conductivity

$$\sigma = \frac{i\,\omega\,e^2 n_f}{m_e\{\omega^2 + i\,(\omega/\tau_f)\}} + \sum_j \frac{-i\,\omega\,e^2 n_j}{m_e\left\{\omega_{0,j}^2 - \omega^2 - i\,(\omega/\tau_j)\right\}} \tag{6.25}$$

and, according to (6.22a), for the permittivity as well

$$\varepsilon = 1 - \frac{\omega_{p,f}^2}{\omega^2 + i\,(\omega/\tau_f)} + \sum_j \frac{\omega_{p,j}^2}{\omega_{0,j}^2 - \omega^2 - i\,(\omega/\tau_j)} \tag{6.26}$$

where we defined $\omega_{p,i}^2 \equiv 4\pi n_i e^2/m_e$, $i = f, j$ (in SI replace 4π by $1/\varepsilon_0$); $\omega_{p,f}$ could be called unreduced plasma frequencies (see below). In the case where both ω^2 and ω/τ_j are much smaller than $\omega_{0,j}^2$, the last sum in (6.26) is just a dimensionless constant to be written as ε_b and the permittivity reduces to

$$\varepsilon = 1 + \varepsilon_b - \frac{\omega_{p,f}^2}{\omega^2 + i\,(\omega/\tau_f)} \tag{6.27}$$

The quantity $\omega_p \equiv \omega_{p,f}/\sqrt{1+\varepsilon_b}$ is the true plasma eigenfrequency at which, in the absence of friction forces ($\tau_f \to \infty$), the permittivity, as given by (6.27), becomes zero. If there are no free electrons, the permittivity in (6.27) becomes a constant

$$\varepsilon = 1 + \varepsilon_b \tag{6.28}$$

which may be quite large if the ratio $\omega_{p,j}^2/\omega_{0,j}^2$ turns out to be large (see Solved Problem 2).

We must mention that the permittivity and the other electrical quantities are in general functions of both the frequency ω and the wavevector **k**. Thc cxprcssions. (6.25) for the conductivity and (6.26) for the permittivity do not show any dependence on k, because for frequencies much larger than $\upsilon_0 k$, the dependence on k is negligible; υ_0 is a characteristic velocity of the electronic system, such as the Fermi velocity $\upsilon_F \equiv p_F/m_e$. Notice that (6.27) describes approximately the high frequency E/M response of metals, while (6.28) is appropriate for the not so high frequency behavior of ε in semiconductors. Keep in mind that the Coulomb interaction between two charges within a semiconductor or insulator is given by

$$\mathcal{V}(r) = \frac{q_1 q_2}{\varepsilon r}, \quad \text{G-CGS}; \qquad \mathcal{V}(r) = \frac{q_1 q_2}{4\pi\varepsilon_0 \varepsilon r}, \quad \text{SI} \tag{6.29}$$

Finally, we mention that the ions in a solid contribute also to the permittivity for non-zero frequency and in some cases even for zero frequency. This contribution must necessarily be included if the sound oscillations are to be found from the zeroes of $\varepsilon(k, \omega)$. This is obvious physically, since sound oscillations in solids propagate by transferring the vibrational motion of an ion to its neighbors and so on.

6.9 Summary of Important Formulae

Total energy of photon gas in equilibrium

$$U = V\varepsilon = V\frac{\pi^2}{15}\frac{(k_B T)^4}{c^3 \hbar^3} \tag{6.30}$$

Energy per unit time and per unit area emitted by a black body

$$I = \frac{1}{4}\varepsilon c = \frac{\pi^2}{60}\frac{(k_B T)^4}{\hbar^3 c^2} \tag{6.31}$$

Frequency distribution of the black body radiation

$$I_\omega = \frac{k_B T\,\omega^2}{4\pi^2 c^2} f\left(\frac{\hbar\,\omega}{k_B T}\right), \tag{6.32}$$

$$\mathrm{f}\left(\frac{\hbar\,\omega}{k_B T}\right) = \frac{\hbar\,\omega}{k_B T}\frac{1}{e^{\hbar\,\omega/k_B T} - 1} \tag{6.33}$$

Energy per unit time emitted by a charged particle and a dipole respectively

$$J = \frac{2}{3} q^2 a^2 / c^3 \tag{6.34}$$

$$J = \frac{2}{3} \boldsymbol{p}^2 \omega^4 / c^3 \tag{6.35}$$

Scattering cross-sections by (a) a charged particle (e.g. an electron) (b) an atom or molecule

$$\text{(a) } \sigma_s = \frac{8\pi}{3}\left(\frac{e^2}{m_e c^2}\right)^2 \equiv \frac{8\pi}{3} r_e^2 \tag{6.36}$$

$$\text{(b) } \sigma_s = \frac{8\pi}{3}\frac{\omega^4}{c^4} a_p^2 \tag{6.37}$$

The mean free path is related to the cross-section (for weak scattering)

$$\frac{1}{\ell} \approx n_s \sigma_s \tag{6.38}$$

Relations between permittivity, susceptibility, and conductivity

$$\varepsilon \equiv 1 + 4\pi\,\chi_e = 1 + \frac{4\pi\,i\,\sigma}{\omega}, \text{G-CGS} \tag{6.39}$$

General expression for the permittivity ($\omega \gg \upsilon_0 k$)

$$\varepsilon = 1 - \frac{\omega_{p,f}^2}{\omega^2 + i\,(\omega/\tau_f)} + \sum_j \frac{\omega_{p,j}^2}{\omega_{0,j}^2 - \omega^2 - i\,(\omega/\tau_j)} \tag{6.40}$$

Coulomb energy between two charges located within a non-conducting material

$$\mathcal{V}(r) = \frac{q_1 q_2}{\varepsilon\, r}, \text{ G-CGS; } \mathcal{V}(r) = \frac{q_1 q_2}{4\pi\varepsilon_0\varepsilon\, r}, \text{ SI} \tag{6.41}$$

6.10 Multiple-Choice Questions/Statements

1. The formula for the pressure of a photon gas in equilibrium is
   (a) $P = \frac{2}{3}(U/V)$, (b) $P = 0$, (c) $P = \frac{1}{3}(U/V)$, (d) $P = \frac{\pi^2}{45}\frac{(k_BT)^3}{(\hbar c)^3}$
2. The formula for the Gibbs free energy of a photon gas in equilibrium is
   (a) $G = \frac{\pi^2}{45}\frac{V(k_BT)^4}{(\hbar c)^3}$, (b) $G = 0$, (c) $G = -PV$, (d) $G = -\frac{\pi^2}{45}\frac{V(k_BT)^4}{(\hbar c)^3}$,
3. The total E/M energy radiated per unit time and per unit area by a black body is
   (a) $I = \frac{\pi^2}{60}\frac{(k_BT)^4}{\hbar^3 c^2}$, (b) $I = c(U/V)$, (c) $I = 3cP$, (d) $I = \frac{\pi^2}{60}\frac{(k_BT)^4}{\hbar^3 c^3}$,
4. The total E/M energy radiated per unit time by a dipole is
   (a) $J = \frac{2}{3}(p\omega^2/c)$, (b) $J = \frac{2}{3}\frac{p^2\omega^3}{c^3}$,
   (c) $J \approx \frac{2}{3}\frac{p^2\omega^4}{c^3}$, (d) $J \approx \frac{2}{3}\frac{p^2\omega^5}{c^3}$
5. The frequency distribution of the black body radiation is proportional to $\omega^3/(e^{\beta\hbar\omega}-1)$, $\beta = (k_BT)^{-1}$. The wavelength distribution is proportional to
   (a) $(1/\lambda^3)\{\exp(2\pi c\beta\hbar/\lambda)-1\}^{-1}$ (b) $(1/\lambda^5)\{\exp(2\pi c\beta\hbar/\lambda)-1\}^{-1}$
   (c) $(1/\lambda^7)\{\exp(2\pi c\beta\hbar/\lambda)-1\}^{-1}$ (d) $(1/\lambda^9)\{\exp(2\pi c\beta\hbar/\lambda)-1\}^{-1}$
6. The maximum of the frequency distribution of the black body radiation appears at
   (a) $\omega_m = \sqrt{3}k_BT/\hbar$ (b) $\omega_m = 5.41k_BT/\hbar$
   (c) $\omega_m = 1.41k_BT/\hbar$ (d) $\omega_m = 2.82k_BT/\hbar$
7. The maximum of the wavelength distribution of the black body radiation appears at
   (a) $\lambda_m = 2.23c\hbar/k_BT$ (b) $\lambda_m = 2.91c\hbar/k_BT$
   (c) $\lambda_m = 1.27c\hbar/k_BT$ (d) $\lambda_m = 2\pi c/\omega_m$
8. The polarizability of an atom or a non-polar molecule of linear dimension r_a, which is defined as the ratio of the induced dipole moment by an electric field over this field, is in the G-CGS system of the form
   (a) const. r_a^4 (b) const. $r_a^3[\omega_0^2/(\omega_0^2-\omega^2-i\omega\gamma)]$
   (c) $r_a^2[\omega_0^2/(\omega_0^2-\omega^2-i\omega\gamma)]$ (d) $r_a^2[\omega_0^2/(\omega_0^2-\omega^2-i\omega\gamma)](c/\omega)$
9. The scattering cross-section of a λ= 600 nm photon by a neutral hydrogen atom is
   (a) $\sigma \approx 0.25\times10^{-20}\,\text{m}^2$ (b) $\sigma \approx 0.75\times10^{-14}\,\text{m}^2$
   (c) $\sigma \approx 0.5\times10^{-26}\,\text{m}^2$ (d) $\sigma \approx 0.64\times10^{-31}\,\text{m}^2$
10. The susceptibility of a gas is related with the polarizability of its molecules and their concentration n as follows:
    (a) $\chi_e = a_p/n$, (b) $\chi_e = a_p$ (c) $\chi_e = a_p n$ (d) $\chi_e = a_p n^2$
11. The relation between mean free path and scattering cross-section is
    (a) $\ell = n_s\sigma^2$ (b) $\ell = n_s\sigma$ (c) $\ell = n_s^2\sigma^{3.5}$ (d) $\ell = (1/n_s\sigma)$. $n_s = N_s/V$
12. The DC $(\omega = 0)$ conductivity of a metal is given by
    (a) $\sigma = e^2 n_f(\tau_f/m_e)$, (b) $\sigma = e^2 n_f\tau_f$,
    (c) $\sigma = e^2(\tau_f/m_e)$, (d) $\sigma = e^2 n_f(m_e/\tau_f)$
13. The maximum ω_m of the frequency distribution and the maximum λ_m of the wavelength distribution of the black body are related as follows
    (a) $\omega_m\lambda_m = 2\pi c$, (b) $\omega_m\lambda_m = 3.1416c$,
    (c) $\omega_m\lambda_m = 3.58c$ (d) $\omega_m\lambda_m = c/2\pi$

6.11 Solved Problems

1. Prove that the mean free path of a photon in a solid is given by the formula $\ell = \lambda/4\pi n_2$, where n_2 is the imaginary part of the refractive index defined in general as $n = n_1 + i n_2 \equiv \sqrt{\varepsilon\mu} \approx \sqrt{\varepsilon}$. In metals and for frequencies below the true plasma frequency, ε is approximately real and negative (see (6.27)). Hence, formally $n \approx i n_2 \approx \sqrt{-|\varepsilon|} = i\sqrt{|\varepsilon|}$.

Solution The propagation of a plane E/M wave within a solid is described by the general relation $E(x,t) = E(0,0)\exp(ikx - i\omega t)$ where E is the electric field and k and ω are connected as follows: $k = \sqrt{\varepsilon}\,\omega/c \approx i\sqrt{|\varepsilon|}\,\omega/c$. Substituting and taking the square of the absolute value of the field we obtain $|E(x,t)|^2 = |E(0,0)|^2\exp\left(-2\sqrt{|\varepsilon|}\,\omega x/c\right)$. Taking into account the definition of the mean free path in Sect. 6.6 and the expression for the wavelength in the vacuum $\lambda = 2\pi c/\omega$ we obtain $\ell = c/(2\omega\sqrt{|\varepsilon|}) = c/(2\omega n_2) = \lambda/(4\pi n_2)$. For a very thin film of dielectric constant ε and thickness d the transmission coefficient is $|t|^2 = 1 - (d/\ell')$, where $\ell' = \lambda/2\pi\varepsilon_2$ and ε_2 is the imaginary part of the dielectric function. This formula for $|t|^2$ is not valid for ideal metals for which ε_2=0; for them $|t|^2 = 1 - (d/\ell'')^2$.

2. Estimate the low frequency permittivity of silicon using formulae (6.28) and (6.26). It is given that the lattice structure of Si is of the diamond type with 8 atoms in the cubic unit cell of length 5.431A. Each atom has 4 "bound" electrons. The average value of $\hbar\omega_{0,j}$ can be estimated as the energy difference between the "lower" part of the empty band of states and the "upper" part of the occupied band of states (see Chap. 12, Sect. 12.9) which is about 5 eV.

Solution It is more convenient to do the calculation in the atomic system of units where $e = 1,\ \hbar = 1,\ m_e = 1$, the unit of length is 0.529 A and the unit of energy is 27.2 eV (see Table I.2 in Appendix I). Thus we have $\omega_{p,j}^2 = 4\pi n e^2/m_e$ $= 4\pi \times 8 \times 4/(5.431/0.529)^3 = 0.37$ a.u. and $\omega_{0,j} = 5/27.2 = 0.184$ a.u. Thus $\varepsilon_b = 0.37/0.184^2 \approx 10.9$ Thus the result for the permittivity, according to (6.28), is $\varepsilon = 11.9$, a value very close to the experimental one.

3. A high-frequency current running along a conducting wire is concentrated mostly near the surface of the wire up to a depth δ (skin-depth). Given that the permeability of the wire is μ and that its conductivity σ is very high $(\mathrm{Re}(\sigma) \gg \omega\,\mathrm{Re}(\varepsilon))$ calculate the skin-depth

Solution By the very statement of the problem it becomes obvious that the skin depth δ must depend on the conductivity σ and on the frequency ω. Moreover, since we are dealing with a high frequency E/M phenomenon, the velocity of light must enter in the formula for δ (in G-CGS). In this system, which we are using extensively in this book, σ has dimensions of 1/time. Hence the most general expression for δ satisfying the dimensional requirements is

$$\delta = \frac{c}{\omega} f\left(\frac{\omega}{\sigma}\right)$$

where f is an arbitrary function of the dimensionless ratio $x \equiv \omega/\sigma$. Let us assume that $f(x) = x^a$.

Then

$$\delta = c_1 \frac{c}{\omega} \frac{\omega^a}{\sigma^a}, \tag{6.42}$$

We know that $\delta \to 0$ when $\sigma \to \infty$. or when $\omega \to \infty$, since then there is no penetration of the E/M field within the metal. This implies that $0 < \alpha < 1$.

Dimensional analysis within the G-CGS system cannot go beyond (6.42). A full theory shows that $f(x) = \sqrt{x/2\pi}$ for all values of x such that $\mathrm{Re}(\sigma) \gg \omega \, \mathrm{Re}(\varepsilon)$. So

$$\delta = c \Big/ \sqrt{2\,\pi\,\sigma\,\omega} \tag{6.43}$$

Let us work in the SI system hoping that it can do better than (6.42) with $0 < \alpha < 1$. In SI besides the conductivity σ and the frequency ω, the permittivity ε and the permeability μ are expected to be present. However in view of (6.22b), ε is not an independent variable, since it is a function of σ and ω. Thus δ ought to be a product of powers of μ, σ, and ω. We need the dimensions of μ and σ. The latter according to the very statement of the problem has dimensions of ε over time. The dimensions of ε are obtained from Coulomb's law: $E = e^2/4\pi\,\varepsilon\, l \Rightarrow [\varepsilon] = [e^2/E\,l]$. The dimensions of μ can be obtained from the magnetostatic formula of the force per unit length F/l of two infinite parallel currents I_1, I_2 at a distance d apart:$F/l = (\mu/2\pi)(I_1 I_2/d)$ $\Rightarrow [\mu] = [E\,t^2/e^2 l]$. It follows that $[\varepsilon\,\mu] = [t^2/l^2]$ and $[\mu/\varepsilon] = [E^2/e^2 I^2] = [V^2/I^2] = [\mathrm{Resistance}^2]$ (actually $\varepsilon_0 \mu_0 = 1/c^2$ and $\mu_0/\varepsilon_0 = Z_0^2$, where c is the velocity of light in vacuum and $Z_0 = 376.6\ \Omega$ is the impendance of the vacuum) Thus $[\sigma\,\mu] = [t/l^2]$ and $[\sigma\,\mu\,\omega] = [1/l^2]$. The final conclusion is that in the SI system the skin depth is determined apart from a constant numerical factor by the relation: $\delta \propto 1/\sqrt{\sigma\mu\omega}$.

6.12 Unsolved Problems

1. Why the sky is blue? Why is getting reddish towards the sunset?
2. Is it possible to see both the Sun and the stars by going above the atmosphere? If yes, why? If no, why?

3. What is the order of the colors in a Rainbow? Is the red or the violet on the top? Can you justify your answer based on the frequency-dependence of index of refraction of water?
4. Obtain the visibility in moist air containing 10^6 droplets of water per cubic meter by calculating the mean free path of light in it. Assume that the average diameter of the droplets is 50 μm and that the permittivity of water at $\lambda = 600$ nm is 1.78.
5. Employing the definitions of polarization P and the electrical current density j, show that $\partial P/\partial t = j$.

Chapter 7
The Other Interactions

Abstract We present in terms of elementary Feynman diagrams the basic processes of strong interactions involving quarks and gluons as well as those of the weak interactions which involve all matter-particles (quarks and leptons) emitting or absorbing vector bosons (Z^0, W^+, W^-). The emission or absorption of a $W^{\pm}$ boson imply a transformation of the *m*-particle involved. Finally, we mention briefly the Newtonian version of gravitational interactions as well as an outline of the formulation of the general theory of relativity developed by Einstein.

7.1 General Remarks

In this chapter we shall employ elementary Feynman diagrams to provide some basic information regarding the Strong and the Weak Interactions at the quantum level where usually an *m*-particle emits or absorbs an ic-particle [1–3]. As we shall see this process of emission or absorption is associated in many cases with a transformation of the *m*-particle. This transformation can be viewed as the annihilation of the incoming *m*-particle and the creation of the outgoing *m*-particle. Depending on the direction of time arrow this same diagram describes other elementary physical processes, such as the annihilation (or the creation) of a particle/antiparticle pair and at the same time the creation (or the annihilation respectively) of an ic-particle. We shall also present diagrams where three or four ic-particles interact directly among themselves. We repeat here that at each vertex the conservation rules presented in Chap. 6 must be obeyed. Depending on the particular interaction, additional conservation rules must also be applied (see Chap. 6 for the additional conservation rules obeyed by the E/M interactions). The conservation rules impose restrictions on which elementary vertices are possible. For example in the elementary Feynman diagram for the E/M interaction presented in Fig. 6.1, the incoming charged particle and the outgoing charged particle must necessarily be the same and the emitted ic-particle must necessarily be a photon. In other words the E/M interaction cannot change any inherent characteristic of any particle.

E.N. Economou, *From Quarks to the Universe*,
DOI 10.1007/978-3-319-20654-7_7

We shall stress once more the importance of Feynman diagrams. At the descriptive level they provide an easy, direct, and vivid physical picture of the ongoing process. They allow also a very precise calculation of measurable quantities, such as the lifetime of a muon, or the magnetic moment of the electron, or the energy levels in a hydrogen atom, etc., by applying certain rules associated with the diagrams. These rules require an advanced background for their understanding and handling and an extensive familiarity with complicated multidimensional integrals for their implementations. In other words the calculational aspects of Feynman diagrams are well beyond the level of this book and thus they are not going to appear in what follows.

At the conclusion of this chapter we shall remind the reader the basic equations of gravitational interactions according to the Newtonian theory. For completeness, we shall also present the basic equations of Einstein's theory of gravity, the so-called general theory of relativity (GTR), and their solutions in some very simple model systems. Both the Newtonian and the Einsteinian versions of the gravitational interactions are classical, i.e. non-quantum. A generally accepted quantum theory of gravity, in spite of many attempts, has not been formulated yet.

7.2 Strong Interactions Involving Quarks and Gluons

In Fig. 7.1 various versions of the elementary Feynman diagram for strong interactions are presented. The quark q_i can be any of the six kinds (usually called *flavors*, $q_i = u, d, c, s, t, b$) and the antiquark $\bar{q}_i$ can be respectively any of $\bar{u}, \bar{d}, \bar{c}, \bar{s}, \bar{t}, \bar{b}$. In process (a) the flavor of the quark remains unchanged and in process (c) the flavor of the antiquark remains also the same. In processes (b) and (d) the pairs of quark/antiquark are of the same flavor, i.e. they can only be one of the following six combinations: $u/\bar{u}$, $d/\bar{d}$, $c/\bar{c}$, $s/\bar{s}$, $t/\bar{t}$, $b/\bar{b}$. We remind the

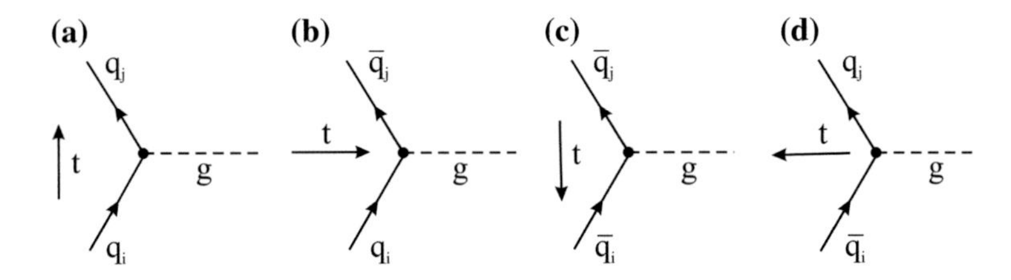

Fig. 7.1 The elementary Feynman diagram for the strong interactions and its four different physical interpretations: **a** a quark emits or absorbs a gluon; the quark either remains unaltered or may change only its color charge depending on the type of gluon involved (see text). **b** A quark and its antiquark (which may be different only in its color anticharge) annihilate and produce a gluon. **c** The same as in physical process **a** but for an antiquark (recall the convention that solid lines running opposite to the time arrow represent antiparticles). **d** A gluon decays giving rise to a pair of quarks/antiquarks; this is the time-reversed process of **b**. In all these elementary processes the color charge is conserved

reader that each quark carries color charge of the red (R) type, or the green (G) type, or the blue (B) type, and each antiquark carries one of the corresponding color anticharge (antired $\bar{R}$, antigreen $\bar{G}$, and antiblue $\bar{B}$). The transformation of the color charge (indicated as c-charge) in quarks involves the exchange of gluons. We have mentioned before in Chap. 2 that there are eight types of gluons:

$$R\bar{G},\ R\bar{B},\ G\bar{R},\ G\bar{B},\ B\bar{R},\ B\bar{G},\ \frac{1}{\sqrt{2}}(R\bar{R}-G\bar{G}),\ \frac{1}{\sqrt{6}}(R\bar{R}+G\bar{G}-2B\bar{B}) \quad (7.1)$$

A gluon $R\bar{G}$ when emitted in the process of Fig. 7.1a will transform an R quark to a G quark, while, if it is absorbed, will transform a G quark to an R quark (the flavor of the quark remains the same). As another example, in the process of Fig. 7.1b a quark of color charge G and an antiquark of color anticharge $\bar{B}$ (both of the same flavor) annihilate to produce a gluon $G\bar{B}$. As a third example the emission or absorption of an $\frac{1}{\sqrt{2}}(R\bar{R}-G\bar{G})$ gluon leaves an R quark or a G quark without any change in their color charge. In all cases the color charge is conserved at each vertex, which means that the total color charge of the ingoing particles (quarks and gluons) must be the same as that of the outgoing particles. Notice that in the process shown in Fig. 7.1a, if any of the first six of the gluons of (7.1) are involved, the c-charge of the incoming quark will change, while, if the last two of (7.1) participate, no change in the c-charge will occur.

Since the gluons carry color charge and anticharge, they also interact directly with each other as shown in the two elementary diagrams of Fig. 7.2. Actual physical processes involving the strong interactions are in general represented by diagrams

combining (repeatedly) the elementary component diagrams shown in Figs. 7.1 and 7.2. The construction of these composite diagrams is not a trivial task, since at each vertex the conservation rules must be obeyed and the energy must be conserved overall. The strong interactions besides the conservation rules presented in Chap. 6, also conserve the parity, the charge conjugation, and the time reversal. Moreover, the strong interactions do not "feel" the electric charge, which means that two quarks differing only in their electric charge are treated as equivalent by the strong interactions. Finally, let us repeat once more that the flavor of a quark or an antiquark cannot change by the action of the strong interactions.

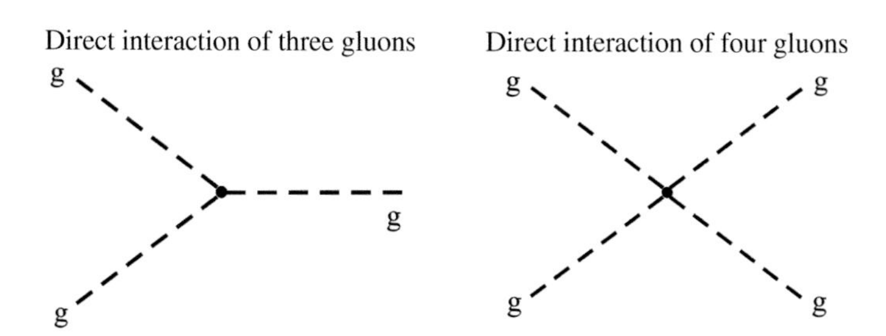

Fig. 7.2 Elementary Feynman diagrams showing the direct (without the intervention of quarks) strong interactions among gluons

We shall conclude this section by mentioning the following:

1. The domain of action of the strong interactions is restricted only to quarks and gluons. They do not act on leptons, or on photons, or on vector bosons.
2. The capability of the strong interactions to transform elementary particles is rather limited, since it cannot change the flavor of quarks and have no influence on leptons.
3. Because they are the strongest and at the same time more complex than the other interactions, as it is evident by the Feynman diagrams of Fig. 7.2, they present serious difficulties in yielding explicit calculational results.
4. The forces they produce among quarks can be either attractive or repulsive depending on the various combinations of color charges.
5. These forces exhibit two very characteristic features: (a) The *asymptotic freedom*, meaning that they become progressively weaker as the quarks come closer together. (b) The *confinement*, meaning that they become increasingly stronger as the distance between quarks becomes larger. As a result, quarks and gluons remain trapped in certain bound states; they cannot be extracted out of them. (See point 7 below).
6. Again as a result of the confinement, the strong interactions appear as being of short range, in spite of the fact that a gluon would produce a long range force, if it could be free, since its zero mass and the relation $r_0 = \hbar/mc$ imply an infinite range.
7. The confinement, which characterizes the strong interactions, can be deduced from the assumption that the lines of (the strong) force connecting a quark to another quark do not stem from the source in a spherically symmetric way as in the EM case. Instead, because of the specific to strong interaction vacuum fluctuations, the lines of force are squeezed within a tube of constant cross-section S connecting the two quarks. Then, the application of Gauss theorem, $S \times F \propto$ c-charge enclosed, produces an attractive force F independent of the distance r from the source and consequently a potential proportional to r: $\mathcal{V}(r) = |F|\, r$. Such a potential obviously accounts for the confinement.

7.3 Weak Interactions

Weak interactions are by far the most efficient ones in producing transformations of all kinds of m-particles and in annihilating (or creating) pairs of particles/antiparticles. Among their three ic-particles, the real transformers are the charged vector bosons, $W^{\pm}$. The neutral one, Z^0, cannot do for charged particles more than the photon and at much weaker strength. However, keep in mind that it is emitted or absorbed by neutral particles as well (any flavor of neutrinos), in contrast to the photon. This is shown in Fig. 7.3.

According to the diagrams in Fig. 7.3 any lepton or any quark can emit or absorb a neutral vector boson without changing any of its inherent characteristics. By

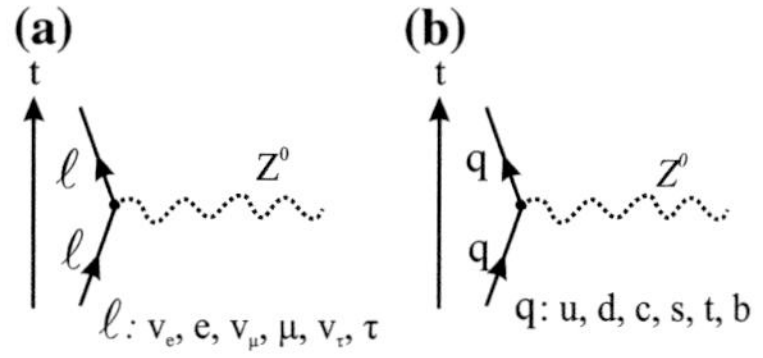

Fig. 7.3 **a** Emission or absorption of a neutral vector boson Z^0 by any lepton of the six species (flavors). The flavor of the lepton remains the same. **b** The same as **a** but for any of the six flavors of quarks. The flavor of the quark remains also the same. Depending on the direction of the time arrow the processes shown in Fig. 7.3 describe also the annihilation or creation of lepton/antilepton or quark/antiquark of the same flavor and of the same c-charge/c-anticharge

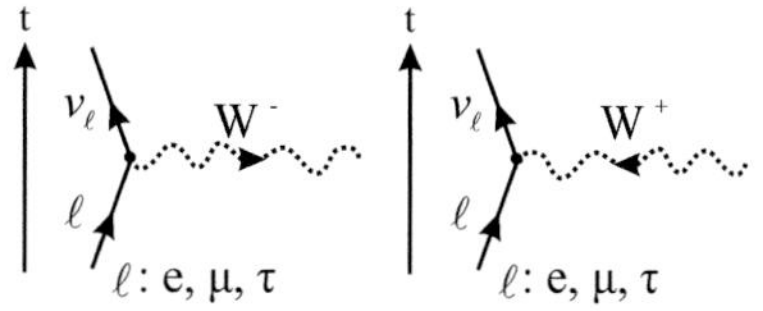

Fig. 7.4 An electrically charged lepton is transformed to the corresponding neutrino by emitting a W^- vector boson (*left*) or by absorbing a W^+ vector boson (*right*)

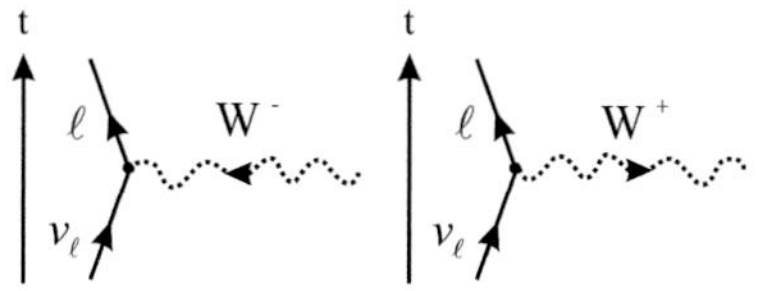

Fig. 7.5 The absorption of a W^- (*left*) and the emission of a W^+ (*right*) by a neutrino is accompanied by its transformation to the corresponding charged lepton

changing the direction of the arrow of time the same diagrams of Fig. 7.3 can describe annihilation (or creation) of a pair lepton/same antilepton or a quark/same antiquark with the creation (or annihilation respectively) of a neutral vector boson.

In Fig. 7.4 a charged lepton emits a negatively charged vector boson W^- or absorbs a positively charged vector boson and at the same time it is transformed to the corresponding neutral lepton, i.e. to the corresponding neutrino. Notice that, because lepton number must be conserved within each family, each charge lepton is associated with its own neutrino. By changing the direction of the arrow of time from left to right, the diagram of Fig. 7.4 on the left describes the annihilation of a charged lepton and its own antineutrino and the creation of a W^-, while the one on the right describes the decay of a W^+ to a pair of neutrino and its own charged antilepton, if the direction of the arrow of time is from the right to the left.

In Fig. 7.5 a neutrino absorbs a W^- vector boson or emits a W^+ vector boson and is transformed to the corresponding charged lepton. The same diagrams can describe

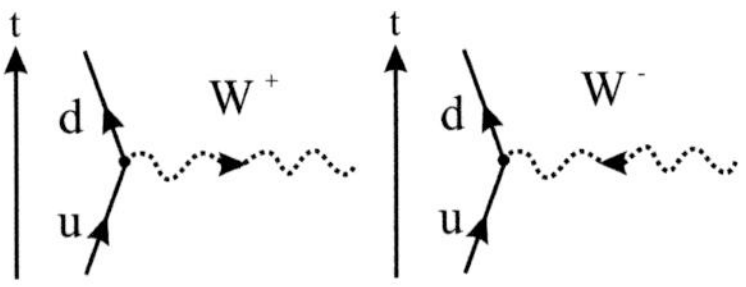

Fig. 7.6 The transformation of a u-quark to a d-quark by the emission of a W^+ (*left*) or by the absorption of a W^- (*right*)

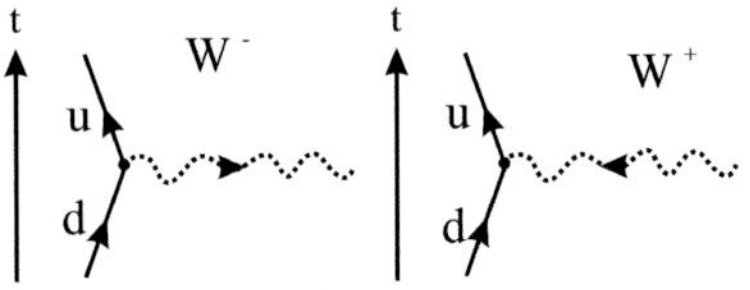

Fig. 7.7 The transformation of a d-quark to a u-quark by the emission of a W^- (*left*) or the by absorption of a W^+ (*right*)

the creation of a charged lepton/corresponding antineutrino pair at the expense of a W^-, or annihilation of a neutrino/corresponding charged antilepton with the creation of a W^+ ic-particle. Notice that by eliminating the arrows on the $W^{\pm}$ ic-particles, the two diagrams in Fig. 7.4 can be merged in one; the same is true for Fig. 7.5.

In Figs. 7.6 and 7.7 we see the transformation capabilities of the $W^{\pm}$ vector bosons on quarks within the same family. In Fig. 7.6 a u-quark emits a W^+ or absorbs a W^- and is transformed to a d-quark. In Fig. 7.7 a d-quark emits a W^- or absorbs a W^+ and is transformed to a u-quark. In Figs. 7.6 and 7.7 the pair of (u, d) quarks can be replaced by the pair (c, s) or the pair (t, b).

The weak interactions through the emission or absorption of the $W^{\pm}$ vector bosons allow transformations between different families of quarks. For example in Fig. 7.6 the u-quark can be replaced by the c-quark or the t-quark. As another example, the d-quark in Fig. 7.7 can be replaced by the s-quark or the b-quark and the u-quark by the c-quark or the t-quark. Examples of such diagrams are shown in Fig. 7.8.

The weak interactions allow also direct processes (without the involvement of m-particles) among the vector bosons themselves but also among the vector bosons and the photon. These processes correspond to Feynman diagrams where three or four wavy lines (corresponding to ic-particles) intersect at a single vertex as shown in Fig. 7.9.

7.4 Gravitational Interactions

We shall not employ Feynman diagrams for the gravitational interactions since an acceptable quantum theory for them has not been formulated yet. In other words the more advanced level of a theory where both the wave and the particle character of

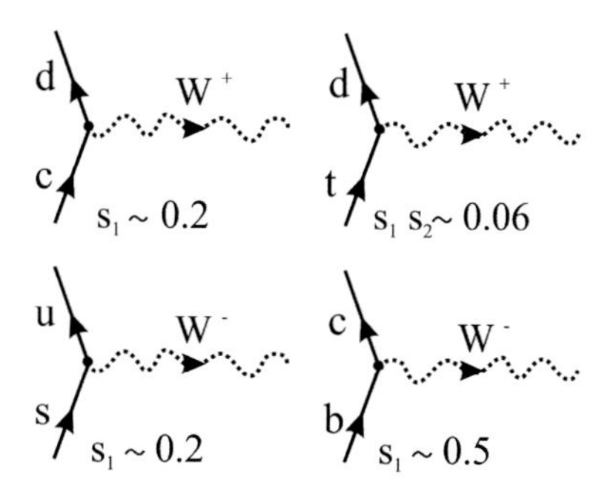

Fig. 7.8 Elementary Feynman diagrams showing the transformation of quarks from one family to another by the emission of a charged vector boson (the charged vector bosons are two of the three ic-particles associated with the weak interactions). These elementary processes have smaller probability amplitude to take place by a factor $s_1 \approx 0.2$ (*the first and the third*), or $s_2 \approx 0.5$ (*the fourth one*), or $s_1 s_2 \approx 0.06$ (*the second one*). These processes do not conserve the property of charmness (*the first one*), or the topness (*the second one*), or the strangeness (*the third one*), or the bottomness (*the fourth one*). Similar transformations take place by the absorption of charged vector bosons

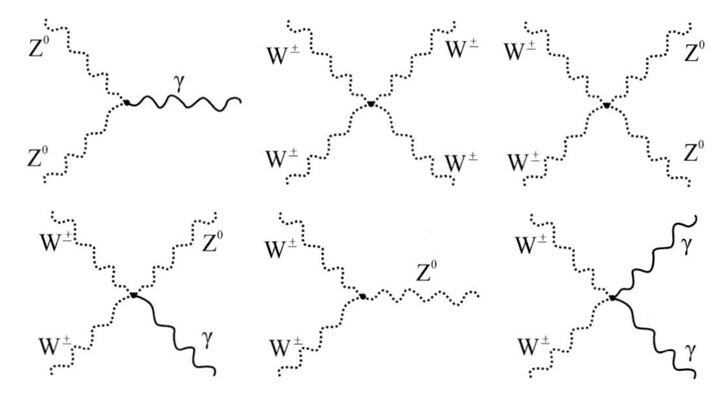

Fig. 7.9 Examples of elementary Feynman diagrams for the direct weak interactions among vector bosons with or without photons. Three or four wavy lines each representing an ic-particle (vector boson or photon) may converge at a single vertex. Charge conservation must be satisfied

reality are taken into account on equal footing, has not been achieved yet for the gravitational interactions, in spite of so many attempts. This is remarkable: The most familiar force, the one experienced continuously by all of us, the first one to be studied quantitatively marking the birth of modern science, turns out to be the less understood at a deep microscopic level.

The next, partially complete level of a theory, which, necessarily includes the wave character of reality, was developed for the gravitational interactions a century ago by a single person, Albert Einstein. This was one of the most (if not the most) remarkable leap and achievement of the human mind. However, due to the extreme weakness of the gravitational interactions, their wave character has not been confirmed yet observationally. The most elementary level for gravitational interactions, which ignores both their particle and their wave character and treats them as a field of infinitely fast transmitted force or potential energy is the one developed in the seventeen century mostly by the work of Newton.

There are three remarkable features of the gravitational interactions: (a) Their extreme weakness, about thirty six orders of magnitude weaker than the E/M interactions. (b) Their absolute universality, meaning that *everything* is subject to them and is α source of them. (c) Their always attractive character, which allows them to become appreciable and eventually to dominate over the other interactions when the number of elementary particles involved is huge (more than 10^{50}).

In the next subsection we shall present the most elementary level of gravitational interactions, which gives the mutual force between two point masses and is known as Newton's law of universal attraction. Actually, we shall present this law in a different but equivalent way, which is better adapted for treating masses of finite extent possessing spherical or cylindrical symmetry.

7.4.1 Newtonian Formulation of Gravitational Interactions

The gravitational field **E** is defined as the ratio of the gravitational force **F** acting on a test point mass m over this mass. The connection of this field to its sources (which are the masses distributed in space) is facilitated by introducing the concept of the *circulation* of the field over an oriented closed line and the concept of the *flux* of the field through an oriented surface. Both of these concepts were introduced originally in the field of hydrodynamics, where the circulation and the flux of the velocity field have a direct and vivid physical meaning. It so happens that the gravitational field as defined above (and the E/M field as described by Maxwell's equations, see Appendix B) have an underlined hydrodynamic character, in the sense that the flux and the circulation of these fields are directly related to their sources, being proportional to them. More explicitly, we have for the gravitational field

$$\text{Circulation of } \mathbf{E} \equiv \oint_c \mathbf{E} \cdot d\mathbf{r} = 0 \tag{7.2}$$

$$\text{Flux of } \mathbf{E} \equiv \oiint_S \mathbf{E} \cdot \mathbf{n}\, dA = -4\pi G M_S \tag{7.3}$$

The symbol c denotes a closed oriented contour and $d\mathbf{r}$ is a tangential infinitesimal vector at each point of c pointing along its positive direction. In (7.2) the circulation of **E** is defined and its value is given; it is zero. The fact that the circulation of **E** along any closed contour is zero means that the line integral of **E** depends only on the end points of the line and that by fixing one of the two points (usually at infinity) we can uniquely define the potential $\tilde{\mathcal{V}}(\mathbf{r})$ (of dimensions energy over mass) at any point **r** as follows:

$$\tilde{\mathcal{V}}(\mathbf{r}) \equiv -\int_{\infty}^{\mathbf{r}} \mathbf{E} \cdot d\mathbf{r}' \tag{7.4}$$

In (7.3) the flux of **E** through a closed surface is defined and its value is connected with the *enclosed* mass within the surface S; **n** is the unit vector perpendicular to the surface at every one of its points and oriented outwards, dA is the infinitesimal area at each point of the surface, G is the gravitational constant (its value is $6.674 \times 10^{-11}\,\mathrm{N\,m^2\,kg^{-2}} = 2.4 \times 10^{-43}$ in atomic units), and M_S is the total mass in the *interior* of the surface S. In other words, masses outside the closed surface do not enter in (7.3). From (7.3) we can easily deduce the law of universal attraction: Consider a point mass M and a concentric sphere of radius r. Because of the spherical symmetry the product $\mathbf{E} \cdot \mathbf{n}$ appearing in (7.3) is constant across the surface of the sphere and can be taken out of the integral; moreover, again because of spherical symmetry, the field **E** is parallel or antiparallel to **n**. Thus $\mathbf{E} \cdot \mathbf{n}\, 4\pi r^2 = -4\pi G M$ or

$$\mathbf{E} = -\frac{GM}{r^2}\mathbf{n} \tag{7.5}$$

which, if multiplied by the test point mass m, becomes identical to the law of universal attraction. In a later chapter we shall use (7.3) to obtain the gravitational field inside the Earth (assumed to be spherical), the acceleration of gravity inside Earth (which is the same as the field), the potential inside Earth (by calculating the integral of the field according to (7.4)), and finally the gravitational self-energy of Earth (by integrating over the volume of Earth half the product of the potential times the mass density).

7.4.2 Gravitational Interactions According to Einstein

This theory, also called general theory of relativity (GTR), is based on the revolutionary idea that gravity modifies the "geometry" of space-time. Space and time are not anymore the familiar frozen, unaltered framework within which events take place. Space and time acquire a plasticity adapting their very structure and modifying their geometrical features (e.g. a straight line may not be the shortest distance) according to the "instructions" of the distribution of masses or energies, which constitute the gravitational sources. Fortunately, mathematicians had already worked out the abstract formalism required to describe quantitatively general classes of non-Euclidean geometries even beyond the ordinary three-dimensional space. Thus the implementation of Einstein's revolutionary idea was facilitated thanks to the intellectually admirable previous work of mathematicians, but still was depending on another crucial step: Which one of the many abstract geometrical quantities associated with the formalism of non-Euclidean geometries are the ones to be connected with the distribution of masses or energies. The successful completion of this step led to the GTR, an impressive edifice of the human mind confirmed by numerous experiments and also considered by many as the most "beautiful" physical theory.

Let us present the above qualitative ideas in an explicitly quantitative framework without any attempt to justify their specific forms. The gravitational sources are connected to the distribution and the motion of energies and are expressed in terms of the so-called energy-momentum tensor

$$T_j^k \equiv (\varepsilon + p)\, u_j u^k - p\, \delta_j^k \tag{7.6}$$

where ε is the energy density, p is the pressure, u_j are the components of velocity divided by c, and summation over repeated indices, if any, is implied. Notice that important details, such as the covariant or the contravariant character of the components of four-vectors, etc., have not been mentioned here, since we do not intend to provide a working knowledge of the GTR, simply to give a first glimpse of its basic structure. The general four-dimensional geometry is defined through the expression of the square of the infinitesimal "length" in the four-dimensional time-space

$$ds^2 = g_{ij} dx^i dx^j, \qquad i, j = 0,\ 1,\ 2,\ 3 \tag{7.7}$$

The tensor g_{ij} defines the geometry of the 4D time-space, where the zero component refers to time and the other three to the 3D space. For our familiar Newtonian conception of time and space, called in the GTR jargon the Minkowski space-time, the components of the tensor g_{ij} are as follows: $g_{00} = 1,\ g_{11} = g_{22} = g_{33} = -1$ with all the other components being zero, $dx^0 = c\, dt$, and the three-dimensional space obeying the ordinary Euclidean geometry. However, in the presence of the energy-momentum tensor, the tensor g_{ij} becomes in general non-diagonal and its components are in general complicated functions of the four space-time variables, $x^i,\ i = 0,\ 1,\ 2,\ 3$. Through the sixteen basic quantities g_{ij}, other tensors, vectors, and scalars are defined, such as the so-called Riemann tensor R_j^k or the Riemann scalar quantity R. In terms of these last quantities the Einstein tensor G_j^k is defined; G_j^k is designed to be connected directly with the gravitational sources, i.e. the energy-momentum tensor:

$$R_j^k - \frac{1}{2} R\, \delta_j^k \equiv G_j^k = -\frac{8\pi G}{c^4} T_j^k - \frac{\Lambda}{c^2} \delta_j^k \tag{7.8}$$

Equation (7.8) is the basic relation of the GTR connecting the geometry of space-time as distilled in G_j^k with the energy-momentum tensor, the source of gravitational interactions; it involves also the last term in the right-hand side of (7.8), which is proportional to the so-called cosmological constant Λ. This term was added by Einstein afterwards in order to obtain time independent cosmological solutions and later was withdrawn by him in view of the discovery of the expansion of the Universe. Recently, the cosmological constant is a quantity of much concern, being an integral part of modern Cosmology, because it seems to be proportional to the so-called dark energy density. The solution of (7.8) must lead to the determination of all the components of the tensor g_{ij}, i.e. to an explicit expression of ds^2 as a function of time and position variables and their differentials.

7.4.3 Two Model Systems Allowing Exact Solutions of the GTR

The first one to be considered here is a model system consisting of a spherical, uncharged, non-rotating body of mass M and radius R.

In this case the exact solution of (7.8), under the conditions $r > R,\ r_S$, gives for ds^2

$$ds^2 = \left(1 - \frac{r_S}{r}\right) c^2 dt^2 - \frac{dr^2}{1 - (r_S/r)} - r^2(d\theta^2 + \sin^2\theta\, d\varphi^2), \quad r > R,\ r_S \tag{7.9}$$

Notice that the space variables have been expressed for convenience in spherical coordinates, i.e. $x = r\sin\theta\cos\varphi$, $y = r\sin\theta\sin\varphi$, $z = r\cos\theta$. The solution (7.9) looks very similar to the Minkowski one; actually along the tangential space direction, the solution is identical to that of the Euclidean geometry. The difference is along the radial space direction and in the time scale because of the factor

$$\{1 - (r_S/r)\}$$

The radius r_S defines the surface of a sphere which is called the event horizon. The quantity r_S is known as the Schwarzschild radius. It is given by the following relation:

$$r_S = \frac{2GM}{c^2} \tag{7.10}$$

Notice that this expression can be obtained (apart from the numerical factor 2) from dimensional analysis, since there is only one combination of the two universal constants G and c and the parameter M which can produce length (apparently, as in the Newtonian case, the radius of the body R does not play any role as long as we are outside the body, $r > R > r_S$). The r_S also happens to be obtained as the radius of a spherical mass M whose escape velocity, according to Newtonian mechanics, is $c: \frac{1}{2}mc^2 = GMm/r_S$.

As $r \to \infty$, (7.9) tends to the Minkowski space-time, i.e. to the familiar Euclidean geometry and to the Newtonian concept of time. However, as we are approaching the body, time runs slower (because of the factor $\{1 - (r_S/r)\}$ multiplying the time differential) and the radial distances become larger (because of the inverse of this factor multiplying the radial differential). Thus the product $4\pi\times$ square of the radial length up to the point r, which in Euclidean geometry is equal to the surface of the corresponding sphere, here in the distorted space is larger than the surface of the sphere.

In the case where $r_S > R$ the event horizon can be physically reached, since the body is not covering it anymore. As the "unprotected" event horizon is approached, the slowing down of time becomes more extreme and at the event horizon, $r = r_S$, time stands still. The case $r_S \geq R$, where the event horizon coincides with or is outside the surface of the body so that it is exposed and can be reached following (7.9) (which is valid for $r \geq R$ and $r \geq r_S$) corresponds to a *black hole*. Its event horizon on which time disappears and the radiant distances tend to blow up separates physically the outside distorted space-time from its interior from which nothing can escape.

The Schwarzschild radius as given by (7.10) is a very small length under ordinary conditions, so that the relation $r_S \geq R$ is far from being realized in the vicinity of our cosmic neighborhood. This becomes clear by substituting numerical values for G, and c in (7.10):

$$r_S\,(\text{in km}) = 1.485 \times 10^{-30} \times M\,(\text{in kg}) \tag{7.11}$$

For the mass of the Earth, $M = 5.972 \times 10^{24}$ kg, $r_S = 0.887$ cm, while for the mass of the Sun, $M = 1.989 \times 10^{30}$ kg, $r_S = 2.95$ km. Thus, these bodies are really very far for satisfying the relation $r_S = R$. However, there are compact dead stars, such as neutron stars, where this relation is satisfied and the dead star becomes a black hole.

The second case, to be mentioned here which allows an exact solution, is the so-called Robertson-Walker model.

This model consists of a uniform, isotropic, and unbounded medium of energy density ε and pressure p. The solution of (7.8) for this model leads to the following explicit expression for the metric tensor g_{ij}:

$$ds^2 = c^2dt^2 - [R(t)]^2\left[\frac{du^2}{1 - \kappa u^2} + u^2(d\theta^2 + \sin^2\theta\, d\varphi^2)\right] \tag{7.12}$$

where the radial variable has been written as $r = R(t)u$. The quantity $R(t)$ is an explicitly time dependent parameter having dimensions of length and determining the size scale of space; u is a dimensionless length variable along the radial direction. It is clear from (7.12) that this solution describes a time dependent geometry of an expanding or a shrinking medium where the distance between two points is increasing or decreasing at a rate which proportional to this distance:

$$\frac{dr}{dt} \equiv \dot{r} = \dot{R}\,u = \dot{R}\,\frac{r}{R} = \frac{\dot{R}}{R}\,r \tag{7.13}$$

The proportionality constant $(\dot{R}/R)$ connecting the rate of change dr/dt of the distance between two points to this distance r is called the *Hubble constant* and is denoted by the symbol H in cosmology. The Hubble constant became a quantity of crucial physical importance, after the discovery by Hubble in the late nineteen-twenties that the Universe really is expanding following the law given in (7.13). Thus the Robertson-Walker model is not just a toy model but is relevant to the behavior of the Universe as a whole.

Two more final remarks regarding the solution (7.12) are in order: First, more information regarding the nature of the energy density and the pressure are needed in order to determine the dependence of R on t. Second, there is an undermined parameter κ in (7.12), which, depending on its values, produces different types of geometries:

- If $\kappa = 0$, the geometry is Euclidean and the space is unbounded extending to infinity in all directions
- if $\kappa = 1$, the space is curved, unbounded, but finite, (The 3D analog of the 2D surface of a sphere)
- if $\kappa = -1$, the space is curved, unbounded, extending to infinity in all directions (The 3D analog of the 2D surface of a hyperboloid of one sheet).

In Chap. 15 of this book dealing with cosmology we shall return to the questions of the time dependence of the scale parameter R and to the value of κ in our Universe.

7.4.4 Universal Physical Constants and the Planck System of Units

The Planck's reduced constant $\hbar$ and the velocity of light in vacuum c are the most "universal" of all the physical constants. This is so because they are not connected to a particular group of elementary m-particles or to a specific one of the four basic interactions. Instead they refer to the physical reality in general by setting some limits. The velocity of light in vacuum is setting the upper limit for the velocity under which mass, or energy, or information can be transmitted. The Planck's constant is setting a lower limit on the size of "length", or "area", or "volume" in phase space; phase space is the product of real × momentum space. More explicitly, the minimum "length", $\delta x \times \delta p_x$, in phase space is equal to $2\pi\hbar = h$, the minimum "area" is $(2\pi\hbar)^2 = h^2$, the minimum "volume" is $(2\pi\hbar)^3 = h^3$, and so on.

We can claim that $\hbar$ is even more universal than c, because it plays a significant role at all scales of matter organization from that of the elementary particles and their interactions, the hadronic matter, the atomic nuclei, the atoms, the molecules, the condensed matter, the living matter, the planets, the stars, the dead stars, and the whole Universe. For the structures of matter, from nuclei all the way to an asteroid, the role of the velocity of light is rather limited (except when E/M external waves are involved). This is why the natural system of units in the range from 10^{-10} to 10^5 m involves mainly $\hbar$ and the constants associated with the electron: e, m_e.

In the Cosmology field, besides $\hbar$, and c, a natural system of units must also include the gravitational constant G, since gravity is the dominant interaction in this regime. A system of units based on these three universal constant is known as the *Planck system*. The formulae and the numerical values of some units in the Planck system are given below.

Length unit: $\ell_P = \sqrt{G\hbar/c^3} = 1.6161 \times 10^{-35}$ m
Time unit: $t_P = \ell_P/c = \sqrt{G\hbar/c^5} = 5.3907 \times 10^{-44}$ s
Mass unit: $m_P = \sqrt{\hbar c/G} = 2.1766 \times 10^{-8}$ kg
Energy unit: $E_P = m_P c^2 = 1.956 \times 10^9$ J
Temperature unit: $T_P = m_P c^2/k_B = 1.4196 \times 10^{32}$ K

7.5 Summary of Important Formulae

$$\text{Flux of } \mathbf{E} \equiv \oiint_S \mathbf{E} \cdot \mathbf{n}\, dA = -4\pi G M_S \tag{7.3}$$

$$\tilde{\mathcal{V}}(\mathbf{r}) \equiv -\int_{\infty}^{\mathbf{r}} \mathbf{E} \cdot d\,\mathbf{r}' \tag{7.4}$$

$$\mathbf{F}_m = -\frac{G M m}{r^2}\mathbf{n} \tag{7.5}$$

$$r_S = \frac{2 G M}{c^2} \tag{7.10}$$

7.6 Multiple-Choice Questions/Statements

1. Which one of the following diagrams describes correctly the decay of μ^-?

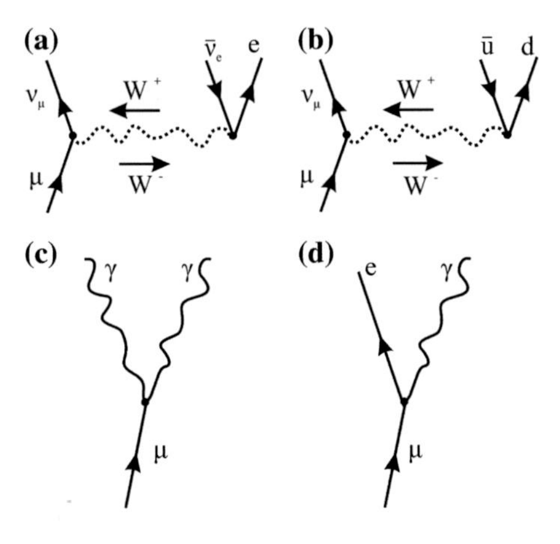

2. Which one of the following diagrams describes correctly the decay of d-quark?

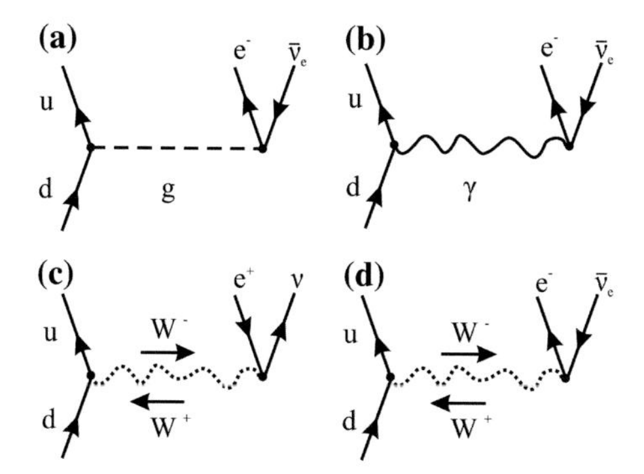

3. Which one of the following elementary diagrams is unphysical?

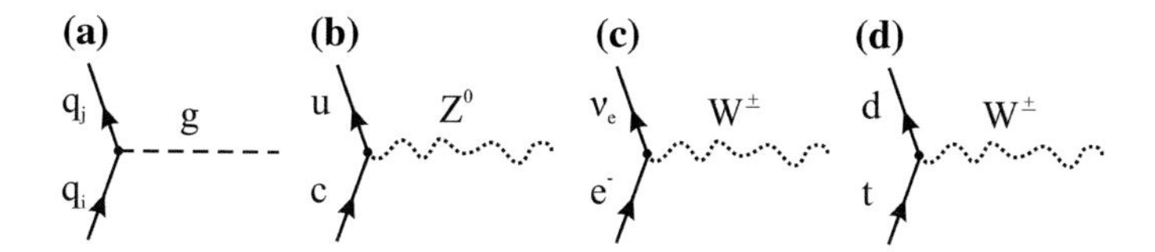

7.7 Problems

Applications of (7.3) and (7.4) will appear in Chaps. 13 and 14 in calculating the acceleration of gravity within a planet or a star, the corresponding potential, the pressure, and the gravitational self-energy.

References

1. D. Perkins, *Introduction to High Energy Physics*, (Addison-Wesley, Reading, 1987)
2. F. Wilczek, *The Lightness of Being: Mass, Ether, and the Unification of Forces* (Basic Books, New York, 2008)
3. W. Cottingham, D. Greenwood, *An Introduction to the Standard Model*, 2nd edn. (Cambridge University Press, Cambridge, 2007)

Part III
Structures Held Together by Strong Interactions

Chapter 8
From Quarks and Gluons to Hadrons

Abstract The first step towards composite structures of matter is the combination of three quarks of total color-charge zero or the combination of a quark and an antiquark of total color-charge zero mainly through the mediation of gluons. Only protons consisting of two up quarks and one down quark are stable. Free neutrons have a mean lifetime of about 15 min and are stabilized only inside non-radioactive nuclei.

8.1 Summary

As was mentioned in Chap. 2, the long journey towards the immense variety of the structures of the World starts with three quarks of the first family combined to form protons and neutrons; protons are the only stable baryons when isolated or as part of non-radioactive nuclei; neutrons, when isolated, are metastable (with mean lifetime[1] of about 887 s) but become stabilized as part of non-radioactive nuclei. Protons consist of two u quarks and one d quark, with all three quarks having different c-charges, so that the total c-charge of each proton to be zero. Neutrons consist of two d quarks and one u quark, with all three quarks having different c-charges, so that the total c-charge of each neutron to be zero. There are many other *baryons* consisting of any three of the six flavors of quarks but always of different c-charge so that the total c-charge to be also zero. All these other baryons are metastable, meaning finite lifetimes ranging from 10^{-24} s (typical when they decay through strong interactions) to about 10^{-10} or even 10^{-8} s (when they decay through the weak interactions).

[1]There are two related times which characterize the rate of exponential decay of metastable structures. If initially ($t = 0$) there were N_0 such structures, their number at a later time t is $N(t) = N_0 \exp(-t/\tau)$, where τ is called their *mean lifetime* or simply lifetime. We define also the *hal-life*, $t_{1/2}$, as the time at which the remaining number $N(t_{1/2})$ is half the initial number N_0; obviously we have $\exp(-t_{1/2}/\tau) = 1/2$, or $t_{1/2} = \tau \ln 2 = 0.693\tau$.

E.N. Economou, *From Quarks to the Universe*,
DOI 10.1007/978-3-319-20654-7_8

There are also combinations of one quark and one antiquark of zero total c-charge called *mesons*; all mesons are metastable. The lower energy mesons made up from quark/antiquark of the first family are called *pions*. Mesons and baryons, collectively called *hadrons*, seem to be the only combinations of quarks in stable or metastable existence. The strong interactions mediated by virtual gluons provide the main mechanism binding together the quarks into baryons and mesons, i.e. hadrons. However, the decay of the latter may involve in general other interactions as well, such as the E/M and the weak ones (see Tables 8.1 and 8.2). Besides protons and neutrons, which make up about 99.97 % of the mass around and inside us, a few other metastable mesons and baryons are reaching the Earth from outer space as cosmic rays or as byproducts of cosmic rays. However, most of the metastable baryons and mesons as well as the metastable leptons and the vector bosons are produced mainly through collisions of particles, such as electrons, protons, positrons, antiprotons, etc. which have been accelerated to reach high kinetic energy by very sophisticated big machines of various types called accelerators or colliders. In the newly operated so-called Large Hadron Collider (LHC) two opposite running beams of protons come to head-on collision; each of these particles will reach a kinetic energy exceeding its rest energy by a factor of about 7500! With the availability of such a huge energy per particle a variety of numerous particles are produced which are detected by big devices, real "miracles" of science and technology. The enormous amount of collected data are analyzed and interpreted through comparison with the results of the established theoretical scheme, called *the Standard Model*. The latter is really an impressive intellectual achievement of many years of collective scientific labor. In this chapter, besides a brief introduction to this extensive and expanding field, an estimate of the masses of proton and neutron is offered based on simple kinetic energy arguments.

Table 8.1 Some of the baryons formed out of three quarks of total c-charge zero

Name/symbol	Composition	Mass/m_e	Size (fm)	Electric charge/$\lvert e \rvert$	Mean lifetime (s)	Main process of decay
Proton/p	uud	1836,15	0.84	1	$>10^{32}$ year	–
Neutron/n	udd	1838,68	0.84	0	889	$\rightarrow pe\bar{\nu}_e$
Δ^{++}	uuu	2411	No accurate values are available	2	5.5×10^{-24}	$\rightarrow p\pi^+$
Λ^o	uds	2183		0	2.63×10^{-10}	$p\pi^-$, $n\pi^o$
Σ^o	uds	2334		0	6×10^{-20}	$\rightarrow \Lambda^o\gamma$
Σ^+	uus	2318		1	0.8×10^{-10}	$\rightarrow p\pi^o$, $n\pi^+$
Σ^-	dds	2343		−1	1.48×10^{-10}	$\rightarrow n\pi^-$
Ξ^o	uss	2573		0	2.9×10^{-10}	$\rightarrow \Lambda^o\pi^o$
Ξ^-	dss	2586		−1	1.64×10^{-10}	$\rightarrow \Lambda^o\pi^-$
Ω^-	sss	3273		−1	0.82×10^{-10}	$\Lambda^o K^-$,$\Xi^o\pi^-$
Λ_c	udc	4471		0	2.1×10^{-13}	$pK^- \pi^+$

Only protons and neutrons are stable (neutrons are stable only inside stable nuclei) [1]

Table 8.2 Some of the mesons formed out of a quark and an antiquark; all of them are metastable [1]

Symbol	Composition	Mass /m$_e$	Electrical charge/ $\lvert e \rvert$	Spin	Mean lifetime(s)	Main process of decay
π^o	$\frac{1}{\sqrt{2}}(u\bar{u}-d\bar{d})$	264	0	0	$0.84\text{x}10^{-16}$	$\gamma\gamma$
π^+	$u\bar{d}$	273	1	0	$2.603\text{x}10^{-8}$	$\bar{\mu}^+\nu_\mu$
π^-	$d\bar{u}$	273	-1	0	$2.603\text{x}10^{-8}$	$\mu^-\bar{\nu}_\mu$
K^o	$d\bar{s}$		0	0	$K_L^o 5.17\times10^{-8}$	$\pi^o\pi^o\pi^o$
$\bar{K}^o$	$s\bar{d}$	970	0	0	$K_s^o 0.892\times10^{-10}$	$\pi^+\pi^-$
K^+	$u\bar{s}$		1	0	$1.237\ \text{x}10^{-8}$	$\mu^+\nu_\mu$
K^-	$s\bar{u}$		-1	0	$1.237\ \text{x}10^{-8}$	$\mu^-\bar{\nu}_\mu$
J/ψ	$c\bar{c}$	6070	0	1	$7.2\text{x}10^{-21}$	Mainly hadrons $e^+e^-, \bar{\mu}^+\mu^-, \eta(\pi^+\pi^-)\pi^o$
D^o	$c\bar{u}$		0	0	$4.15\text{x}10^{-13}$	$K^-\pi^+\pi^o$ κ.α.
$\bar{D}_o$	$u\bar{c}$	3650	0	0		
D^+	$c\bar{d}$		1	0	$1.04\ \text{x}10^{-12}$	
D^-	$d\bar{c}$		-1	0		
B^+	$u\bar{b}$		+1		1.62×10^{-12}	
B^-	$b\bar{u}$		-1			
B_d^o	$d\bar{b}$	10315	0			
$\bar{B}_d^o$	$b\bar{d}$		0		$1.56\text{x}10^{-12}$	
B_s^o	$s\bar{b}$		0			
$\bar{B}_s^o$	$b\bar{s}$		0			

8.2 Baryons and Mesons

The basic elementary process accounting for the formation of baryons and mesons out of quarks was shown schematically in the diagram of Fig. 7.1. In Fig. 7.2 the direct interaction of three or four gluons was also shown. By properly combining elementary diagrams, such as those of Figs. 7.1 and 7.2 and employing the rules associated with those diagrams, one can in principle calculate theoretically explicit values for experimentally determined quantities, such as the rest mass and the magnetic moment of protons and neutrons, the mean lifetime of an isolated neutron, etc. Before even attempting to calculate such specific quantities, one has to answer first the following general qualitative questions:

(a) Why do the bound combinations of quarks, stable or metastable, have to be colorless (i.e., of zero total c-charge)?
(b) Why do only bound combinations of three quarks or bound quark/antiquark pairs exist? In other words, why stable (or metastable) zero total c-charge bound combinations of 3n quarks, or stable (or metastable) bound combinations of n quarks/n antiquarks do not exist?
(c) Could real gluons (as opposed to virtual ones) be constituents of stable (or metastable) bound clusters?

We know that the answers to such questions could in principle be obtained by the general idea already presented in Chap. 4. We have to calculate the total energy of the observed composite particles, i.e. the baryons and the mesons, and show that it is considerably lower than the energy of the other hypothetical particles mentioned above. The implementation of this obvious approach meets serious practical difficulties due to the high value of the strength a_s of the strong interactions. As a result of a_s being of the order of unity, each of a huge number of diagrams of ever increasing number of vertices and complexity contributes to the total energy an amount of the same order of magnitude. This makes the diagram-based explicit calculations practically impossible [1]. To understand this difficulty, it may be of help to compare the theory of electron-photon E/M interaction (named quantum electrodynamics, QED) with the theory of quark-gluon strong interaction (named quantum chromodynamics, QCD [2, 3]). In QED there are no diagrams similar to those shown in Fig. 7.2, since the photons do not have e-charge and, hence do not interact directly with each other; as a result, the diagrams in QED are much simpler. Moreover, the dimensionless coupling constant α in QED is only 1/137 (see Table 2.2), in contrast to $a_s \approx 1$ in QCD; as a result, and in contrast to QCD, there is no need in QED to keep diagrams with more than a few vertices.[2] The contributions of a few simple diagrams in QED suffice to give theoretical results in impressive agreement with the experimental data. On the other hand, in QCD the strength of the coupling constant $a_s \approx 1$ and the extra complication introduced by the elementary diagrams of Fig. 7.2 prevent us from repeating the phenomenally successful calculations of QED. Nevertheless, a special very laborious computational scheme has been developed which predicts the masses of several baryons and mesons in reasonable agreement with the corresponding experimental data [2]. There is another issue where the comparison with QED is helpful: We mentioned before, Sect. 2.3 that the range r_o of each interaction is related to the mass m of the

[2]Each additional vertex in a diagram introduces a factor of $\alpha = 1/137$ in QED, while the corresponding factor in QCD is a_s, i.e., approximately equal to one.

corresponding ic-particle through the formula $r_o = \hbar/mc$. That's why the range of Coulomb interaction energy between two e-charged particles mediated by photon exchange is infinite. By analogy we should expect that the range of the strong interaction energy between two quarks mediated by gluons, having zero mass, to be infinite. However, we know experimentally that this is not true: The range of the strong interactions is quite small of the order of 10^{-15} m. How does QCD account for this apparent discrepancy? There are strong indications, although no rigorous proof, within the framework of QCD that the interaction energy between two quarks (or between two c-charged particles) tends to infinity as their separation tends to infinity (see point 7 of Sect. 7.2). Thus, it seems that the energy cost to completely isolate a quark or a c-charged particle is infinite. In other words, it is forbidden by energy considerations to have isolated quarks, or other entities carrying a non-zero c-charge. This argument provides an answer to question (a) above and at the same time explains why the range of the strong forces is so short: Quarks or gluons or any other c-charged particles are always locked-up within c-neutral structures such as protons, neutrons, pions, etc. This *confinement* explains automatically the effectively short-range nature of the strong interaction, since isolated gluons are always locked up in c-neutral structures. What is going to happen if we try to extract a quark from a c-neutral structure by supplying a lot of energy? The system may be deformed and stretched in complicated ways until enough energy is provided to one of the virtual gluons keeping the quarks together; then the gluon may break to a quark/antiquark pair as shown in Fig. 7.1d. The newly created quark will stay within the initial system, while the newly created antiquark will accompany the initial quark in its exit. Thus the end result will be to extract a meson, i.e. a c-neutral structure of non-zero mass and leave behind another c-neutral structure containing the same number of quarks as the initial one. A similar mechanism is at work when two c-neutral composite particles (e.g., a neutron and a proton, collectively called *nucleons*) come close together; their interaction, instead of being described through gluon exchange, may be described in a simpler, but essentially equivalent way, as the creation and exchange of virtual pions and possibly other mesons between the two c-neutral particles. This meson mediated description of the interaction of proton/proton, proton/neutron, and neutron/neutron is shown in an oversimplified form in Fig. 8.1; it will be used in the next chapter to account for the formation of atomic nuclei out of protons and neutrons. This meson mediated description may break down when the two c-neutral composite particles are forced to come so close together as to almost merge into one. In this case the more fundamental description in terms of quarks and gluons may be necessary. In any case it is shown experimentally that the merging of the proton/proton pair (or the proton/neutron, or the neutron/neutron pair) into a single six-quark composite particle is not energetically favorable and does not occur spontaneously. This feature is exhibited in Fig. 8.1 which shows a very strong repulsion when the distance between the centers of two nucleons is approximately around or below 0. 5 fm. This strong repulsion prevents the merging of two nucleons into a single six-quark particle.

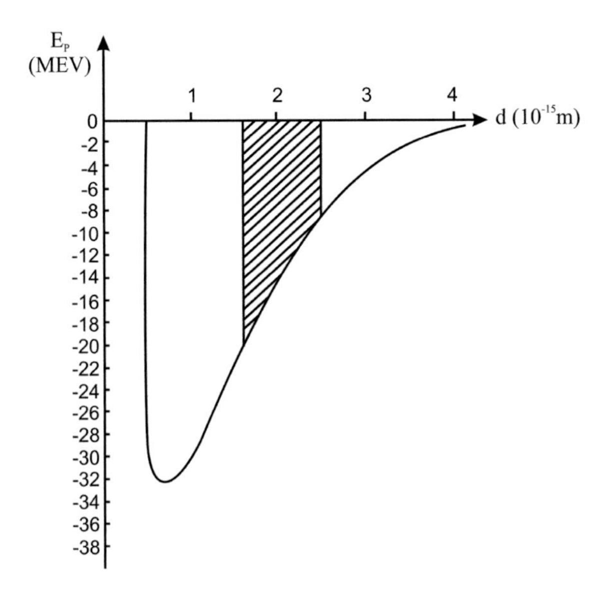

Fig. 8.1 An oversimplified form of the average interaction energy E_P between two nucleons as a function of the distance d between their centers. There is a strong attraction for 1 fm $\leq d \leq$ 4 fm. For larger distances the interaction energy is approaching exponentially zero, which is consistent with the picture of pion mediated short range interaction. For $d \leq 0.5$ fm a very strong repulsion appears indicating the impossibility of merging two nucleons into a single particle. The *shaded area* indicates the range of d between two nearest neighbor nucleons in a nucleus

On the other hand, the possibility of a large number of quarks and gluons merging together to form a dense quark/gluon "soup", called *quark/gluon plasma*, under conditions of high temperature and concentration, has been studied extensively.

It must be pointed out that the theoretical scheme of QCD, as summarized by the elementary diagrams shown in Figs. 7.1 and 7.2, shows that the coupling "constant" a_s is not so constant: It does depend on the energy or, equivalently, on the distance between two quarks. As the energy increases or the distance decreases the coupling "constant" is reduced and tends to a very small value. This is the feature of *asymptotic freedom* [2] for the discovery of which the 2004 Nobel prize in Physics was awarded. It means that the theory of QCD at say 10^9 eV with the corresponding value of the coupling "constant" a_s is equivalent to the very same theory at 10^{10} eV but with a smaller value of a_s. This length (or energy) dependence of the QCD coupling "constant" is reminiscent of the electrostatic screening of the interaction between two electrons in a metal by the dielectric "constant" due to all the other electrons. The inverse of the dielectric "constant" times e^2 which is the analog of the coupling "constant" in QCD takes its largest value ($=e^2$) as the distance of the two electrons becomes very small, and it tends to zero as this distance becomes very large. In QCD, a_s behaves in exactly the opposite way: The coupling "constant", a_s, takes larger and larger values as the distance increases and, hence, it prevents the extraction of a c-charged entity; and it takes very low values for short distances (high energies). Another difference between the electrostatic screening and the QCD asymptotic freedom is that in the latter the length dependence of the coupling "constant" is taking place in the vacuum as a result of quantum vacuum fluctuations, while the reduction of Coulomb interaction between two extra charges in a metal is due to the screening response of the free electrons of the metal, especially those lying between the two extra charges.

8.3 Estimating the Rest Energy of Proton or Neutron

We shall conclude this very short chapter by attempting to estimate the rest energy of the proton (or the neutron) by assuming asymptotic freedom at its extreme, i.e., zero coupling constant for the quarks inside the proton (or the neutron). Then, the rest energy of the proton (or neutron) equals the rest energy of the three quarks plus their kinetic energy due to the confinement within the volume of the proton (or neutron). The center of mass is free to move everywhere; so its three degrees of freedom (corresponding to a single particle in three dimensions) are not subject to confinement. The latter concerns only the relative motion of the quarks i.e., the remaining six degrees of freedom (approximately corresponding to two particles in three dimensions). Thus, we can approximately write for the average energy of a proton (or a neutron) whose center of mass is at rest:

$$\langle \varepsilon \rangle \approx 2\left\langle \sqrt{m^2c^4 + c^2p^2} \right\rangle \approx 2\langle cp \rangle \tag{8.1}$$

where m is an appropriate linear combinations of the rest masses, m_1, m_2, m_3 of the quarks. Notice that we are at the extreme relativistic limit, $cp \gg mc^2$, since $\langle \varepsilon \rangle$ is of the order of the rest energy of proton or neutron, i.e. about 1000 MeV, while the rest energy mc^2 of a quark is less than 10 MeV (see Table 2.2). The formula for the quantity $\langle cp \rangle$ has been obtained in Chap. 3, (3.14): $\langle cp \rangle \approx 1.875\,\hbar\, c/R$. In the present case the radius of the proton (or neutron) is about $0.8\,\text{fm} = 1.51 \times 10^{-5}$ atomic units; then the numerical value of $\langle cp \rangle$ is 463 MeV. Ignoring the Coulomb interactions of the three quarks in the proton or the neutron we obtain for the proton and neutron mass:

$$m_p c^2 \approx m_n c^2 \approx 926\,\text{MeV} \tag{8.2}$$

surprisingly close to the experimental values of 938.27 and 939.57 MeV respectively.

8.4 Multiple-Choice Questions/Statements

1. Which one of the following Feynman diagrams describes correctly the decay of π^0 at rest? Explain your choice.

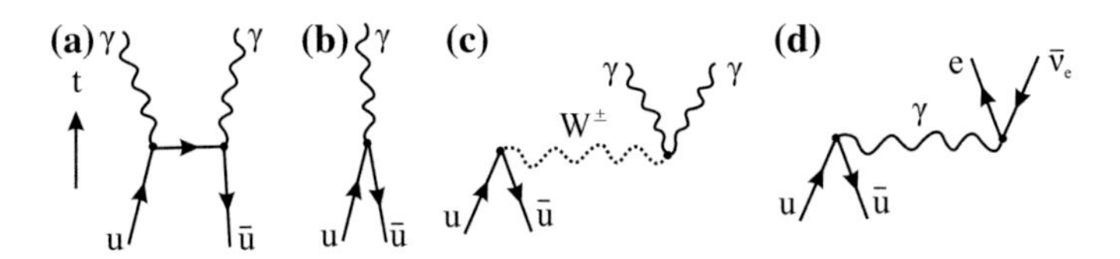

2. Which one of the following Feynman diagrams describes correctly the decay of π^+ at rest? Explain your choice.

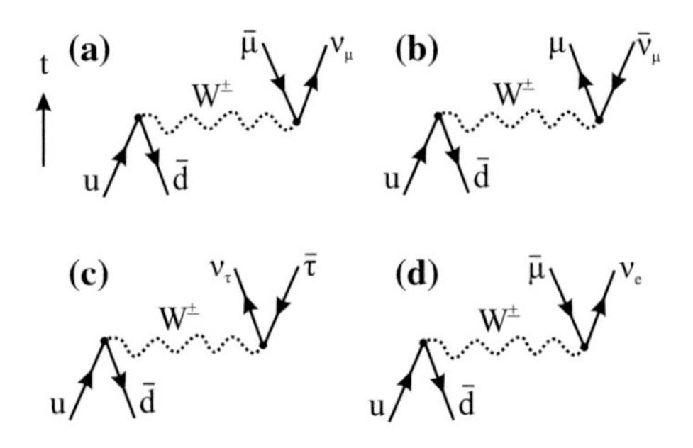

8.5 Solved Problems

1. *Which quarks make up the following baryons: p, n,* $\Lambda^0, \sum^0, \sum^+, \Omega^-$?
   For the answer see Table 8.1
2. *What combination of quark/antiquark make up the following mesons*: π^-, K^0, K^-, J/ψ, B^-?
   For the answer see Table 8.2
3. Which ones of the following reactions are taking place and which are not?

$$K^- + p \to \Sigma^- + \pi^+, \qquad K^+ + p \to K^+ + p + n,$$

$$K^- + p \to \Lambda^0 + \pi^0, \qquad \Lambda^0 \to \pi^- + p,$$

$$K^- + p \to \Lambda^0 + K^0 \qquad \pi^0 \to \mu^- + e^+ + \nu_e$$

$$\mu^- \to e + \gamma \qquad p + p \to p + p + n$$

$$\gamma + p \to n + \pi^0$$

(See the answers at the Book. See Appendix H)

4. *Draw the Feynman diagrams describing one of the main decay processes of the following baryons:* Σ^0, Δ^{++}, Ξ^0. *Could you estimate their lifetime?*

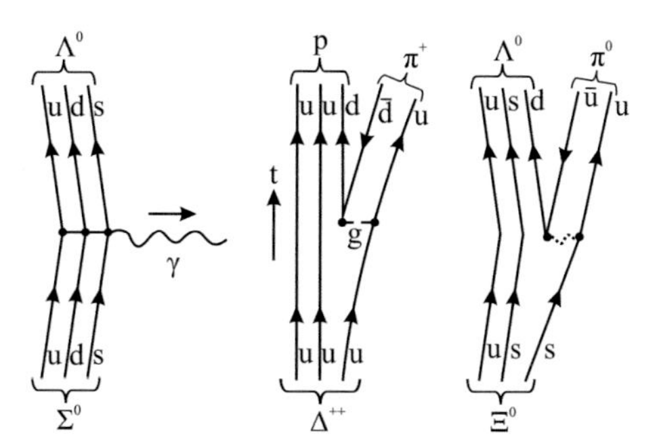

5. *Draw the Feynman diagrams describing one of the main decay processes of the following mesons:* K^+, J/ψ, K^0, $\bar{K}^0$. *Could you estimate their lifetime?*

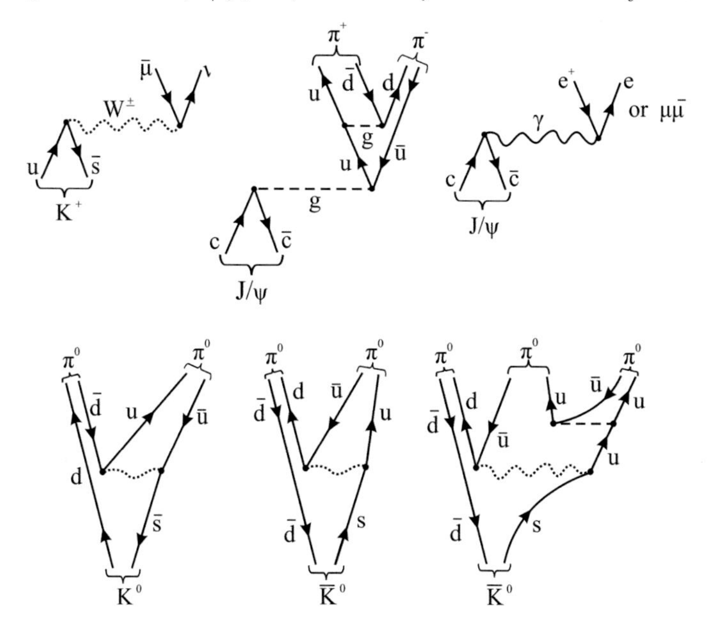

References

1. D. Perkins, *Introduction to High Energy Physics* (Addison-Wesley, Reading, 1987)
2. F. Wilczek, *The Lightness of Being: Mass, Ether, and the Unification of Forces* (Basic Books, NY, 2008)
3. W. Cottingham, D. Greenwood, *An Introduction to the Standard Model*, 2nd edn. (Cambridge University Press, Cambridge, 2007)

Chapter 9
From Protons and Neutrons to Nuclei

Abstract Protons and neutrons combine to form stable and metastable nuclei. The total energy of the nuclei depends on the number of protons and neutrons involved. Studying this dependence of the total energy allows us to account for many properties of the various nuclei.

9.1 Summary

The energy E of a nucleus consisting of Z protons and N neutrons, i.e. of $A \equiv Z + N$ nucleons,[1] is equal to $Zm_pc^2 + Nm_nc^2 - B$, where B, a positive quantity, is the binding energy of all the A nucleons. The formula for $-B$, a negative quantity, is the following:

$$\boxed{-B = -\alpha A + \beta A^{2/3} + \gamma \frac{Z(Z-1)}{A^{1/3}} + \alpha_K \frac{(A-2Z)^2}{A} - \delta; \quad A \equiv Z + N} \tag{9.1}$$

The first term in the right hand side of (9.1) is due to the difference of two contributions: (a) The (residual) attractive strong interactions among nearest neighbor nucleons, if all of them were in the interior of the nucleus. (b) The quantum kinetic energy of the nucleons under the assumption of $Z = N = A/2$. The second term is a correction due to the fact that the nucleons located at the surface of the nucleus have fewer nearest neighbors than the ones in the bulk; as a result their binding energy per nucleon is lower than those in the bulk. The third term is due to the *long range* repulsive Coulomb interactions among pairs of protons. The fourth term is due to the *excess* kinetic energy occurring when $|N - Z| \equiv |A - 2Z| \neq 0$.

[1]*Nucleon* is the common name used for both the proton and the neutron.

E.N. Economou, *From Quarks to the Universe*,
DOI 10.1007/978-3-319-20654-7_9

The last term δ is due to the discrete nature of the energy levels occupied by neutrons and protons in the nucleus; if both N and Z are even (a situation to be denoted as [e,e]), a more efficient exploitation of these energy levels is obtained resulting to a lowering of the total energy relative to the case where one of N, Z is odd and the other is even (i.e. $N + Z = \text{odd}$); in contrast, if both N and Z are odd (a situation to be denoted as [o,o]) an increase of the total energy is expected to occur relative to the case [e,o]. An empirical formula describing approximately this situation is the following

$$\delta = \pm \frac{34}{A^{3/4}} \quad \text{in MeV} \tag{9.2}$$

where the plus sign holds for the [e,e] case, and the minus for the [o,o] case; if A is odd, then $\delta = 0$.

The radius R of a nucleus is given by

$$R = c_s A^{1/3},$$

where the "constant" c_s is between 1.18 fm (for nuclei of large A) and 1.3 fm (for nuclei of small A) with a typical value $c_s \approx 1.24$ fm.

Having an expression of the total energy E as a function of Z and A, we minimize it with respect to Z keeping A constant and therefore setting $N = A - Z$; thus we obtain an estimate of the equilibrium percentage of protons (and hence of neutrons) in a nucleus of A nucleons; the result is

$$\frac{Z}{A} \approx \frac{1 + \frac{(m_n - m_p)\, c^2}{4\,\alpha_K} + \frac{\gamma A^{-1/3}}{4\,\alpha_K}}{2 + \frac{\gamma A^{2/3}}{2\,\alpha_K}} \tag{9.3}$$

The last two terms in the numerator of (9.3) are very small compared to one. Hence the ratio Z/A is about 1/2 for small A and tends to decrease as A increases because of the second term in the denominator of (9.3).

In this chapter we derive (9.1) and (9.3) by employing an additional simplifying approximation for the (residual) strong interaction between two nearest neighbor nucleons. Then, combining (9.1) and (9.3), we are in a position to study the stability of various nuclei i.e., to predict how many and which stable nuclei exist, how many and which metastable (i.e., radioactive) nuclei exist, and which combinations of Z protons and N neutrons are unstable and, hence, do not exist. We can also provide quantitative answers to various questions such as:

How can one extract energy from nuclei? Why the byproducts of a nuclear reactor are radioactive? Why is uranium out of all elements of such critical importance in the extraction of nuclear energy? Why is uranium-235 fissionable, while uranium-238 is not? etc.

9.2 Calculating the Total Energy

We already mentioned in Chap. 8 that there is an attractive (residual) strong interaction between two neighboring nucleons; this attraction appears when their distance[2] d is in the range of about 1 fm to about 4 fm. For smaller distances the interaction becomes strongly repulsive as a result of the fact that the merging of two nucleons into a single six-quark particle is energetically unfavorable. For distances larger than 4 fm the residual interaction soon becomes negligible, since beyond 4 fm it decays exponentially as

$$\frac{1}{d}\exp\left(-\frac{d}{d_o}\right)$$

where $d_o = \hbar/c\,m_\pi \approx 1.47$ fm is related to the mass m_π of the pions (see Sect. 8.2). On the basis of these remarks we expect that the mean distance between neighboring nucleons in a nucleus to be somewhere between 1 and 4 fm but closer[3] to the 1 fm. Actually this mean distance is about 2 fm, which implies that the percentage of the volume of the nucleus occupied by the nucleons is on the average approximately equal to $r_o^3/1^3\,\text{fm}^3 = 0.59$, where $r_o = 0.84\,\text{fm}$ is the radius of each nucleon. If we include the *non-available* empty space among randomly placed *touching* spheres of equal size this 0.59 is decreased to become about 0.50. Thus the volume V' of the actually available empty space for the motion of the nucleons is approximately 50 % of the volume V of the nucleus. These numbers could be obtained theoretically, if the total energy of a nucleus were expressed in terms of the mean distance d between neighboring nucleons. However, in reality the attractive (residual) strong interaction between neighboring nucleons is a very complicated function not only of d but also of the relative orientation of their spins, of their relative motion, etc. As a result, in what follows we will accept the empirical value of $d \approx 2$ fm and the empirical *mean* value of the attractive interaction $\mathcal{V}_s \approx -11.4$ MeV corresponding to $d \approx 2$ fm according to Fig. 8.1. We are now in a position to estimate the total energy E_t of a nucleus of Z protons and N neutrons. Besides the rest energy of the Z protons and the N neutrons, there are three contributions to E_t:

(a) The (residual) attractive strong interaction gives a contribution E_s equal to $N_{\text{pairs}}\mathcal{V}_s$ where N_{pairs} is the number of pairs of interacting nucleons. Because of the short range character of the residual strong interaction each nucleon interacts appreciably only with its nearest neighbor nucleons. As we mentioned in Sect. 9.1, the number N_B of nearest neighbors of a nucleon well inside the nucleus is larger than the corresponding number N_S of a nucleon

---

[2] d is the distance between the centers of the two nucleons.

[3] Being closer to 1 fm, it fully exploits the attraction but it increases the quantum kinetic energy.

located at the surface of the nucleus. Actually for a nucleon in the bulk a reasonable estimate for the average number of its nearest neighbors is $N_B \approx 8$, while for a nucleon at the surface a reasonable estimate is $N_S \approx 5$. Therefore the pairs of interacting nucleons is

$$N_{\text{pairs}} = \frac{1}{2} A_B N_B + \frac{1}{2} A_S N_S$$

where A_B and A_S are the number of nucleons in the bulk (i.e., in the interior of the nucleus) and at the surface respectively. The 1/2 factors correct for the double counting of nucleons (once as a member of A_i and once as a member of $N_i,\ i = B,\ S$). The number of nucleons at the surface is obviously proportional to the area of the surface $4\pi R^2$: $A_S \sim R^2 \sim A^{2/3}$ (in the last relation we took into account that $R \sim A^{1/3}$); moreover, if we assume that the nucleons at the surface of the nucleus are locally arranged in a square lattice with a lattice constant twice that in the bulk (2×2.1 fm) we obtain the proportionality factor to be about one, $A_S \approx A^{2/3}$, and therefore $A_B = A - A_S$. Hence, recalling that $\mathcal{V}_s \approx -11.4$ MeV, and taking into account the above estimates for $A_i,\ N_i$ we find that the total (residual) strong interaction is:

$$\begin{aligned} E_s &= \left\{ \frac{1}{2}(A - A^{2/3}) \times 8 + \frac{1}{2} A^{2/3} \times 5 \right\} \times (-11.4) \\ &= -45.6A + 17.1A^{2/3} \text{ in MeV} \end{aligned} \tag{9.4}$$

(b) The second contribution to $-B$ is due to the quantum kinetic energy of the protons plus the quantum kinetic energy of the neutrons; it is of repulsive nature, and it is given (according to (3.22b)) by

$$E_K = 2.87 \frac{\hbar^2}{mV'^{2/3}} \left(Z^{5/3} + N^{5/3} \right)$$

Z and N can be written as follows: $Z = \frac{1}{2}\{A - (N - Z)\},\ N = \frac{1}{2}\{A + (N - Z)\}$. By replacing these equivalent expressions of Z and N respectively in $Z^{5/3} + N^{5/3}$ and by expanding it in powers of $N - Z$ to obtain

$$Z^{5/3} + N^{5/3} \approx \frac{1}{2^{2/3}} \left[A^{5/3} + \frac{5}{9} \frac{(N - Z)^2}{A^{1/3}} \right]$$

Substituting the above expression and the numerical values $R \approx 1.24A^{1/3}$ (in fm) $\approx 2.34 \times 10^{-5} A^{1/3}$ a.u., $V' = 0.5V = (0.5)\, 4\pi R^3/3 \approx 4A\ \text{fm}^3 \approx 2.68 \times 10^{-14} A$ a.u. and $m = 1837.5$ a.u., we have for E_K the following result:

$$E_K \approx 29.78A + 16.54 \frac{(N - Z)^2}{A} \text{ in MeV} \tag{9.5}$$

We see from (9.5) that the minimum value of the kinetic energy is obtained when $N = Z = A/2$

(c) The Coulomb repulsion among protons is

$$E_c = \frac{1}{2}\sum_{i,j=1}^{Z}\frac{e^2}{r_{ij}} = \frac{Z(Z-1)}{2}\frac{e^2}{r},$$

where $Z(Z - 1)/2$ is the number of proton pairs and e^2/r is the average Coulomb interaction for a pair of protons. The value of r can be obtained by a procedure quite analogous to the one described in the text just below (7.5); if one assumes that the protons are distributed within the nucleus in such a way as to create a uniform electric charge density, the resulting value of r is (5/6) R (see unsolved problem 5). The value of R, as was mentioned before, is proportional to $A^{1/3}$. Taking into account that the proportionality factor is about 1.24 fm so that $R \approx 2.34 \times 10^{-5}A^{1/3}$a.u., we obtain

$$E_C \approx 0.70\frac{Z(Z-1)}{A^{1/3}} \text{ MeV} \tag{9.6}$$

As was mentioned in Sect. 9.1, the final contribution to the total energy of a nucleus has to do with the discrete nature of the energy levels that a nucleon can occupy. In view of Pauli's exclusion principle, each such (non-degenerate) level, can be occupied by up to four nucleons (two protons with opposite spin and two neutrons with opposite spin). Thus in an [e,e] nucleus, under conditions of minimum total energy, the highest occupied level will be occupied by exactly two protons and two neutrons (if the Coulomb interactions are ignored). Let us now change one of the neutrons to a proton transforming the nucleus to an [o,o] one; then one of the resulting three protons has to move to the next higher level, increasing thus the total energy by $\mathcal{E}_{n+1} - \mathcal{E}_n$ which is equal to the difference of the δ-contributions $\delta_{[o,o]} - \delta_{[e,e]} = 2\delta$. The quantities $\mathcal{E}_n$, $\mathcal{E}_{n+1}$ are the highest occupied levels for the [e,e] and the [o,o] case respectively. In the case of odd A the highest occupied level will have, let us say, two neutrons and one proton. Changing one neutron to proton will leave the highest occupied level still having three nucleons; thus no reoccupation of the levels will take place and, hence, no change in energy will occur. These arguments account for the appearance of the last term in (9.1). Adding together (9.4), (9.5), and (9.6) and including the δ term we end up with (9.1); the values obtained for the coefficient in (9.1) according to the previous estimates are as follows:

$$\begin{aligned} &\alpha \approx 45.6 - 29.78 = 15.82\,, \ \beta \approx 17.1, \ \gamma \approx 0.70, \text{ and} \\ &\alpha_K \approx 16.54 \text{ all in MeV} \end{aligned} \tag{9.7}$$

Notice that the value of α is coming by subtracting the coefficient of the first term in (9.4) from the coefficient of the first term in (9.5). The values of the coefficients in (9.1) obtained by the best fit to the experimental data are [1]:

$$\alpha = 15.75\,,\ \beta \approx 17.8,\ \gamma \approx 0.71,\ \text{and}\ \alpha_K \approx 23.69\ \text{all in MeV} \qquad (9.8)$$

Our simplistic approach produced results unexpectedly close to the best fit data (with the exception of α_K).

9.3 Minimizing the Total Energy

Having the total energy

$$E_t = Zm_pc^2 + Nm_nc^2 - B \qquad (9.9)$$

we can minimize it with respect to the various available free parameters. These are the following:

(a) The percentage of protons of or neutrons for a given number of nucleons in a nucleus. This is a free parameter since the weak interactions allow neutrons to change to protons and vice versa until the optimum ratio Z/A, as given by (9.3) is achieved (see Figs. 2.4 and 2.6 and recall that a proton consists of u, u, d quarks, while a neutron of u, d, d quarks). Keep in mind that the actual ratio of Z/A (or N/A) corresponds to the minimum of the *total* energy of the nucleus (see also solved problem 1). In applying formula (9.3) take into account that it is an approximate relation so that a degree of flexibility is required, since both Z and N ought to be integers and that [e,e] nuclei have an energy advantage over the [o,o] nuclei.

(b) The number of nucleons. This can possibly change by spontaneously breaking the original nucleus to two or more daughter nuclei or even expelling a neutron or a proton:

$$A \to A_1 + A_2 + \cdots,\ \text{where}\, A_1 + A_2 + \cdots = A$$

It is quite common for large nuclei to break into two pieces one of which is the nucleus of helium-4 ($Z_1 = N_1 = 2$) and the other has $Z_2 = Z - 2,\ N_2 = N - 2$. This is known as *alpha-decay*. It is so common, because the helium-4 nucleus has very low total energy per nucleon (see Fig. 9.1) given its size, while at the same time its small mass makes it easier to escape by overcoming a potential barrier through quantum mechanical tunneling. A very large nucleus can also break into two usually unequal but of comparable size daughter nuclei plus one or more isolated neutrons. This is known as *fission*. It occurs in nuclear reactors not spontaneously but with the incorporation of an external neutron to the nucleus of uranium-235 or plutonium-239. In the case of high concentration of nuclei the possibility of combining two nuclei into a single one (*fusion*) is present as it will be discussed below.

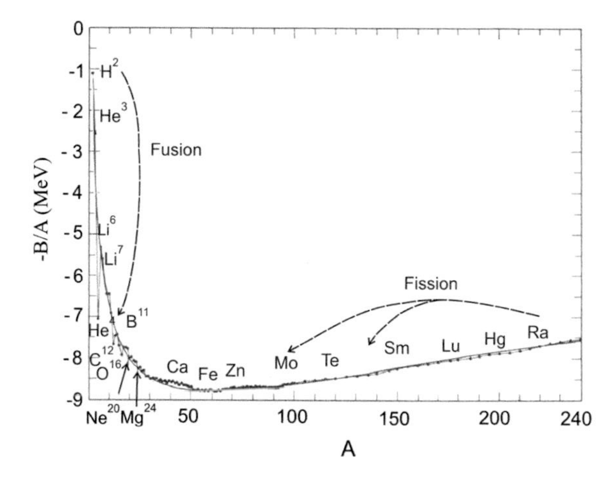

Fig. 9.1 The energy reduction per nucleon, $-B/A$, (as a result of the nucleus formation) versus the number of nucleons A. The continuous line is based on (9.1) without the δ term and on (9.3), (9.8). The *dots* are the experimental data. The discrepancies between the continuous line and the experimental data almost disappear with the inclusion of the δ term (check the extreme case of helium-4, the experimental value of which for $-B/A$ is -7.074 MeV). The minimum of $-B/A \approx 8.8$ MeV occurs for iron-56. The rise of the curve for large A is due to the repulsive Coulomb forces. This rise makes possible the extraction of energy by *fission* (i.e., by the splitting of an initial large nucleus to two smaller ones of comparable size); it is also responsible for the non-existence of nuclei with A larger than about 240 (because, then, the nucleus becomes unstable either by the emission of a He^4 nucleus or by fission). The rise of the *curve* for small A is due to the increased percentage of nucleons at the surface where the exploitation of residual strong nuclear force is less effective, since the number of nearest neighbors is smaller. This rise is responsible for the fact that energy can be extracted also by *fusion*, i.e., by merging together two small nuclei into a single one, as shown in the figure; this process takes place in stars and in the thermonuclear bombs. Notice that the dependence of $-B/A$ on A, for $A > 100$, is to a good approximation linear: $-B/A = -9.45 + 0.008A$ MeV, $A > 100$

(c) The radius R of a nucleus for a given A. In the present approach we fixed this parameter by using experimental results and not by minimizing the total energy with respect to R. The reason is the non-availability of a simple analytical form of the residual strong interaction between nucleons.
To minimize the total energy with respect the parameter Z for fixed A is very simple, since it is enough to set the derivative of the total energy with respect to Z (after replacing N by $A - Z$ as required by the constancy of A) equal to zero. By doing this simple operation, we obtain (9.3); substituting in (9.3) the values of the coefficients as given by (9.8), we find

$$\frac{Z}{A} \approx \frac{1.014 + 0.0075\,A^{-1/3}}{2 + 0.015\,A^{2/3}} \tag{9.10}$$

The 0.014 is the contribution of the second term in the numerator of (9.3), while the third term makes a contribution to the numerator much less than 1 %. Thus (9.10) can be simplified further

$$\boxed{\frac{Z}{A} \approx \frac{1}{2 + 0.015\,A^{2/3}}} \tag{9.11}$$

We repeat that weak interactions allow a nucleus to change neutrons to protons and vice versa until the optimum ratio Z/A, as given by (9.3) or approximately by (9.10), or (9.11) is achieved.
It is clear from (9.5) that the quantum kinetic energy becomes minimum, when $Z = N = A/2$ (if A is odd, $Z = \frac{1}{2}(A-1)$ and $N = \frac{1}{2}(A+1)$), while the Coulomb energy becomes minimum, when $Z = 0$ or $Z = 1$; also protons are slightly favored because they have a smaller rest energy. All these conflicting requirements, each according to its own "weight", determine the optimum ratio Z/A which minimizes the total energy, as shown in (9.3). Indeed the term $(m_n - m_p)\,c^2/4\,\alpha_K \approx 0.014$ is there because of the difference in the rest masses of a proton and a neutron and it tends to increase very slightly the percentage of protons. The other terms on the right hand side (r.h.s.) of (9.3) are proportional to the ratio γ/a_K, i.e. the ratio of the coefficient of the Coulomb term over the coefficient of the excess kinetic energy and are due to the competition between the kinetic energy which dictates $Z = N = A/2$ and the Coulomb repulsion among protons which is minimized when $Z = 0$ or $Z = 1$. If the Coulomb repulsion were absent, the ratio Z/A would be a constant equal to $0.507 \approx 1/2$. In the presence of the Coulomb repulsion this ratio decreases monotonically with increasing A starting with a value about equal to 0.50 for small A and reaching about 0.39 for large A, such as $A = 238$ corresponding to uranium. This behavior of Z/A is due to the fact that the "weight" of the Coulomb contribution relative to that of the excess quantum kinetic energy is almost negligible for small A, but it is increasing faster with A than that of the excess kinetic energy[4] and, as a result, the balance tips to more neutrons than protons. Substituting (9.3) and (9.8) in (9.1) we obtain $-B$ as a function of A. In Fig. 9.1 we plot the ratio $-B/A$ versus A. It is preferable to plot $-B/A$ rather than $+B/A$, as several books do, because the former gives essentially the total energy per nucleon in the nucleus (apart from the almost constant quantity $(Zm_pc^2 + Nm_nc^2)/A \approx mc^2$, where $m \approx \frac{1}{2}(m_p + m_n)$).

9.4 Questions and Answers

The formula (9.9) for the total energy of a nucleus allows us to give answers to some reasonable questions:

[4]The excess kinetic energy term increases linearly with A, while the Coulomb term increases almost proportionally to $A^{5/3}$ (see (9.5) and (9.6)). Thus the ratio of the two terms is proportional to $\gamma A^{5/3}/a_K A = \gamma A^{2/3}/a_K$ as it appears in the denominator of (9.3).

1. *Why is the atomic nucleus such an enormous giant in energy?*

Because it is such a minute dwarf in size. Its energy scale per nucleon is determined by the quantum kinetic energy given by (3.11), $\mathcal{E}_K \approx \hbar^2/m_p r_o^2 \approx 40\,\text{MeV}$ for $r_o \approx 1\,\text{fm}$.

2. *Why are there no nuclei of A larger than about* 240?

Because it is energetically favorable and feasible for such a nucleus to break into two pieces. Let us examine the case where one of the two pieces is a He^4 nucleus $(Z=N=2)$. For this process, known as *α-decay*, to happen we must have the inequality $E(A) \geq E(A-4) + E(4)$ or, equivalently, $B(A) \leq B(A-4) + B(4)$. As shown in Fig. 9.1, $B(4) \approx 4 \times 7.074$ and the $-B/A$ can be approximated for large A $(A > 100)$ by a straight line, $-B/A \approx -9.45 + 0.008\,A$. Hence the inequality becomes $9.45\,A - 0.008\,A^2 \leq 9.45\,(A-4) - 0.008(A-4)^2 + 4 \times 7.074$ or $A \geq 150.5$

However, the requirement that the rest energy of the final state to be lower than that of the initial is not enough for the reaction to proceed; an energy barrier may stand between the initial and the final state preventing the realization of the process. If the barrier is not very high and wide and the reduced mass of the two fragments is not very large, quantum mechanical tunneling may spontaneously allow the penetration of the barrier to take place; otherwise, an external supply of energy may be needed in order to take the initial system over the barrier. In either case, our previous theoretical result shows that α-decay could not take place in nuclei with A smaller than about 150, while it may take place for A larger than 150. The experimental data are in good agreement with these conclusions: No nucleus with $A < 144$ exhibits α-decay; between $A = 144$ and $A = 151$ there only five nuclei undergoing α-decay, but with extremely long lifetimes (of the order of 10^{16} years!). For A between 152 and 208 there are a few nuclei undergoing α-decay and many others which are stable. For $A > 208$ all nuclei are metastable. For example, U^{238} through a series of α-decays (and β-decays) ends up as Pb^{206}.

Let us examine also the case where a nucleus may break spontaneously into two fragments of comparable size. For simplicity we shall take the two fragments to be equal, although unequal fragments would produce smaller reduced mass and, hence, higher probability for tunneling; unequal fragments will release slightly less energy as inspection of Fig. 9.1 indicates. Using (9.1) and (9.8) we find that the rest energy difference between the initial and the final state equals to $\Delta E \approx 0.2627 Z^2/A^{1/3} - 4.628 A^{2/3}$ (β-decay of the daughter nuclei is not included); this is positive, if $(Z^2/A) \geq 17.62$ which corresponds to zirconium with $Z = 40$, $A = 91$, and $Z^2/A = 17.58$; however, since the reduced mass of the fragments in this case is quite large, $A\,m_p/2$, it is expected that the potential barrier, if any, would be impenetrable. Thus we have to calculate the height of the energy barrier, E_B, and find when it would become zero. As it is shown in Fig. 9.2, $E_B \approx E_C - \Delta E$, where E_C is the maximum value of the Coulomb repulsion as the two fragments are brought back towards the initial state, but before the (residual) strong interaction between them would become appreciable: $E_C \approx (Z_1 Z_2 e^2/d_i)$, where $d_i \approx R_1 + R_2 + 3.5\,\text{fm} \approx$

$1.57\,(A_1^{1/3}+A_2^{1/3})$ fm for $A \approx 300$. In the present calculation we have $Z_1 = Z_2 = Z/2$ and $A_1 = A_2 = A/2$. Substituting these values in the expression for E_C, we obtain $E_C \approx 0.14Z^2A^{-1/3}$. Combining this value of E_C with ΔE as obtained above, we have for E_B : $E_B = 4.628A^{2/3} - 0.1127\,Z^2/A^{1/3}$ which becomes zero when $Z^2/A \approx 38$. Keep in mind that Uranium-238, which is at the limit of stability against spontaneous fission has $Z^2/A \approx 35.56$ with $-B/A \approx -7.566$ MeV. However, (9.10), for $A \approx 300$, $Z \approx 114$ gives $Z^2/A \approx 43$. Moreover, metastable, but extremely short-lived nuclei of $Z = 116$, $A = 292$ ($Z^2/A \approx 46$) and $Z = 114$, $A = 298$ ($Z^2/A \approx 43.6$) have been made artificially. The conclusion is that our simple theory for the upper limit of the stability of nuclei is close to reality, although the latter shows that the onset of instability does not depend only on the ratio of Z^2/A but on other factors as well, (such as the energy levels, etc.).

It is worthwhile to calculate the quantities E_C and ΔE for the important case of U^{238}, using the above formulae: We obtain ΔE = 181.05 MeV, while the corresponding experimental value is 180.9 MeV, and E_C = 191.21 MeV, while the corresponding experimental value is 186.9 MeV. The experimental value for the height of the energy barrier in U^{238} is about 6.2 MeV.

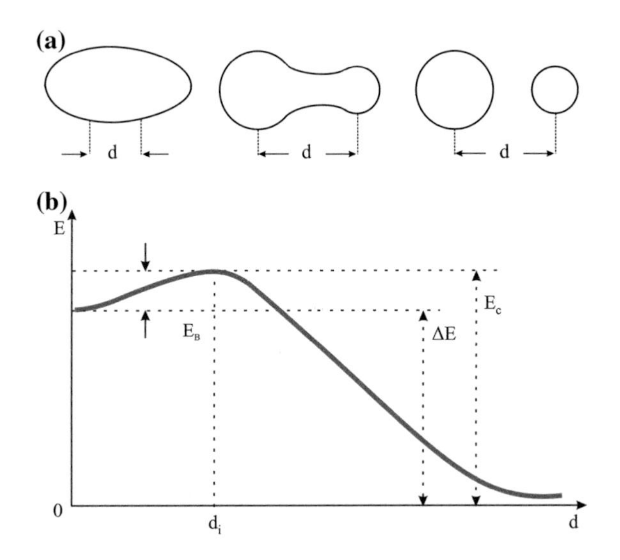

Fig. 9.2 **a** Fission of a large nucleus, such as U^{235}, proceeds usually through the absorption of an external neutron accompanied by the elongation of the nucleus; next a neck is formed and finally the nucleus breaks into two fragments plus 2 or 3 isolated neutrons; these neutrons may cause further fissions, leading thus to the so-called nuclear chain reaction. **b** The continuous *red curve* gives the total energy of the system shown in **a** as a function of the separation distance d; the zero of energy has been chosen to be the sum of the rest energies of the two fragments as $d \to \infty$ but before any β-decay occurs. $\Delta E \approx 0.2627Z^2/A^{1/3} - 4.628A^{2/3}$ is the energy difference between the initial and the final rest energies; E_c is the maximum energy calculated by fictitiously reversing the process of fission and using Coulomb repulsion up to a separation distance d_i just before the strong interactions become appreciable; $E_B \approx E_C - \Delta$E is the height of the energy barrier(colour online)

3. *Why are the fission fragments of* U^{238} *or* U^{235} *undergoing* β*-decay?*

In the fission fragments the percentage of neutrons is almost[5] the same as in the parent nucleus, i.e., about 60.3 %. Let us assume that the smaller fragment has A_1 = 106 and the larger has A_2 = 130. According to (9.10), the equilibrium percentage of neutrons for the smaller and the larger fragments are equal to 56.5 and 57.5 % respectively. Thus both fragments have more neutrons than the equilibrium number corresponding to their size. As a result, one or two neutrons in each fragment will change to protons by emitting an electron and an antineutrino per decaying neutron (see Fig. 2.4), i.e., a β-decay will take place.[6]

4. *How many stable nuclei exist?*

Since for each A there is one optimum Z which minimizes the total energy, it is reasonable to expect that we must have as many stable nuclei as the maximum number of A corresponding to a stable nucleus. The only exception to this conclusion may occur, if the optimum Z is exactly in the middle between two successive integers; in this case it is reasonable to expect that either of these two integers would give the same total energy. Such an improbable coincidence almost appears for four odd nuclei: $A = 87$ and $Z = 38$ or 37; $A = 113$ and $Z = 49$ or 48; $A = 115$ and $Z = 50$ or 49; $A = 123$ and $Z = 51$ or 52. The above argument does not apply to nuclei with even A. To see why consider three such nuclei of the same even A: (Z, N), $(Z - 1, N + 1)$, $(Z - 2, N + 2)$ with Z being even. Let us assume that the nucleus (Z, N) has the lowest energy among the three. Because of the δ term in the total energy, it is quite possible that the [o,o] nucleus $(Z - 1, N + 1)$ has higher energy than any of the other two; then the [e,e] nucleus $(Z - 2, N + 2)$, although of higher energy than the (Z, N), is stable, because to end up in the latter, it has to go through the intermediate step of becoming a $(Z - 1, N + 1)$ nucleus by a β-decay; but this β-decay is not allowed, since the $(Z - 1, N + 1)$ nucleus has higher energy than the $(Z - 2, N + 2)$ one [1, 2]. Thus for the same even A there may be two or even three stable [e,e] nuclei; as a result there are in total about 270 stable nuclei corresponding to values of A from 1 to 208 and values of Z from 1 to 82 excluding the $Z = 43$ (Tc) and $Z = 61$ (Pm) which are artificial radioactive nuclei.

5. *Why* U^{235} *is fissionable, while* U^{238} *is not?*

Spontaneous fission of a uranium nucleus is a very rare event: There is only one spontaneous fission for every 50 million α-decays. Neutron assisted fission is quite another story: Indeed, if an external neutron is incorporated in a uranium nucleus, let us say (A, Z), a new uranium nucleus, $(A + 1, Z)$, will result. This new nucleus is not in its ground state but in an excited state above the ground one by an amount E_{exc}. If this energy amount E_{exc} exceeds the potential barrier E_B shown in Fig. 9.2 an immediate fission will occur. Thus we have to estimate the magnitude of E_{exc}.

[5]During the fission 2–3 isolated neutrons are released.

[6]The nucleus resulting after a β-decay is usually in an excited state; it returns to its ground state by emitting a high energy photon in the γ range; this process is known as γ*-radioactivity*.

Let the ground state energy of the (*A*, *Z*) be E_0 and the ground state energy of the (*A* + 1, *Z*) be $E_{0,+}$. The total energy of the initial configuration before the incorporation of the neutron is $E_0 + \mathcal{E}_n + \mathcal{E}_K$ where $\mathcal{E}_n$ is the rest energy of the neutron, and $\mathcal{E}_K$ is its kinetic energy. The energy $E^*_{0,+}$ of the nucleus (*A* + 1, *Z*) after the neutron incorporation is (by conservation of energy) equal to $E_0 + \mathcal{E}_n + \mathcal{E}_K$. Thus the quantity $E_{exc} \equiv E^*_{0,+} - E_{0,+} = (E_0 + \mathcal{E}_n + \mathcal{E}_K) - E_{0,+} = A\mathcal{E}_n - B_0 - (A+1)\mathcal{E}_n + B_0^+ + \mathcal{E}_n + \mathcal{E}_K = B_{0,+} - B_0 + \mathcal{E}_K$. We have mentioned before that $B(A) \approx 9.45A - 0.008A^2 + \delta$ (see the caption of Fig. 9.1), where for odd A $\delta = 0$, while for even A $\delta = 34/A^{3/4}$. Let us assume that *A* is odd as in the case of U-235. Then $B_{0,+} - B_0 + \mathcal{E}_K = 9.45(A+1) - 0.008(A+1)^2 + |34/A^{3/4}| - 9.45A + 0.008A^2 + \mathcal{E}_K$ which is equal to $9.45 - 0.008(2A+1) + (34/A^{3/4}) + \mathcal{E}_K = 6.25$ MeV + $\mathcal{E}_K$, for $A = 235$, and 6.207 MeV + $\mathcal{E}_K$ for $A = 239$. Let us consider now the case of even *A* as in U-238. Then $E^*_{0,+} - E_{0,+} = 9.45(A+1) - 0.008(A+1)^2 - 9.45A + 0.008A^2 - |34/A^{3/4}| + \mathcal{E}_K$ or equals to $9.45 - 0.008(2A+1) - (34/A^{3/4}) + \mathcal{E}_K =$ 5.073 MeV $+\mathcal{E}_K$. Thus the excitation energy $E_{\text{exc}} \equiv E^*_{0,+} - E_{0,+}$ (which is how much above its ground state energy is the nucleus (*A* + 1, *Z*) after the incorporation of the neutron) is at least 6.25 MeV for U-235, while it is only 5.073 MeV for U-238 (if the kinetic energy of the neutron was zero). This difference of about 1.2 MeV between the two isotopes is due to the fact that in the U-235 case the term $34/A^{3/4} \approx 0.56$ MeV adds to this excitation energy, while for the U-238 case it subtracts. This difference between U-235 and U-238 is due to that the former isotope, by incorporating a neutron, is transformed from an [e,o] nucleus to an [e,e] one, while the latter isotope is transformed from an [e,e] nucleus to an [e,o] one.

We have mentioned before that the barrier height for the fission of large nuclei (*A* of the order of 240) is about 6.2 MeV. Hence for U-235 or Pu-239 the incorporation of a neutron of zero kinetic energy excites the resulting nucleus U-236 (or Pu-240) sufficiently above its ground state energy as to overcome the potential barrier and to undergo immediate fission. In contrast, for U-238 the excitation energy due to the incorporation of a neutron of zero kinetic energy falls short by about 1.1 MeV for driving this nucleus above the potential barrier; the incorporated neutron had to have a kinetic energy (before the incorporation) of at least 1.1 MeV to induce the immediate fission of U-238. This crucial for humanity difference (as far as the easy availability of nuclear weapons is concerned) between the abundant isotope U-238 (comprising 99.3 % of natural uranium) and the very hard to get U-235 (only 0.7 %) is due to the δ term in (9.1) which in turn stems from the discrete nature of the energy levels. We conclude that U^{235} is *fissionable*, i.e., the incorporation of a neutron of even zero kinetic energy does lead to its immediate fission, while U^{238} is not fissionable because the incorporation of a neutron of zero kinetic energy *does not* lead to its immediate fission. Thus U^{238} cannot sustain by itself a chain reaction. This is so, because even if the neutron has enough kinetic energy to lead to fission (i.e. more than 1.1 MeV) no fission will occur *if the neutron is not incorporated*; however, the probability of incorporation decreases with increasing kinetic energy of the neutron. As a result the probability of incorporation of an 1.1 MeV in U-238 is insufficient for sustaining a chain reaction.

Besides U^{235}, which is the only naturally occurring fissionable isotope (only 0.7 % of natural uranium is U^{235}), other fissionable nuclei are: Plutonium-239(Pu^{239}) and Uranium-233 (U^{233}); both of them are produced artificially by neutron incorporation to the non-fissionable nuclei U^{238} and Th^{232} respectively and by subsequent β-decays.

6. *How much energy is released by the fission of* U^{235}

About 211 MeV per nucleus. The formula $\Delta E \approx 0.2627\, Z^2/A^{1/3} - 4.628A^{2/3}$ obtained in the answer of question no. 2 (see Sect. 9.4) gives ΔE = 181.05 versus 180.9 MeV for the experimental value. To that value one has to add 30.4 MeV which are released as a result of β-decay and γ-decay of the daughter nuclei. From this last figure a part equal to 21.6 MeV remains in the reactor. The rest 30.4 − 21.6 = 8.8 MeV is the energy carried by the antineutrinos created during the β-decays; these 8.8 MeV per fissioned U^{235} nucleus is transferred to the Universe.

We can try to estimate theoretically the total energy released (including the β-decays and γ-decays of the daughter nuclei by employing the formula $-B = -9.45A + 0.008A^2 - \delta$. The quantity B for the initial state of A = 235 is −1778.95 MeV. The quantity B for the final state of daughter nuclei with $A_1 = 100$, $A_2 = 134$ and two free neutrons is $-865 \; - (34/100^{3/4}) - 1122.65 - (34/134^{3/4}) = -1989.59$ MeV. Thus the estimated total energy released from the fission of uranium-235 including the energy released later on by the β-decays and γ-decays of the daughter nuclei is 210.64 MeV in amazing agreement with the corresponding experimental value of 211.3 MeV.

9.5 Summary of Important Formulae and Related Comments

The formation of a nucleus by Z protons and N neutrons coming together is accompanied by a reduction, denoted by $-B$, of the total energy relative to its value when the protons and the neutrons are at their ground state energy and at infinite distance from each other. Thus the total energy of the nucleus is

$$E_t = Zm_pc^2 + Nm_nc^2 - B \tag{9.9}$$

where m_p, m_n are the rest masses of protons and neutrons respectively and the reduction $-B$ in the total energy is given by

$$-B = -\alpha A + \beta A^{2/3} + \gamma\,\frac{Z(Z-1)}{A^{1/3}} + \alpha_K\,\frac{(A-2Z)^2}{A} - \delta; \quad A \equiv Z + N \tag{9.1}$$

The first term in the right hand side (r.h.s.) of (9.1) is due to the attractive (residual) strong interaction (as given approximately in Fig. 8.1) of each nucleon with its nearest neighbors and the quantum kinetic energy (of repulsive character) under the condition $Z = N = A/2$.

The second term in the right hand side (r.h.s.) of (9.1) takes into account that the nucleons located at the surface have fewer nearest neighbors and, hence, less attractive interactions.

The third term is due to the long range Coulomb repulsion among protons; it is the term responsible for setting an upper limit to the size A of the nuclei and making possible the α-decay and the fission of large nuclei and the resulting release of energy.

The fourth term is the excess quantum kinetic energy when $N - Z = A - 2Z \neq 0$. The competition between the 3rd and the 4th term is the main factor determining the percentage of protons or neutrons for a given A.

The last term in (9.1) is due to the discrete nature of the energy levels and their more efficient occupation by nucleons when both N and Z are even numbers, relative to when only one is even number or to when both are odd numbers

$$\delta = \pm \frac{34}{A^{3/4}} \text{ in MeV} \tag{9.2}$$

where the upper sign holds for both even, the lower for both odd, while, if only one is even, then $\delta = 0$.

By setting the derivative of (9.9) with respect to Z (under A = constant) equal to zero we obtain the equilibrium percentage of protons (and, hence, of neutrons) for a nucleus of given A

$$\frac{Z}{A} \approx \frac{1 + \frac{(m_n - m_p)\, c^2}{4\, \alpha_K} + \frac{\gamma A^{-1/3}}{4\, \alpha_K}}{2 + \frac{\gamma A^{2/3}}{2\, \alpha_K}} \tag{9.3}$$

The coefficients in (9.1) and (9.3) are

$$\alpha = 15.75\,,\ \beta \approx 17.8,\ \gamma \approx 0.71,\ \text{and}\ \alpha_K \approx 23.69 \text{ all in MeV} \tag{9.8}$$

Thus (9.3) becomes

$$\frac{Z}{A} = \frac{1.014 + 0.0075\, A^{-1/3}}{2 + 0.015\, A^{2/3}} \tag{9.10}$$

For A larger than about 90 the binding energy B can be approximated by the simple formula

$$B(A) \approx 9.45A - 0.008\, A^2 + \delta \text{ MeV} \tag{9.12}$$

9.6 Multiple-Choice Questions/Statements

1. Why are there no nuclei with A larger than about 240 ?
(a) Because their binding energy B becomes negative.
(b) Because the kinetic energy forces them to break to two pieces.
(c) Because it is both favorable and feasible energetically to break to two pieces
(d) Because it so happened that the conditions for their formation were never realized

2. Why do nuclei with large A have smaller binding energy $|B/A|$ per nucleon?
(a) Because the average number of nearest neighbors is smaller
(b) Because the Coulomb repulsion is relatively larger
(c) Because the kinetic energy is larger
(d) Because the kinetic energy is smaller

3. Estimate the order of magnitude of nuclear energy per nucleon from its kinetic energy knowing that the volume per nucleon is $4\pi r^3/3$, $r = 1\text{fm}$. The result is about
(a) 4GeV, (b) 400MeV (c) 40MeV (d) 40 keV

4. Why do nuclei with very small A have smaller binding energy per nucleon?
(a) Because their kinetic energy is smaller
(b) Because their kinetic energy is larger
(c) Because their Coulomb repulsion is larger
(d) Because the average number of nearest neighbors per nucleon is smaller

5. The binding energy B/A per nucleon for uranium-238 is approximately
(a) 3.6MeV (b) 5.6 MeV (c) 7.6MeV (d) 9.6MeV

6. The binding energy B/A per nucleon for iron-56 is approximately
(a) 9MeV (b) 7MeV (c) 5MeV (d) 3MeV

7. The binding energy B/A per nucleon for helium-4 is approximately
(a) 4.1MeV (b) 5.1MeV (c) 6.1MeV (d) 7.1MeV

8. The percentage of protons in a nucleus as a function of A is as follows:

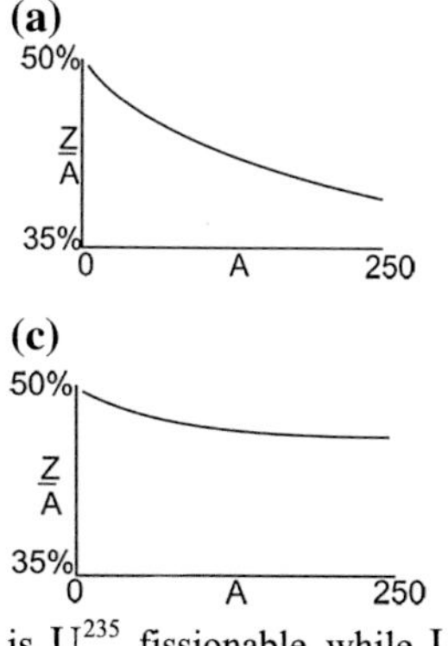

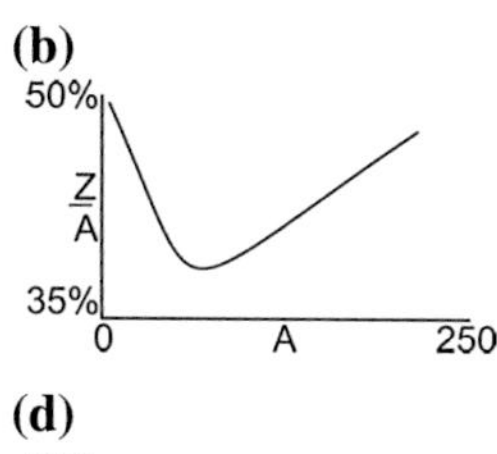

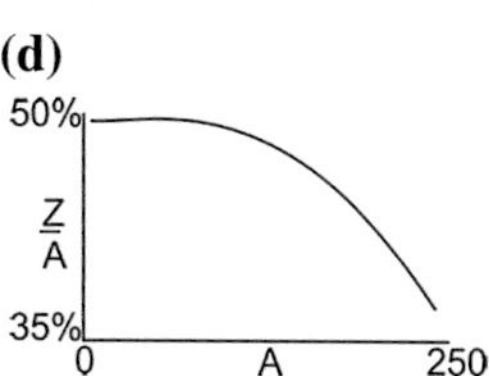

9. Why is U^{235} fissionable, while U^{238} is not ?
(a) Because U^{235} has fewer neutrons. As a result the incorporation of an additional neutron offers enough energy to overcome the potential barrier. U^{238} has too many neutrons for this to happen.

(b) The binding energy per nucleon *B*/*A* is larger for U^{235} than that for U^{238}. As a result the potential barrier is lower for the former isotope than for the latter.
(c) The energy of U^{236} after the reaction $n+U^{235} \rightarrow U^{236}$ is about 6.25MeV above the ground state energy of U^{236}, i.e. higher than the potential barrier of 6.2MeV, while the energy of U^{239} after the reaction $n+U^{238} \rightarrow U^{239}$ is only 5.1MeV above the ground state energy of U^{239}, i.e. 1.1MeV below the top of the barrier. This difference between 6.25MeV and 5.1MeV is due to the smaller number on neutrons in U^{236} than in U^{239}.
(d) The energy of U^{236} after the reaction $n+U^{235} \rightarrow U^{236}$ is about 6.25MeV above the ground state energy of U^{236}, i.e above the potential barrier of 6.2MeV, while the energy of U^{239} after the reaction $n+U^{238} \rightarrow U^{239}$ is only 5.1MeV above the ground state energy of U^{239}, i.e. 1.1MeV below the top of the barrier. This difference between 6.25MeV and 5.1MeV is due to the fact that in the case of U^{235} the incorporation of neutron transforms an [e,o] nucleus to an [e,e] one, while, in contrast, in the case of U^{238} the transformation is from an [e,e] nucleus to an [e,o] one.

10. Why does the fission of U^{235} in a nuclear reactor produces radioactive nuclei ?
(a) Because each fragment has a lot of kinetic energy (about 90MeV)
(b) Because the two fragments have in general unequal number of neutrons which tends to become equal by β-decay
(c) Because the percentage of neutrons in the fragments is almost equal to that of U^{235} which is higher than the one which corresponds to equilibrium for their size
(d) Because they collide violently with other nuclei and tend to break

9.7 Solved Problems

1. *An isolated neutron breaks down according to the exothermic reaction* $n \rightarrow p + e + \bar{\nu}_e + 0.78$ *MeV. Why all neutrons in a nucleus do not undergo this reaction?*

Solution As it was mentioned before, the equilibrium is established when the *total* energy of an isolated system (in the present case the nucleus and not an individual nucleon) is minimized. Even if we consider the individual reaction $n \rightarrow p + e + \bar{\nu}_e$ taking place within the nucleus, the gain in energy will not necessarily be 0.78 MeV; as a matter of fact it may be negative as well depending on the percentage of protons or neutrons in the nucleus. The reason is that the proton produced by this reaction, which remains within the nucleus, cannot go to levels already fully occupied by the other protons. Thus the levels available to the produced proton may be of energy higher than the energy the proton would have if the reaction occurred in vacuum plus 0.78 MeV.

2. *The composition of natural uranium is* 99.3 % *U-238 and* 0.7 % *U-235. Their half-lifes are* 4.51×10^9 s and 7.1×10^8 s *respectively. Obtain limits for the age of our planetary system and the age of the Universe.*

Solution We assume that both U^{235} and U^{238} were created during a violent supernova explosion practically at the same time, $t = 0$, and of equal number N_o. Thus their number $N_1(t)$ and $N_2(t)$ at any subsequent time t is

$$N_1(t) = N_o e^{-t/\tau_1} \quad \text{for } U^{235}$$

$$N_2(t) = N_o e^{-t/\tau_2} \quad \text{for } U^{238}$$

Hence at present time t

$$\ln\frac{N_2(t)}{N_1(t)} = t\left(\frac{1}{\tau_1} - \frac{1}{\tau_2}\right) = 0.693t\left(\frac{1}{t_1} - \frac{1}{t_2}\right)$$

or

$$0.693\left(1.421 \times 10^{-9} - 2.238 \times 10^{-10}\right)t = \ln\frac{99.284}{0.716}$$

or

$$t = 5.94 \times 10^9\,\text{yr} = 5.94\,\text{Gyr}$$

It follows that the age of our planetary system t_p must be smaller than the age of the uranium nuclei, while the age of the Universe t_U is obviously larger than t:

$$t_p < 5.94\,\text{byr} < t_U$$

3. *The distribution of the mass number of the fragments of the neutron induced fission of U-235 in a nuclear reactor exhibits a double peak at* A = 92 *and* A = 140. *(See p.* 654 *of Eisberg-Resnick, Quantum physics* [3].*) On the contrary, the fragments of a fission bomb exhibit a broad single maximum at* A ≈ 116. *What may be the reason for this disparity, which is also a useful tracer of nuclear bomb testing?*

Solution At least two factors influence the probability of each specific fission: (a) The rest energy difference between the initial and the final state; as a general rule, the larger this difference the more probable the reaction is. If both fragments had rest energies per nucleon lying on exactly the same straight line fitting the $-B/$

A versus A for large A ($A > 100$), then this difference would be exactly the same no matter what the size of the two fragments would be; in reality this factor very slightly favors equal size fragments because the $-B/A$ versus A curve does not follow this straight line as the value of A becomes smaller, but it bends to a small degree above this straight line (see Fig. 9.1). (b) If quantum mechanical tunneling is involved in the fission process, even in a minute degree, then the reduced mass of the fragments will enter in the exponent favoring unequal fragments which produce smaller reduced mass. In the fission reactors the fission is achieved by neutrons of almost zero kinetic energy (in order to increase the probability of neutron capture by the uranium-235 nuclei); hence, there, the tunneling is expected to play a significant role strongly favoring unequal fragments. On the other hand, the nuclear explosion in fission bombs occurs by very energetic neutrons for which tunneling plays a small role, if at all. Thus in bombs the equal size fragments are greatly enhanced relative to those in nuclear fission reactors.

9.8 Unsolved Problems

1. Why is helium-4 so much off the continuous line in Fig. 9.1?
2. Which nuclear reaction is the first step in the thermonuclear process from hydrogen to helium occurring in active stars? Estimate the minimum required temperature for the initiation of this reaction. Which is the rate of "burning" hydrogen to helium in the Sun? (see Chap. 14)
3. Why nuclear reactors employ a material as a moderator? A moderator slows down fast neutrons released during the fission so that they reach thermal kinetic energies i.e. a fraction of an eV. Why heavy water is the best moderator?
4. The nucleus of uranium-238 through a series of α- and β-decays ends up as a nucleus of lead-206. How much energy is released during this whole process from U^{238} to Pb^{206}? (It is given that the difference $M c^2 - A m_u c^2$ is 116.6 MeV, 36.15 MeV, and 3.607 MeV for U^{238}, Pb^{206}, and He^4 respectively)
5. What is the so-called depleted uranium? How one can separate the two isotopes of natural uranium? By chemical or physical methods? Assume that the protons are arrange within the nucleus as to create a uniform positive electric charge. Show then that

$$E_c = \frac{1}{2}\sum_{i,j=1}^{Z}\frac{e^2}{r_{ij}} = \frac{Z(Z-1)}{2}\frac{e^2}{r}, \quad \text{with } r = \frac{5}{6}R$$

References

1. R. Leighton, *Principles of Modern Physics* (McGraw-Hill, NY, 1959)
2. A. Das, T. Ferbel, *Introduction to Nuclear and Particle Physics*, 2nd edn. (World Scientific, Singapore, 2003)
3. R. Eisberg, R. Resnick, *Quantum Physics*, 2nd edn. (J. Wiley, NY, 1985)

Part IV
Structures Held Together by Electromagnetic Interactions

General Remarks

The structures of matter from the level of atom up to the level of an asteroid, i.e. for 15 orders of magnitude from about 10^{-10} m to about 10^5 m are dominated by only one of the four interactions: The electromagnetic one whose strength is determined by the charge e (of the proton). The quantum kinetic energy, which counterbalances this force (for equilibrium to be established), depends (according to (3.11)) on the ratio $\hbar^2/m$. The dominant kinetic energy will be the one with the smaller mass in the denominator, i.e. the mass of the electron, m_e.

Thus the properties of all these structures, which "live" in the realm of electromagnetic forces, must necessarily depend at least on the three universal constants $e, \hbar, m_e$. The three quantities $e, \hbar, m_e$ define a system of units called atomic system of units (a.s.u.). In this system, the units of mass, length, and time are the following:

- unit of mass: m_e, the rest mass of electron $\approx 9.109 \times 10^{-31}\,kg$
- unit of length: Bohr radius, $a_B \equiv \hbar^2/m_e e^2 \approx 0.529\,\text{A}$
- unit of time: $t_o \equiv \hbar^3/m_e e^4 \approx 2.42 \times 10^{-17}$ s.

The expressions for a_B and t_0 in SI are obtained from the above expressions in G-CGS by replacing e^2 by $e^2/4\pi\,\varepsilon_0$: $a_B \equiv 4\pi\,\varepsilon_0\hbar^2/m_e e^2 \approx 0.529\,\text{A}$ and $t_o = (4\pi\,\varepsilon_0)^2\hbar^3/m_e e^4 \approx 2.42 \times 10^{-17}$ s.

The units of various other physical quantities such as velocity, pressure, temperature, etc. in the a.s.u are given in Appendix, where we find it more convenient to use the equivalent triad $a_B, \hbar, m_e$, rather than $e, \hbar, m_e$. In estimating the values of various quantities, we shall often use $r \equiv \bar{r} a_B$ instead of Bohr's radius a_B, where

r is the radius of the relevant atom and $\bar{r}$ is the radius of the relevant atom measured in units of a_B.

The preceding remarks allow us to write a general expression for any physical quantity X referring to a system held together only by electromagnetic interactions

$$\boxed{X = X_0 f\left(Z, \frac{m_a(Z)}{m_e}, \frac{T}{T_0}, \frac{P}{P_0}, \frac{c}{\upsilon_0}\right)}$$

where

$X_0 = \hbar^\alpha m_e^\beta a_B^\gamma$ with α, β, γ are such that $[X_0] = [X]$

f is a function (of the indicated dimensionless independent variables); this function cannot be determined by dimensional analysis,
Z is the relevant atomic number(s),
m_a is the mass of the relevant atom(s),
T is the temperature and $T_0 = \hbar^2/k_B m_e a_B^2 = 315628\,\text{K}$ is its a.s.u. unit.
P is the pressure and $P_0 = \hbar^2/m_e a_B^5 = 2.94 \times 10^{13}$ Pa is its a.s.u unit.
c is the velocity of light in vacuum and $\upsilon_0 = \hbar/m_e a_B = c/137 = 2.189\,\text{m/s}$ is the a.s.u. unit of velocity.

The above expression for X as a general rule becomes more accurate if everywhere a_B is replaced by $r = \bar{r} a_B$ or by a relevant length (see Chaps. 11 and 12).

Chapter 10
From Nuclei and Electrons to Atoms

Abstract In this chapter the main properties of the electrons trapped around each nucleus to form atoms are presented. Of special interest is the size of the atoms as determined by the outer electrons as well as the main features of the motion of the latter and their energy levels. The connection of these properties with the Periodic Table of the elements is examined.

10.1 Summary

A nucleus carrying a positive electric charge Ze attracts, mainly through the Coulomb potential, and binds around it Z electrons, each of which carry negative charge $-e$, to form electrically neutral structures called *atoms*.[1] The simpler atom is that of hydrogen consisting of a proton and an electron; the latter is trapped around the proton at a distance[2] $r = a_B$. We should stress that the electron is not at a fixed distance from the proton. Instead, it can be found at any distance between zero and several times the Bohr radius with a probability distribution; the latter in the ground state of the hydrogen atom exhibits a maximum at a distance equal to a_B. *Thus, when we are referring to the radius of any atom in its ground state, we mean the distance at which the probability to find the most external electron is maximum.*

---

[1]It is not uncommon to have $Z - \zeta$ electrons around the nucleus; then the structure carries overall a positive charge ζe and it is called a ζ^+ *cation*. It is also possible to have $Z + \xi$ electrons trapped around the nucleus, creating thus a structure of total negative charge equal to $-\xi e$ called a ξ^- *anion*. Cations and anions (collectively called *ions*) are metastable structures which tend to combine with each other or with electrons to form electrically neutral structures.

[2]Instead of a proton we may have a deuteron, i.e., a proton/neutron nucleus. As far as the electron is concerned, the only change this will bring, it will be a minute increase in the reduced mass of the system from $m_e m_p/(m_e + m_p) = 0.9995\, m_e \approx m_e$ to $0.9997\, m_e \approx m_e$. The reduced mass replaces the electronic mass in the formula for the atomic radius and other electronic properties.

E.N. Economou, *From Quarks to the Universe*,
DOI 10.1007/978-3-319-20654-7_10

The electron in the hydrogen atom, being confined within a finite volume, exhibits discrete energy levels, as in Figs. 3.2 and 3.3(d′), line 01′, given by [1, 2]

$$\boxed{\varepsilon_n \approx -\frac{e^2}{2a_B}\frac{1}{n^2} \approx -13.6\frac{1}{n^2}\,\text{eV}, \quad n = 1, 2, 3, \ldots} \tag{10.1}$$

The $n = 1$ level corresponds to the ground state. (If instead of a proton of charge e we had a nucleus of charge Ze, then the radius of a single electron trapped around it would be $r = a_B/Z$ and the energy levels of (10.1) would be multiplied by Z^2). There are as many different atoms as the existing stable (or long lived metastable) nuclei, i.e., about 300. However, as far as the electronic properties of atoms are concerned, the number of neutrons N in the nucleus plays a very minute role (see Footnote 2). Thus we count as different atoms those with different Z which are about 90 and which are arranged in a table called the *periodic table of the elements*. Atoms having a nucleus of the same Z but different N are termed *isotopes* of the same element. The most important property of an atom is the so-called first *ionization energy*, I_P, defined as the minimum energy required to extract an electron from it, when this atom is in its ground state. The ionization energy of the negative ion with an extra electron is called the *electron affinity*, E_A, of the atom. The quantities I_P and E_A give the energy scale of chemical reactions. The radius, r_a, of each atom in its ground state, as defined before, although not easily determined experimentally, is also of high interest because to a large extent determines the bond length of molecules, solids, and liquids.

In this chapter we shall obtain an order of magnitude estimate (or even better than that) of the quantities I_p, E_A, and r_a by dimensional considerations. We shall also determine the angular dependence of the electronic stationary waves (see Appendix A), called *atomic orbitals*, associated with each of the discrete single electron levels (such as those in Table 10.1). To a single electron level may correspond more than one atomic orbitals. The energy ordering of the atomic orbitals given in Table 10.1 allows the construction of the periodic table of the elements (see Table I.3 in Appendix I), which constitutes the foundation of Chemistry.

10.2 Size and Relevant Energy of Atoms

In the study of atoms the relevant universal physical constants, as it was mentioned before, are the following three,[3] e, $\hbar$, m_e; the number Z of protons in the nucleus, called *atomic number*, is also important. Hence, by dimensional considerations, we can conclude that the radius, r_a, of any atom is given by the formula,

[3]The velocity of light c may also play a role, usually small; however, some properties, such as the magnetic properties, depend critically on c, and disappear in the electrostatic limit, $c \to \infty$.

$$r_a = \bar{r}_a(Z)\, a_B \tag{10.2}$$

since $a_B \equiv \hbar^2/m_e e^2$, the Bohr radius, is the only combination of e, $\hbar$, m_e with dimensions of length. As shown in Fig. 10.1, it turns out that the numerical factor $\bar{r}_a(Z)$ is in a range between about 1 and 5. A qualitative explanation of the saw-like shape of the curve in Fig. 10.1 shall be provided later on when the stationary wavefunctions (also called eigenfunctions or atomic orbitals) are examined. Again, by dimensional considerations, we obtain the following result for the first ionization energy:

$$I_P = \bar{I}'_P(Z)\frac{e^2}{a_B} = \bar{I}_P(Z)\frac{e^2}{r_a}; \qquad \frac{e^2}{a_B} \equiv \frac{\hbar^2}{m_e\, a_B^2} \equiv \frac{e^4 m_e}{\hbar^2} = 27.2\,\text{eV} \tag{10.3}$$

since $e^4 m_e/\hbar^2$ is the only combination of e, $\hbar$, m_e with dimensions of energy. The formulae in (10.3) are in G-CGS. To change in SI replace e^2 by $e^2/4\pi\varepsilon_0$. In Fig. 10.2a I_P is plotted versus Z. The numerical factor $\bar{I}'_P(Z)$ is obtained by dividing the values of I_P in eV by 27.2; it varies between 0.14 (for Cs) and 0.9 (for He). If we replace a_B in the denominator of (10.3) by the actual radius r_a of each atom, the resulting numerical factor $\bar{I}_P$ exhibits a smoother dependence on Z and it varies in a more narrow range of values, as shown in Fig. 10.2b. This shows that there is a strong correlation between I_P and r_a: the smaller the radius the larger the ionization potential and vice versa, as expected in view of the Coulomb potential which is inversely proportional to the length r.

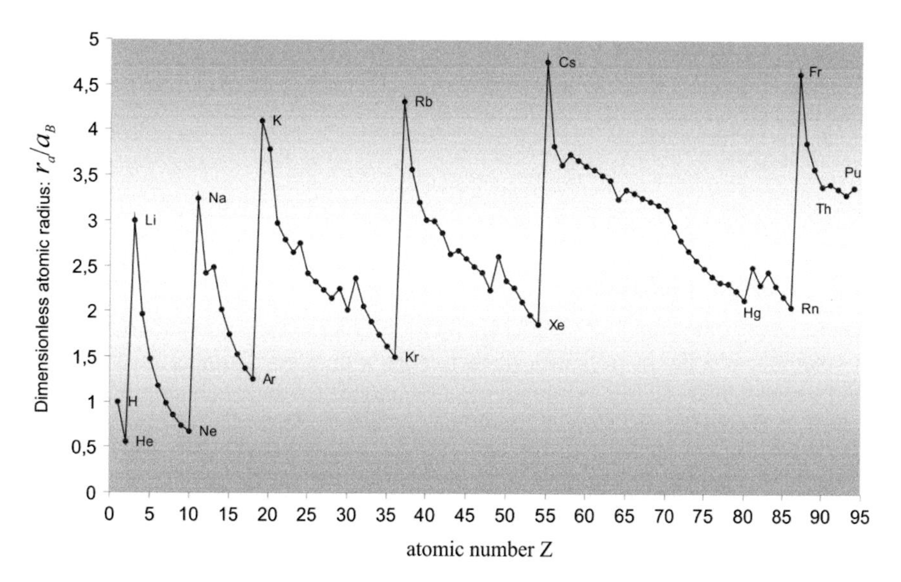

Fig. 10.1 Dimensionless radius of atoms, r_a/a_B, versus their atomic number, Z. Local minima are associated with atoms of noble gases and local maxima with atoms of alkalis

Table 10.1 Electronic energy levels in non-hydrogenic atoms

ENERGY ↑				Number of electrons in the fully occupied shell	Total number of electrons up to this shell	
7s	7p	6d	5f	32	32+86=118	7th shell, 7th line
6s	6p	5d	4f	32	32+54=86	6th shell, 6th line
5s	5p	4d		18	18+36=54	5th shell, 5th line
4s	4p	3d		18	18+18=36	4th shell, 4th line
3s	3p			8	8+10=18	3rd shell, 3rd line
2s	2p			8	8+2=10	2nd shell, 2nd line
1s				2	2	1st shell, 1st line in the P.T.E
$s \Leftrightarrow \ell = 0$	$p \Leftrightarrow \ell = 1$	$d \Leftrightarrow \ell = 2$	$f \Leftrightarrow \ell = 3$			
2	6	10	14	⇐ Maximum number of electrons for each value of ℓ including spin		

The ordering of these levels and Pauli's principle determines the structure of the periodic table of the elements (PTE)

10.3 Atomic Orbitals

To obtain the electronic properties of each atom with atomic number Z we have to find first the single electron stationary wavefunctions, the so-called atomic orbitals, as well as their corresponding discrete energy levels. Then we have to populate each atomic orbital with two electrons of opposite spin starting with the orbital of the lowest energy and continuing up to the next ones in energy until all Z electrons are exhausted. This way we obtain the lowest total energy of the atom, the so-called ground state energy, consistent with Pauli's exclusion principle. Excited states of the atom can be obtained by redistributing the electrons among the atomic orbitals as to populate some that were empty in the ground state.

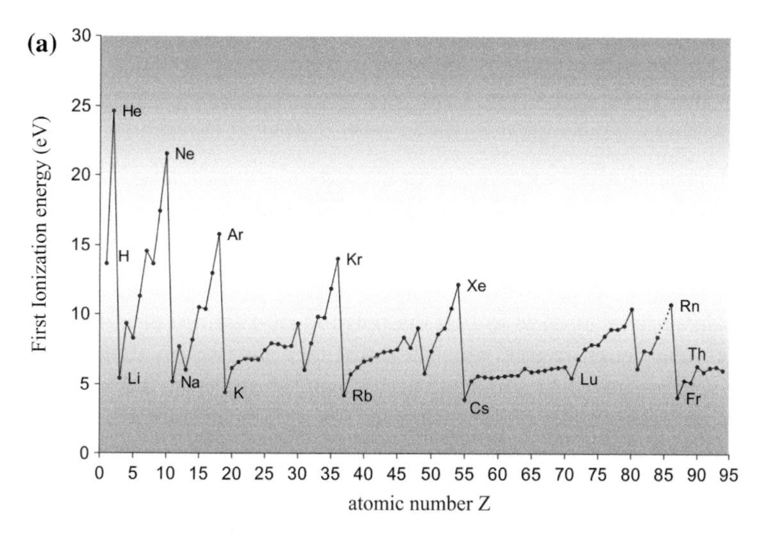

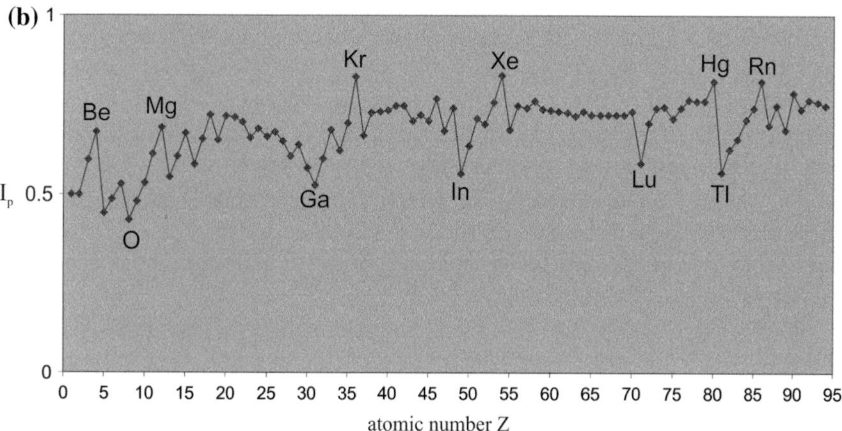

Fig. 10.2 **a** First Ionization Energy versus the atomic number Z. The local maxima are associated with completed shells, as in noble gases, and subshells. The local minima are associated mainly with alkalis and the Boron column of the periodic table. **b** The dimensionless first ionization energy $\bar{I}_P(Z) = I_P/(e^2/r_a)$ versus Z (see (10.3)) shows a smoother dependence on Z than I_P or $\bar{I}'_P(Z)$ with an average value of $\bar{I}'_P \approx 0.26$

Each atomic orbital is fully determined by a wavefunction $\psi(\mathbf{r})$ which is a stationary solution (i.e., one of fixed energy) of the wave equation known as *Schrödinger's equation* (see (3.4)); in the latter enters the Coulomb potential energy felt by an electron at position $\mathbf{r}$ due not only to the nucleus but to *all the other electrons*. This potential energy, $\mathcal{V}(r)$, as a result of the overall spherical symmetry of the atom, depends only on the distance r from the nucleus and not on the direction of $\mathbf{r}$. On the other hand, each individual stationary wavefunction, $\psi(\mathbf{r})$, may depend not only on the distance r but also on the angles $\theta, and\ \varphi$ which determine the direction of $\mathbf{r}$. Because $\mathcal{V}(r)$ depends only on r, we can write $\psi(\mathbf{r})$ as a product, $\psi(\mathbf{r}) = R(r)Y(\theta, \varphi)$, with the angular part $Y(\theta, \varphi)$ being independent of the form and the values of $\mathcal{V}(r)$. Exploiting this independence we choose, $\mathcal{V}(r) = E$, so that Schrödinger's equation reduces to the so-called Laplace equation,

$$\frac{\partial^2 \psi}{\partial x^2} + \frac{\partial^2 \psi}{\partial y^2} + \frac{\partial^2 \psi}{\partial z^2} = 0, \tag{10.4}$$

the angular part of which is the same as the angular part of Schrödinger's equation. Hence, all we have to do is to find the polynomial solutions of (10.4) and extract from them their angular part, which is the same as that of the atomic orbitals.

It is very easy to find polynomial solutions of (10.4) of degree ℓ. One such solution is a polynomial of zero degree, $\ell = 0$, i.e., a constant whose angular dependence is also a constant. The solutions of Schrödinger's equation (i.e., the atomic orbitals) with no angular dependence corresponding to $\ell = 0$, are denoted by the letter s.

Next, we consider polynomial solutions of (10.4) of degree $\ell = 1$, denoted by the letter p; there are exactly three *independent* such solutions: $\psi \sim x$ with angular dependence $p_x = x/r = \sin\theta\cos\varphi$; $\psi \sim y$ with angular dependence $p_y = y/r = \sin\theta\sin\varphi$; and $\psi \sim z$ with angular dependence $p_z = z/r = \cos\theta$.

Next, we consider polynomial solutions of (10.4) of degree $\ell = 2$, denoted by the letter d. There are exactly five *independent* such solutions: $\psi \sim xy,\ yz,\ zx,\ x^2 - y^2$, and $y^2 - z^2$. Continuing like this we find that there are exactly seven independent solutions corresponding to $\ell = 3$ denoted by the letter f. In general, there are $2\ell + 1$ independent polynomial solutions of (10.4) of degree ℓ with all their terms being of the same degree ℓ.

We conclude that the angular part of the atomic orbitals are characterized by two integer numbers: the degree ℓ of the polynomial solution of (10.4) and a number m identifying each one of the $2\ell + 1$ angular parts belonging to the same ℓ. To fully characterize an atomic orbital we need a third integer, denoted by n_r, $n_r = 0, 1, 2, \ldots$, and called the radial quantum number; n_r determines the radial part $R(r)$. Usually, instead of n_r, another integer is used, called *principal quantum number*, appearing in (10.1), defined as $n \equiv n_r + \ell + 1$, and therefore taking the values $n = \ell + 1,\ \ell + 2,\ \ell + 3, \ldots$. In Fig. 10.3 we present a few atomic orbitals to indicate their angular dependence and to show how their relative size depends on the quantum numbers n and ℓ.

10.4 Energy Ordering of the Atomic Orbitals ψ_{nlm}

We have already stated that the energy levels of the hydrogen atom depend only on the principal quantum number n (see (10.1)). This is a peculiarity of the $1/r$ potential appropriate for the hydrogen atom. For all other atoms, which have more

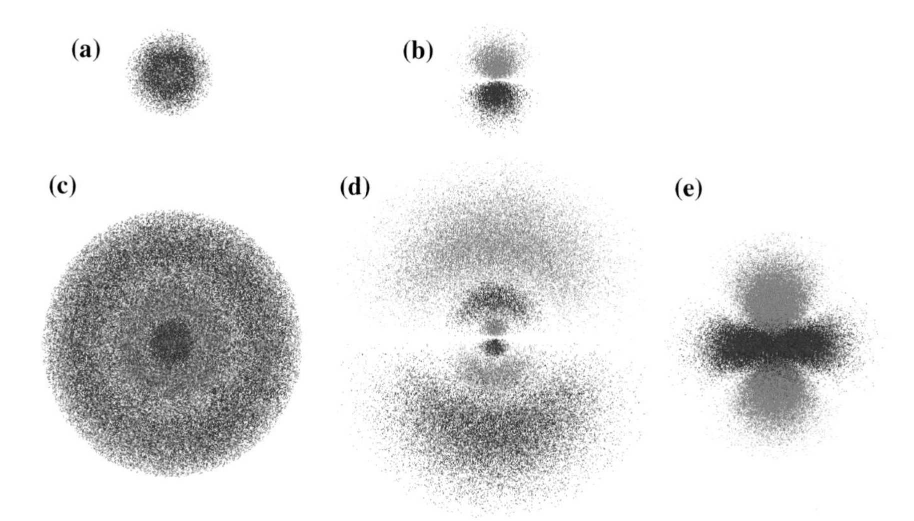

Fig. 10.3 Various atomic orbitals. *Orange color* shows positive values of ψ and blue negative. In the following notation the *number* indicates the principal quantum number n and the *letter* indicates the angular dependence. **a** $2s$; **b** $2p_z$; **c** $4s$; **d** $4p_z$; **e** $3d_{z^2-\frac{1}{3}r^2}$. The (**c**) and the (**e**) orbitals, in spite of their difference in size, correspond to comparable energies. Notice the increase in the size of the orbitals with n.

than one electron, the potential $\mathcal{V}(r)$ is more complicated and, as a result, the energy level corresponding to the atomic orbital ψ_{nlm} depends on both n and ℓ, but not on m. This last non-dependence is a consequence of the fact that the rigid rotation of an atomic orbital around an axis passing through the nucleus does not change its energy, since there is no preferred direction in an isolated atom. If a preferred direction is introduced by placing, e.g. the atom in an external field, then a dependence on m as well is expected to appear. The energy ordering of the atomic levels is very important, because it allows the identification of which orbitals are occupied in the ground state of the atom and which is the highest energy occupied orbital. The latter is of special importance in chemistry, because usually it is the most external one and, hence, the first one to come in contact with another atom (or atoms) to form molecules.

The rules of energy ordering of the atomic orbitals are relatively simple:

(1) For the same principal quantum number n the higher ℓ has the higher energy: $\varepsilon_{n\ell} < \varepsilon_{n\ell'}$, if $\ell < \ell'$. This follows from the observation that higher ℓ (for the same n) is associated with the electron being further away from the nucleus (compare c and d in Fig. 10.3) and, hence, weaker attraction.
(2) The orbitals (n + 2, s), (n + 1, d), and (n, f), have atomic levels of comparable energy. The energy ordering among these orbitals depends on their relative

occupation by electrons. As a result, the ground state occupation of the atomic orbitals for the transition and the rare earth elements, which do not have fully occupied d and f orbitals, is not always a priori obvious.

(3) The level of the orbital $(n + 2, p)$ is always higher than $(n + 1, d)$, and (n, f), but lower than the orbital $(n + 3, s)$.

These rules allow us to energy-order the various atomic orbitals (with the uncertainties mentioned in point (10.2) above), as shown in Table 10.1 and to construct line by line and shell by shell the periodic table of the elements (PTE).

10.5 The Structure of the Periodic Table of the Elements

Let us present, as an example, the construction of the fourth line of the PTE, from Z = 19 (K) till Z = 36 (Kr) based on Table 10.1: The first two electrons go to the orbital 4s completing thus the s-subshell at the element Ca (Z = 20). The next three electrons go to 3d orbitals indicating that up to this occupation the 4s orbital is lower in energy than the 3d; we have this way the elements: Z = 21, Sc, with [Ar]$4s^2 3d^1$; Z = 22, Ti, with [Ar]$4s^2 3d^2$; and Z = 23, V with [Ar]$4s^2 3d^3$. The exponents indicate the number of electrons in the corresponding orbital and the symbol [Ar] presents in a condensed way the configuration $1s^2 2s^2 2p^6 3s^2 3p^6$ which is that of Ar. In the next element Z = 24, Cr, the first reversal of the ordering appears, since the configuration is not [Ar]$4s^2 3d^4$, but [Ar]$4s^1 3d^5$ indicating that at this stage of occupation, the 3d orbital is lower in energy than the 4s. In the next four elements, Z = 25, Mn, Z = 26, Fe, Z = 27, Co, and Z = 28, Ni the original ordering returns: [Ar]$4s^2 3d^5$, [Ar]$4s^2 3d^6$, [Ar]$4s^2 3d^7$, and [Ar]$4s^2 3d^8$ respectively. In the next element Z = 29, Cu, the anomaly returns: the configuration is [Ar]$4s^1 3d^{10}$. With the next element, Z = 30, Zn, the s/d subshell is completed. The next six elements, from Z = 31, Ga to Z = 36, Kr, correspond to the gradual occupation of the three 4p orbitals.

Now, let us go back to Figs. 10.1 and 10.2 and examine how the atomic radius and the first ionization energy vary as we go along the elements of the fourth line of the PTE. The first observation is the rather systematic drop of the atomic radius and the corresponding increase of the ionization energy. This trend is due to the fact that the first electron of a new shell is loosely bound, while all the electrons of a completed shell are strongly bound, since a large energy difference separates the orbital n, p from the $n + 1$, s. Notice though the non-monotonic behavior appearing at Z = 30 and Z = 31. The element Z = 30 behaves somehow similar to that of Z = 36 in the sense that it exhibits a local minimum of the atomic radius and a local maximum of the I_P; this is because the s/d subshell is completed for Z = 30.

In contrast, the element $Z = 31$, exhibits a local maximum of the atomic radius and a local minimum of the I_P. In this sense it is similar to Z = 19, K, because it has only one electron in a new subshell. The non-monotonic behavior of the atomic radius appearing between Z = 23 and Z = 24 and between Z = 28 and Z = 29 is due to the anomaly of the orderings between the 4s and 3d orbitals.

10.6 Summary of Important Relations

Radius of an atom

$$r_a = \bar{r}_a(Z)\, a_B, \quad 1 < \bar{r}_a(Z) < 5; \qquad a_B = \frac{\hbar^2}{m_e e^2} = 0.529\,A \tag{10.5}$$

Energy levels in the hydrogen atom

$$\varepsilon_n \approx -\frac{e^2}{2a_B}\frac{1}{n^2} \approx -13.6\frac{1}{n^2}\,\text{eV}, \; n = 1, 2, 3, \ldots$$

Energy ordering of levels in atoms other than the hydrogen

$$\varepsilon_{n\,\ell} < \varepsilon_{n\,\ell'}, \text{ if } \ell < \ell', n = 2, 3, 4, \ldots \tag{10.6}$$

$$\varepsilon_{n+2,s} \approx \varepsilon_{n+1,d} \approx \varepsilon_{n,f}, \; < \varepsilon_{n+2,p}, \quad n = 2, 3, 4, \ldots \tag{10.7}$$

$$\varepsilon_{n,s} < \varepsilon_{n,p} < \varepsilon_{n+1,s}, \qquad n = 2, 3, 4, \ldots \tag{10.8}$$

First Ionization energy

$$I_P = \bar{I}'_P(Z)\frac{e^2}{a_B}, \; 0.14 < \bar{I}'_P(Z) < 0.9; \qquad \frac{e^2}{a_B} \equiv \frac{\hbar^2}{m_e\, a_B^2} \equiv \frac{e^4 m_e}{\hbar^2} = 27.2\,\text{eV} \tag{10.9}$$

Electron affinity of an atom is the first ionization potential of its -1 anion (if such an anion is formed). The concept is most relevant for halogens. For hydrogen it is about 0.75 eV, which means that the proton can bind two electrons; to extract one of them a minimum of 0.75 eV is needed, while to extract the one left behind a minimum of 13.6 eV is needed.

10.7 Multiple-Choice Questions/Statements

1. One of the following formulae for the ground state energy of the hydrogen atom is wrong. Which one?
(a) $-e^2/2a_B$ (b) $-e^4/2m_e\hbar^2$ (c) $-e^4m_e/2\hbar^2$ (d) $-\hbar^2/2m_ea_B^2$

2. The average value of the kinetic energy $\langle n,l,m|(p^2/2m_e)|n,l,m\rangle$ in the hydrogen atom is: (*We clarify that* $\langle n,l,m|$ *and* $|n,l,m\rangle$ *are alternative ways of writing* ψ^*_{nlm} *and* ψ_{nlm} *respectively;* $\langle n,l,m|A|n,l,m\rangle$ *is an alternative way of writing the integral* $\int\psi^*_{nlm}(A\psi_{nlm})d^3r$, *where A is the operator corresponding to the physical quantity A*).
(a) $\hbar^2/2m_ea_B^2n^2$ (b) e^2/a_Bn^2 (c) $e^4m_e/4\hbar^2n^2$ (d) $e^4m_e/\hbar^2n^2$

3. The average value of the total energy for the ground state of the positronium atom (e, e^+) is:
(a) $-\hbar^2/2m_ea_B^2$ (b) $-e^2/2a_B$ (c) $-e^4m_e/\hbar^2$ (d) $-\hbar^2/4m_ea_B^2$;
$a_B\equiv\hbar^2/m_ee^2$

4. The average value of the potential energy in the ground state of the (p, μ^-) atom is: (The mass of μ^- equals $207\,m_e$, $a_B\equiv\hbar^2/m_ee^2$)
(a) $-207e^2/a_B$ (b) $-207\hbar^2/2m_ea_B^2$ (c) $-186e^4m_e/\hbar^2$ (d) $-186e^2/2a_B$

5. The first ionization energy of the (p, μ^-) atom is: (The mass of μ^- equals $207\,m_e$, $a_B\equiv\hbar^2/m_ee^2$)
(a) $93e^4m_e/\hbar^2$ (b) $186e^2/a_B$ (c) $207\hbar^2/2m_ea_B^2$ (d) $207e^2/2a_B$

6. The first ionization energy of the He atom is approximately (in eV):
(a) 13.6 (b) 24 (c) 27.2 (δ) 54.4

7. The first ionization energy of the Li atom is approximately (in eV):
(a) 13.6 (b) 5 (c) 10 (d) 14

8. The electron affinity of the Cl atom is approximately (in eV):
(a) 13.6 (b) 27.2 (c) 3.6 (d) 54.4

9. The electronic configuration of the C atom is:
(a) $1s^22s^22p^2$ (b) $2s^22p^4$ (c) $1s^12s^12p^13s^13p^13d^1$ (d) $1s^22p^4$

10. The electronic configuration of the O atom is:
(a) $2s^22p^6$ (b) $1s^41p^4$ (c) $1s^22p^6$ (d) $1s^22s^22p^4$

11. The electronic configuration of the Cu is:
(a) $[\mathrm{Ar}]3d^{10}4s^1$ (b) $[\mathrm{Ar}]3d^94s^14p^1$ (c) $[\mathrm{Ar}]3d^84s^14p^2$ (d) $[\mathrm{Ar}]3d^{11}$

12. The angular dependence of the p_x orbital is:

(a) $\sin^2\theta$ (b) $\sin\theta\sin\phi$ (c) $\sin\theta\cos\varphi$ (d) $\cos\theta$

13. The angular dependence of the d_{zx} orbital is:

(a) $\cos^2\theta$ (b) $\cos\theta\sin\theta\cos\varphi$ (c) $\sin^2\varphi\sin\theta$ (d) $\cos\theta\sin\theta\sin\varphi$

14. The sixth ionization energy of the C atom is about (in eV):

(a) 13.6 (b) 27.2 (c) 81.6 (d) 489.6

15. The fourth line of the periodic table of the elements (PTE) has

(a) 8 (b) 18 (c) 32 (d) 60 elements

16. The element located at the fifth line and the first column of the PTE has

(a) 19 (b) 27 (c) 35 (d) 37 electrons

17. The average value $\langle n,l,m,|r^{-1}|n,l,m\rangle$ of $1/r$ in hydrogen atom is

(a) $1/na_B$ (b) $1/n^2a_B$ (c) $1/2n^2a_B$ (d) $1/2na_B$

10.8 Solved Problems

1. *Obtain the second and the third ionization energy of Li. It is given that* $\langle 1/r_{12}\rangle = 5/(8a)$, *where* a *is the radius of the 1s orbital of the system of a Li nucleus plus two electrons in the corresponding 1s orbital.*

Solution The Hamiltonian of the Li^+ cation consists of the kinetic energy of the two electrons plus the Coulomb interactions. Since the electrons are in a 1s orbital with a radius a (to be determined) we have for the average value of the Hamiltonian

$$\langle H\rangle = \frac{\hbar^2}{2m_e a^2} + \frac{\hbar^2}{2m_e a^2} - \frac{Ze^2}{a} - \frac{Ze^2}{a} + \frac{5e^2}{8a}, \quad Z = 3 \tag{10.10}$$

We use the atomic system of units ($\hbar = m_e = e = 1$) and we set $a = 1/x$ so we have

$$\langle H\rangle = x^2 - \left(6 - \frac{5}{8}\right)x \tag{10.11}$$

By minimizing (10.12) we respect to x we obtain $x = 43/16$ and $\langle H\rangle = -(43/16)^2$ or $|\langle H\rangle| = (43/16)^2 \times 27.2 = 196.46$ eV. This is the minimum energy needed to extract the two electrons from the Li^+ cation, i.e. the sum of the second and the third ionization energy of lithium. The third ionization energy is 13.6 $\times\, 3^2 = 122.4$ eV. Thus the second ionization energy of lithium is 196.46 − 122.4 = 74.06 eV. The experimental values for the second and the third ionization energy are 75.64 and 122.45 eV respectively.

2. *Estimate the velocity of an electron being in the orbital 1s of the uranium atom.*

Solution The kinetic energy $\frac{1}{2}m_e\upsilon^2$ of the electron at the $1s$ orbital of uranium is approximately Z^2 times that of hydrogen, $\hbar^2 Z^2/2m_e a_B^2$, (the Coulomb repulsion of the two electrons occupying the $1s$ level has been omitted, see problem 1). Thus $\upsilon = Z\hbar/m_e a_B = Zc/137 = (92/137)c$. Actually, given that our estimate of this velocity came so close to the velocity of light, we need the relativistic quantum equation of Dirac and not that of Schrödinger (on which our previous estimate was based) to really determine this velocity.

3. *Obtain the electron affinity of hydrogen. It is given that* $\langle 1/r_{12}\rangle = 35/(64a)$ *where a is the radius of the 1s orbital of the system of a proton plus two electrons.*

Solution The electron affinity of the hydrogen atom is by definition the first ionization energy of the anion of hydrogen. We shall work as in the case of Li^+ cation. The average Hamiltonian is, assuming that both electrons of H^- occupy the $1s$ orbital with a radius $a = 1/x$,

$$\langle H\rangle = x^2 - \left(2 - \frac{35}{64}\right)x \tag{10.12}$$

Thus, by minimizing, we have $x = 93/128$, $|\langle H\rangle| = (93/128)^2 \times 27.2 = 14.358$ eV, and the electron affinity is $14.358 - 13.6 = 0.758$ versus 0.7542 eV for the experimental value.

10.9 Unsolved Problems

1. How much would the density of the human body change if the electron mass were half its present value and everything else remain unchanged?
2. Consider two neutral hydrogen atoms at a distance R between their protons, 1 fm $< R <$ 20 A, 1 $A = 10^{-10}$ m. Plot the energy of this system versus R in a log-log graph omitting the kinetic energy of the protons. Mark the characteristic energies and distances.
3. The experimental value of the 20th ionization energy of calcium is 5469.86 eV. Is your estimate for this ionization potential consistent with the experimental value? If not, what is your explanation for the observed discrepancy?

References

1. R. Feynman, R. Leighton, M. Sands, *The Feynman Lectures on Physics* (Addison-Wesley, Reading, 1964)
2. R. Shankar, *Principles of Quantum Mechanics*, 2nd edn. (Springer, NY, 1994)

Chapter 11
From Atoms to Molecules

Abstract Atoms attract each other to form molecules. For each molecule we would like to know the relative positions of its atoms, their vibration properties, and the changes in the electronic states induced by molecule formation. Dimensional analysis and various computational methods such as the so-called LCAO are useful tools in obtaining this information.

11.1 Introduction

The atoms come close together to form molecules, because by doing so they lower their energy, as shown in Fig. 11.1. There is an unlimited number of different molecules starting from simple diatomic molecules, such as N_2 or NaCl, and reaching the huge biomolecules consisting of millions of atoms. The so-called *bond length*, that is the equilibrium distance d between the nuclei of neighboring atoms in a molecule, is of the order of the sum of the atomic radii, $d \approx r_{a1} + r_{a2}$, since neighboring atoms in a molecule essentially touch each other (atoms of noble gases are exceptions). For each molecule we would like to explain and/or predict a number of properties such as its stereochemistry (i.e., the relative position of every one of its atoms in space), the spectrum of the discrete frequencies and the corresponding stationary states of ionic vibrations around the ionic equilibrium positions,[1] as well as the electronic stationary states (i.e., the molecular orbitals, occupied and unoccupied, when the molecule is in its ground electronic state).

In this chapter we shall first explain, by means of the important plot 11.1, why atoms form molecules. Then we shall employ dimensional considerations to

---

[1]Each atom in a molecule is divided into the external valence electrons responsible for the bonding and the remaining ion, the electronic states of which are considered unaffected by the molecule formation.

E.N. Economou, *From Quarks to the Universe*,
DOI 10.1007/978-3-319-20654-7_11

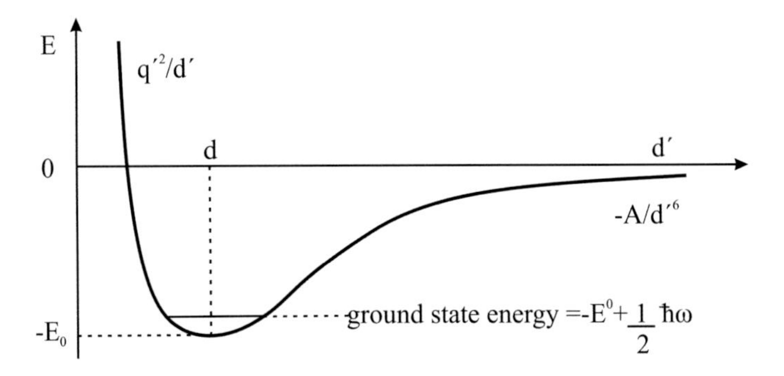

Fig. 11.1 Schematic plot of the variation of the total energy E of two neutral atoms versus the distance d' between their nuclei. The ground state energy of the two atoms when $d' = \infty$ is chosen as the zero of energy. If the zero point motion of the ions is taken into account, the energy increases at every d'; this increase at the equilibrium distance d (corresponding to the minimum energy) is $\frac{1}{2}\hbar\omega$

estimate the values of various quantities pertaining to molecules. A method known as Linear Combination of Atomic Orbitals (LCAO), expressing each molecular orbital as a linear combination of atomic orbitals, will be introduced. It will be used in order to make an educated guess regarding the form of the molecular orbital, and to estimate the corresponding molecular electronic levels, as well as the stereochemistry of at least some simple molecules. Finally, the idea of hybridization of atomic orbitals, i.e. the predetermined linear combinations of atomic orbitals of the same atom as a preparation for molecular formation, will be presented.

11.2 The Residual Electric Interaction Between Two Atoms

Let us consider a system of two atoms at a distance d' between their nuclei. We would like to obtain the interaction energy, E, of this system as a function of d', when the latter varies between, let us say, 20 and 0.001 A. This energy includes the Coulomb interactions among all charged particles of both atoms and the kinetic energy of all electrons; the latter are assumed to be in their lowest total energy for every distance d'. The kinetic energy of the two nuclei is omitted for the time being. The dependence of E on d' is shown schematically in Fig. 11.1 and its gross characteristics can be justified as follows: At a distance d' much larger than the equilibrium distance d (at which E reaches its minimum value $-E_0$), one can show that E is inversely proportional to the sixth power[2] of d':

$$\boxed{E = -\frac{A}{d'^6}, \quad \text{with } A > 0, \ d' \gg d; \quad \text{van der Waals interaction}} \tag{11.1}$$

On the other hand, at d' much smaller than d, the Coulomb repulsion between the two nuclei will dominate[3] giving a contribution to E as shown in Fig. 11.1 with q'^2 approaching $Z_1Z_2e^2$ as $d' \to 0$. Since the E versus d' curve starts by being negative and decreasing with decreasing d' and ends up by being positive and increasing, it must have at least one negative minimum for some intermediate value of d'. It turns out that there is only one minimum occurring at the distance where the highest occupied orbitals of each atom have a small overlap. It has been found empirically that this curve can be approximated around the minimum $(-1 \leq x \leq 5)$ as follows:

$$E(d') = -E_0\varphi(x), \quad \varphi(x) = e^{-x}(1 + x + 0.05x^2), \quad x \equiv \frac{d' - d}{l}, \quad l \approx 0.2d \tag{11.2}$$

If the two atoms are squeezed together beyond the equilibrium point, a repulsion develops partly due to the increased overlap of the occupied atomic orbitals (and the subsequent increase of the electronic kinetic energy) and partly due to the Coulomb repulsion of the two nuclei.

Actually what happens as the two atoms start approaching each other from large distance, is that there is first a slight reorganization of the external electronic states within each atom giving rise to induced polarizations and the resulting van der Waals attraction. When the atoms start touching each other the reorganization is

---

[2] In electrostatics the interaction energy E of two neutral, initially unpolarized, systems at a distance d' much larger than their size is proportional to $-E_1(d') \cdot p_2$, where $E_1(d') \propto p_1/d'^3$ is the electric field created by the dipole moment p_1 at a distance d'; $p_2 = \alpha_2 \cdot E_1(d')$ is the dipole moment of the system 2 induced by the field $E_1(d')$ and α_2 is the polarizability of the system 2. (See (6.12)). For dimensional reasons, α_2 is proportional to the volume r_{a2}^3, where r_{a2} is the radius of system 2; this volume can be written with the help of (10.3) and $p_2^2 \propto e^2 \cdot r_{a2}^2$ as $r_{a2}^3 = r_{a2}^2 r_{a2} = r_{a2}^2 c_{p2} e^2/I_{P2} \propto p_2^2/I_{P2}$. Substituting $E_1(d')$, p_2, and α_2 in $E = -E_1(d') \cdot p_2$ we obtain $E = -A/d'^6$, where $A \propto p_1^2 p_2^2/I_{P2}$. If we had started with the equivalent relation $-E_2(d') \cdot p_1$ for E instead of $-E_1(d') \cdot p_2$, the result for A would have been $A \propto p_1^2 p_2^2/I_{P1}$; it is not unreasonable to assume, for symmetry reasons, that the correct expression for A is the average of the two: $A = c_W p_1^2 p_2^2 (\frac{1}{I_{P1}} + \frac{1}{I_{P2}})$. For the hydrogen-hydrogen case, where $p_1^2 = p_2^2 = 3\,e^2 a_B^2$, the numerical factor c_W is equal to 0.72.

[3] In the case $d' \ll d$, the electrons will not be squeezed between the two nuclei as to screen their repulsion; on the contrary, as $d' \to 0$, they will approach the ground state configuration of an atom of atomic number $Z_1 + Z_2$, where Z_1, Z_2 are the atomic numbers of the two atoms under consideration.

more drastic and involves the spreading of the external electrons to the other atom, i.e., a process that transforms the atomic orbitals to molecular orbitals.[4]

The E versus d' curve allows us to extract valuable information regarding diatomic molecules: One is the equilibrium distance d, which is identical with the so-called bond length; actually there is a fluctuation Δd of this length, since otherwise Heisenberg's relation $\Delta d \cdot \Delta p_d \geq \hbar/2$ would be violated. This fluctuation gives rise to an ionic kinetic energy $\hbar^2/8\,m_r \Delta d^2$ ($m_r \equiv m_{a1} m_{a2}/(m_{a1} + m_{a2})$, where m_{a1}, m_{a2} are the masses of the two atoms), and an additional potential energy equal to $\frac{1}{2}\kappa\,\Delta d^2$ where $\kappa = \partial^2 E/\partial d'^2)_{d'=d}$. Minimizing with respect to Δd the sum ε_{vo} of these two energies (due to the minimum relative motion of the two ions also known as *zero point motion*) we obtain that $\Delta d^2 = \hbar/2\,m_r \omega$, where $\omega \equiv \sqrt{\kappa/m_r}$. Substituting this value of Δd to the sum of the kinetic and the potential energies we find that $\varepsilon_{vo} = \frac{1}{2}\hbar\omega$.

To summarize: From the E versus d' dependence we can extract the following information: (1) The bond length d and, hence, the rotational spectrum $\varepsilon_{r\ell} = \hbar^2 \ell(\ell+1)/2I$, where $\hbar^2 \ell(\ell+1)$ is the square of the angular momentum, $\ell = 0, 1, 2, \ldots$, and $I = m_r d^2$ is the moment of inertia of the diatomic molecule. (2) The fluctuation in the bond length Δd. (3) The natural frequency of oscillation $\omega \equiv \sqrt{\kappa/m_r}$ and, hence, the vibrational spectrum $\varepsilon_{vn} = (n + \frac{1}{2})\,\hbar\omega$, $n = 0, 1, 2, \ldots$. (4) The dissociation energy (i.e., the minimum energy needed to separate the molecule to the two neutral atoms), $D = |E_o| - \frac{1}{2}\hbar\omega$.

11.3 Estimates Based on Dimensional Analysis

Some of the molecular quantities mentioned in the previous section can be estimated by dimensional analysis as follows:

(1) *Bond length d* (see Fig. 11.1 and footnote 4 in this chapter)

$$d \approx (r_{a1} + r_{a2}) = (c_{a1} + c_{a2})\,a_B \approx 0.5(c_{a1} + c_{a2})\,\text{A}, \quad 1\,\text{A} = 10^{-10}\,\text{m} \qquad (11.3)$$

The argument in support of this relation was presented in the previous section. The proportionality factors c_{a1}, c_{a2} can be estimated from Fig. 10.1.

[4]Atoms of the noble gases are an exception: Because of their fully completed outer shells and the large energy separation of the next empty level, no overlap of atomic orbitals belonging to different atoms is tolerated, and the curve E versus d' (see Fig. 11.1) starts moving upwards before any overlap occurs and before any molecular orbital is formed. As a result, the equilibrium distance d is considerably larger than the sum $r_{a1} + r_{a2}$ and the energy gain $|E_o|$ is much smaller than usually.

(2) *Dissociation energy D*

$$D = c_d \frac{\hbar^2}{m_e d^2} = c_d \frac{\hbar^2}{m_e a_B^2} \frac{a_B^2}{d^2} \approx c_d \, 27.2 \frac{a_B^2}{d^2} \text{ eV} \approx 27.2 \frac{1}{d^2}, \quad c_d \approx 1 \tag{11.4}$$

It is expected that the order of magnitude of the dissociation energy must be given by the basic unit of energy, $\hbar^2/m_e a_B^2$. Moreover, it is reasonable to expect that the replacement of a_B by the characteristic length d of the molecule will make our estimate molecule-specific, the same way that the replacement $a_B \to r_a$ made our estimate of I_P atom-specific (see Fig. 10.2). The experimental value for the quantity D for the molecules O_2, Rb_2, AgI is 5.17, 0.51, 2.42 eV respectively. To obtain these values from (11.4) and the corresponding experimental data for d, $d =$ 1.207, 4.4, 2.54 A (1 A $= 10^{-10}$m), we must choose c_d equal to 1.028, 1.29, 2.05 respectively. Notice that (11.3) and Fig. 10.1 give d = 0.9, 4.49, 2.64 A respectively.

(3) *Natural vibrational frequency* ω. As explained in Chap. 4, the vibrational frequency ω must be proportional to ω_o, where $\omega_o = e^2/a_B\hbar$ or, equivalently, $\hbar/m_e a_B^2$ is the only quantity with dimensions of frequency made out of e, m_e, $\hbar$. However, the vibrating masses are those of the ions; thus the reduced mass of the ions $m_r = A_{Wr} u = 1823 A_{Wr} m_e$ must enter the formula for ω and it must enter in a dimensionless way. Hence the most general formula for ω is of the following form: $\omega = \omega_o f(m_r/m_e)$, where f is an arbitrary function of m_r/m_e. We demonstrate in Appendix A that in an oscillation, the oscillating mass enters as the inverse square root. Therefore, by implementing also the replacement $a_B \to d$, the formula for ω becomes

$$\hbar\omega = c_v \frac{e^2}{d} \sqrt{\frac{m_e}{m_r}} \approx \frac{637}{d\sqrt{A_{Wr}}} \text{ meV}, \; c_v \approx 1 \tag{11.5a}$$

or, if we start from the formula $\omega_0 = \hbar/m_e a_B^2$,

$$\hbar\omega = c_v' \frac{\hbar^2}{m_e d^2} \sqrt{\frac{m_e}{m_r}} \approx \frac{3026}{d^2\sqrt{A_{Wr}}} \text{ meV} \tag{11.5b}$$

Notice that, as a result of the replacement $a_B \to d$, the two expressions (11.5a, b) are no longer equivalent. The second one, which is proportional to $1/d^2$, is more reasonable, since it is obtained by taking into account (11.2), (11.4), and that $\omega = \sqrt{\kappa/m_r}$ with $\kappa = (\partial^2 E/\partial d'^2)_{d'=d}$. Following this procedure we obtain $c_v' \approx$ 4.75 except for the special case of the hydrogen molecule where this value is about 1.9 instead of 4.75. Choosing the experimental values for d, (11.5b) gives that $\hbar\omega$

Table 11.1 Comparison of experimental data for *d, D,* and $\hbar\omega$ of some typical diatomic molecules with the corresponding estimates based on dimensional analysis (11.4 with $c_d = 1$, (11.5b), upper row of values, (11.5a) lower row)

	H_2		N_2		O_2		Na_2		AgI	
	exp.	est.	exp.	est.	exp.	est.	exp.	est.	exp.	est.
d(A)	0.74	1.06	1.098	1.04	1.21	0.90	3.08	3.426	2.54	2.33
A_{Wr}	0.5		7		8		11.5		58.3	
D (eV)	4.48	13.9	9.79	6.31	5.165	5.2	0.775	0.8	2.42	1.18
$\hbar\omega$ (meV)	516	875	292	265	196	204	19.7	26	25.6	18
		647		117		99		32		17

equals 875, 265, 204, and 26 meV for H_2, N_2, O_2, and Na_2 respectively, while the corresponding experimental values are 516, 292, 196, and 19.7 respectively.

(4) *The rotational spectrum.* As it was mentioned in the previous section, the quantum of the rotational spectrum of a diatomic molecule is equal to

$$\frac{\hbar^2}{I} = \frac{\hbar^2}{m_r d^2} = \frac{\hbar^2}{m_e a_B^2} \frac{m_e}{m_r} \frac{a_B^2}{d^2} = 27.2 \frac{m_e}{m_r} \frac{a_B^2}{d^2} \text{ eV} \tag{11.6}$$

For O_2 the quantity $\hbar^2/I$ is equal to 0.36 meV. In general the rotational quantum (such a quantum is what we called $\delta\varepsilon$ in Chap. 3) is smaller than the typical dissociation energy D, by a factor (m_e/m_r), as shown in (11.6). Its range of values is about a fraction of a meV which corresponds to a few degrees Kelvin. On the other hand, the vibrational quantum $\hbar\omega$ is smaller than the typical electronic energy by a factor $\sqrt{m_e/m_r}$ which brings it down from the typical electronic energy of a few eV to the few tens (even hundreds) of meVs (corresponding to hundreds or even thousands of degrees Kelvin). We conclude that at room temperature the rotational degrees of freedom of a diatomic molecule are fully excited, while the vibrational ones are usually frozen, at least for small, light molecules.

Out of the $3N_a$ ionic degrees of freedom of a non-linear molecule consisting of N_a atoms, 3 degrees of freedom are associated with the motion of the center of mass of the molecule and another 3 are associated with its rotations as a rigid body.[5] The remaining $3N_a - 6$ ionic degrees of freedom are associated with stationary oscillations (or waves, see Fig. A.1c in Appendix A) to each one of which corresponds a natural frequency common to all participating ions (Table 11.1).

[5] If the molecule is linear, there are only two rotational degrees of freedom and $3N_a - 5$ vibrational ones.

11.4 Linear Combination of Atomic Orbitals (LCAO)

As we mentioned before when two atoms approach each other at a distance comparable to the sum of their radii, the highest occupied atomic orbital of atom 1 is spread to the space occupied by atom 2 and vice versa. Thus these two atomic orbitals, ψ_1, ψ_2 are transformed to two molecular orbitals, ϕ_a, ϕ_b, one of them with energy ε_b below the lowest one of the two atomic levels and the other with energy ε_a above the highest one of the two atomic levels, as shown in the diagram below.

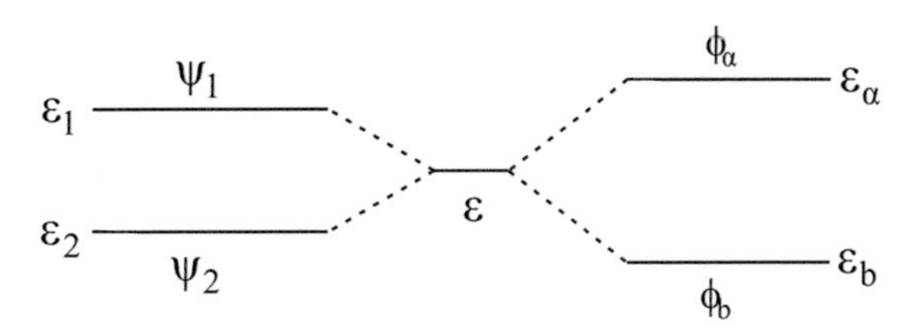

If each atomic orbital was initially occupied by two electrons (as in noble gases), the four electrons will fully occupy both the resulting molecular orbitals and as a consequence no energy gain will be obtained by the attempted chemical bond which, therefore, will not be materialized. On the contrary, if the total number of electrons in these two atomic orbitals is between one and three (0 + 1, as in $H^+ + H \rightarrow H_2^+$, (1 + 1, as in Na + Na $\rightarrow$ Na_2, 0 + 2, as in one of the triple bond in CO, 1 + 2, as in H + Mg $\rightarrow$ HMg) the possibility arises of lowering the total energy and hence creating a bond. For example, if initially the total number of electrons were 2 (0 + 2 or 1 + 1) both of these electrons will occupy the low lying molecular orbital and an energy gain is expected to result.

Since a molecular orbital (MO) involves the spreading of the atomic orbital to the nearby atom, it is only natural to write the MO as linear combination of atomic orbitals (LCAO) of both atoms (see the first few pages of [1]). The LCAO method ignores the fact that the spreading of the wavefunction is also associated with a reorganization of the part that remains in the initial atom and the part that goes to the nearby atom. Nevertheless the LCAO has been proven a reasonable approximation. To be specific we shall consider first the simple molecule AgI assuming that only the two partly occupied atomic orbitals, the 5s of Ag (denoted by ψ_1), and the 5p_x of I (denoted by ψ_2), are relevant for the formation of the molecule. This assumption, although oversimplified, is not unreasonable, since these two atomic orbitals are the highest singly occupied ones (each of them has just one electron). Then, according to the LCAO method, and the above simplifying assumption, the molecular orbital ϕ is given by:

$$\phi = c_1 \psi_1 + c_2 \psi_2 \tag{11.7}$$

The physical meaning of the coefficients c_1, c_2 (assumed to be real numbers) is the following: If an electron is in the molecular orbital ϕ, the probability p_1 to find it in the atomic orbital ψ_1 is given by $p_1 = c_1^2/(c_1^2 + c_2^2)$, and the probability to find it in the atomic orbital ψ_2 is given by $p_2 = c_2^2/(c_1^2 + c_2^2)$ assuming that ψ_1, ψ_2 are orthogonal. The average value of the energy of an electron being in the orbital ϕ can be obtained from the following general formula[6]:

$$\varepsilon_\phi = \frac{\langle\phi|\hat{H}|\phi\rangle}{\langle\phi|\phi\rangle} = \frac{c_1^2\varepsilon_1 + c_2^2\varepsilon_2 + 2c_1c_2V_2}{c_1^2 + c_2^2} \tag{11.8}$$

where $\varepsilon_1 \equiv \langle\psi_1|\hat{H}|\psi_1\rangle$, $\varepsilon_2 \equiv \langle\psi_2|\hat{H}|\psi_2\rangle$, $V_2 \equiv \langle\psi_1|\hat{H}|\psi_2\rangle$ and $\hat{H} = (\hat{p}^2/2m_e) + \mathcal{V}(r)$. We can determine c_1 and c_2 by minimizing ε_ϕ. More specifically, we determine the extrema of ε_ϕ by setting the partial derivatives $\partial\varepsilon_\phi/\partial c_1$, $\partial\varepsilon_\phi/\partial c_2$ equal to zero. We thus find the following homogeneous linear system of two equations with two unknown quantities:

$$(\varepsilon_1 - \varepsilon_\phi)\, c_1 + V_2 c_2 = 0, \quad (\varepsilon_2 - \varepsilon_\phi)\, c_2 + V_2 c_1 = 0 \tag{11.9}$$

which has two solutions, only if the determinant is set equal to zero. By doing so we find the lower eigenenergy which is

$$\varepsilon_\phi = \varepsilon_b = \varepsilon - \sqrt{V_2^2 + V_3^2}, \quad \text{where } \varepsilon \equiv \frac{\varepsilon_1 + \varepsilon_2}{2}, \quad \text{and } V_3 \equiv \frac{\varepsilon_1 - \varepsilon_2}{2} > 0 \tag{11.10}$$

The reader may verify that the values of c_1, c_2 of the orbital ϕ_b corresponding to ε_b are such that

$$p_{1b} = \frac{c_1^2}{c_1^2 + c_2^2} = \frac{1 - a_p}{2} \quad \text{and} \quad p_{2b} = \frac{c_2^2}{c_1^2 + c_2^2} = \frac{1 + a_p}{2} \tag{11.11}$$

where the so-called *polarity index* a_p is defined as follows:

$$\boxed{a_p \equiv \frac{V_3}{\sqrt{V_2^2 + V_3^2}} > 0} \tag{11.12}$$

If we place the two valence electrons in the molecular orbital ϕ_b with opposite spins, their total energy will be $2\varepsilon_b = 2\varepsilon - 2\sqrt{V_2^2 + V_3^2}$, while before the molecule formation the corresponding energy was $\varepsilon_1 + \varepsilon_2 \equiv 2\varepsilon$. Hence, there is a lowering of the total energy due to the molecule formation equal to $2\sqrt{V_2^2 + V_3^2}$. Actually the decrease is $2\sqrt{V_2^2 + V_3^2} - U$, because the presence of the two electrons in the same molecular orbital leads to an increase U of the Coulomb repulsive energy relative to the initial state where the two electrons were in different atoms. We conclude that

[6] If a waveparticle is in a state ϕ, the average value of a physical quantity A is obtained by the following formula: $\langle A\rangle = \langle\phi|\hat{A}|\phi\rangle/\langle\phi|\phi\rangle$ where $\hat{A}$ is the operator corresponding to A and by definition we have $\langle\phi|\hat{A}|\phi\rangle \equiv \int\phi\,(\hat{A}\,\phi\,)d^3r$ and $\langle\phi|\phi\rangle \equiv \int\phi\,\phi\,d^3r$; ϕ, c_1, c_2 are assumed to be real.

placing the two electrons to the orbital ϕ_b lowers the total energy by $D = 2\sqrt{V_2^2 + V_3^2} - U - \frac{1}{2}\hbar\omega$, and, hence, bonds the two atoms together as to form a diatomic molecule. For this reason, ϕ_b is called *bonding molecular orbital* and the level ε_b is called bonding molecular level. Equation (11.11) shows that out of the two electrons populating the orbital ϕ_b, $2p_{2b} = 1 + a_p$ electrons are on the average in the atom 2 and $2p_{1b} = 1 - a_p$ electrons are in the atom 1. Therefore, the molecule formation is accompanied by a net transfer of electron equal to a_p from atom 1 (with the higher ε_1) to atom 2 (with the lower ε_2); this justifies the name polarity index for a_p.

Let us now examine the other solution of the system (11.9). The resulting molecular orbital, denoted by ϕ_a, is given by

$$\phi_a = \frac{1}{\sqrt{2}}\left(\sqrt{1 + a_p}\,\psi_1 - \sqrt{1 - a_p}\,\psi_2\right) \tag{11.13}$$

and the corresponding molecular level is given by

$$\varepsilon_\phi = \varepsilon_a = \varepsilon + \sqrt{V_2^2 + V_3^2} \tag{11.14}$$

Placing two electrons in the orbital ϕ_a will increase now the energy by an amount equal to $2\sqrt{V_2^2 + V_3^2} + U'$. Hence, the molecular orbital ϕ_a works in the opposite direction than that of molecule formation and for this reason is called the *antibonding molecular orbital*; ε_a is called the antibonding molecular level.

What would have happened, if the atomic levels ψ_1, ψ_2 were fully occupied with two electrons each? In that case, we would be forced to place two electrons in the bonding molecular orbital, ϕ_b, and two in the antibonding ϕ_a; the result would be to increase the energy by an amount about equal to $U + U'$, instead of lowering it. We conclude that combining two completely occupied atomic orbitals does not favor molecule formation through molecular orbitals and, as a first approximation, can be ignored (however, see footnote 4). The ones that favor molecule formation are those which are singly occupied each or, more generally, those where the number of electrons satisfy the relation $1 \leq n_1 + n_2 \leq 3$.

Up to now the presentation of the LCAO method was kept at a qualitative level. To reach a semiquantitative level we need to know the levels ε_1 and ε_2 and the value of V_2, as well as an estimate of U. The levels ε_1 and ε_2 are given approximately in Table 11.2, while for V_2 the approximate expression proposed by Harrison [2] is used. According to this:

$$V_2 = c_2 \frac{\hbar^2}{m_e d^2} \tag{11.15}$$

with $c_2 = -1.32,\ 1.42,\ 2.22,\ -0.63$ for s/s, s/p_x, p_x/p_x, p_y/p_y atomic orbitals respectively, where the x-axis is taken along the axis of the molecule (see Fig. 11.2). U is usually taken as $c_U\sqrt{V_2^2 + V_3^2}$ with c_U usually in the range between about 1 and

Table 11.2 Calculated values (according to the Hartree-Fock method of the energy levels $-\mathrm{E}_{n's}$, $-\mathrm{E}_{n'p}$, $-\mathrm{E}_{n'-1,d}$, and $V_1 \equiv \left(\mathrm{E}_{n'p} - E_{n's}\right)/4$ for the highest partially or fully occupied subshell (From Harrison' s book [2])

H																	***He*** 24,97 -	
Li 5,34 - -	***Be*** 8,41 - -	Non-transition elements					Transition elements					***B*** 13,46 8,43 1,26	***C*** 19,37 11,07 2,08	***N*** 26,22 13,84 3,10	***O*** 34,02 16,72 4,33	***F*** 42,78 19,86 5,73	***Ne*** 52,51 23,13 7,35	$n' = 2$
Na 4,95 6,88 is Mg -	***Mg*** 6,88 - -	***C*** 19.37 ← $-E_{n's}$ 11.07 ← $-E_{n'p}$ 2.08 ← V_1					***Fe*** 7.08 ← $-E_{n's}$ 16.54 ← $-E_{n'-1d}$					***Al*** 10,70 5,71 1,25	***Si*** 14,79 7,58 1,80	***P*** 19,22 9,54 2,42	***S*** 24,01 11,60 3,10	***Cl*** 29,19 13,78 3,85	***Ar*** 34,75 16,08 4,67	$n' = 3$
K 4,01 - -	***Ca*** 5,32 - -	***Sc*** 5,72 9.35	***Ti*** 6,04 11.04	***V*** 6,32 12.55	***Cr*** 6,59 13.94	***Mn*** 6,84 15.27	***Fe*** 7,08 16.54	***Co*** 7,31 17.77	***Ni*** 7,52 18.96	***Cu*** 6,49 13.36	***Zn*** 7,96 4 0.99	***Ga*** 11,55 5,67 1.47	***Ge*** 15,15 7,33 1.96	***As*** 18,91 8,98 2.48	***Se*** 22,86 10,68 3.05	***Br*** 27,00 12,43 3.64	***Kr*** 31,37 14,26 4.28	$n' = 4$
Rb 3,75 - -	***Sr*** 4,85 - -	***Y*** 5,34 6,80	***Zr*** 5,68 8,46	***Nb*** 5,95 10,03	***Mo*** 6,19 11,56	***Tc*** 6,39 13,08	***Ru*** 6,58 14,59	***Rh*** 6,75 16,16	***Pd*** 6,91 17,66	***Ag*** 5,98 14,62	***Cd*** 7,21 4 0,803	***In*** 10,14 5,37 1,19	***Sn*** 13,04 6,76 1,57	***Sb*** 16,02 8,14 1,97	***Te*** 19,12 9,54 2,40	***I*** 22,34 10,97 2,84	***Xe*** 25,69 12,44 3,31	$n' = 5$
Cs 3,36 - -	***Ba*** 4,29 - -	***La*** 4,35 -	***Hf*** 5,72 8,14	***Ta*** 5,98 9,57	***W*** 6,19 10,96	***Re*** 6,38 12,35	***Os*** 6,52 13,73	***Ir*** 6,71 15,13	***Pt*** 6,85 16,55	***Au*** 6,01 14,17	***Hg(α)*** 7,10 4,11 0,75	***Tl*** 9,82 5,23 1,15	***Pb*** 12,48 6,53 1,49	***Bi*** 15,19 7,79 1,85	***Po*** 17,96 9,05 2.23	***At*** 20,82 10,33 2.62	***Rn*** 23,78 11,64 3.04	$n' = 6$
Fr	***Ra***	***Ac***																
			Ce	***Pr***	***Nd***	***Pm***	***Sm***	***Eu***	***Gd***	***Tb***	***Dy***	***Ho***	***Er***	***Tm***	***Yb***	***Lu***		
			Th	***Pa***	***U***	***Np***	***Pu***	***Am***	***Cm***	***Bk***	***Cf***	***Es***	***Fm***	***Md***	***No***	***Lr***		

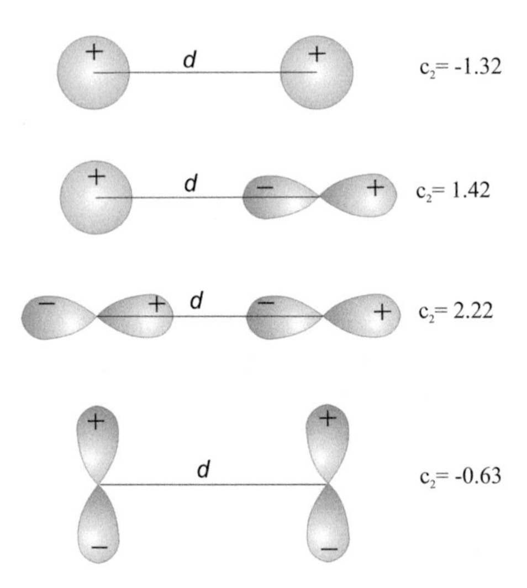

Fig. 11.2 The non-diagonal matrix elements V_2 of the Hamiltonian used in the LCAO method is given by the formula $V_2 = c_2\hbar^2/m_e d^2$, where d is the bond length and the numerical factor c_2 depends on the type of atomic orbitals involved, as shown above, and the orientation of the p (or the p's) relative to the direction of the bond

1.7; the smaller values for single long bonds and the larger for triple short bonds. For the molecule AgI we have the following values $\varepsilon_1 = -5.98\,\text{eV}$, $\varepsilon_2 = -10.97\,\text{eV}$, $c_U = 1$, $V_2 = 1.42\hbar^2/(m_e d^2) = 1.675$ eV and $D = (2-1)\sqrt{V_2^2 + V_3^2}$ eV $= \sqrt{1.675^2 + 2.495^2} = 3$ eV versus $D = 2.42 \pm 0.3\,\text{eV}$ for the experimental value.

For diatomic molecules consisting of identical atoms, the above formulae become simpler, because then $V_3 = a_p = 0$ and $c_1^2/(c_1^2 + c_2^2) = c_2^2/(c_1^2 + c_2^2) = 1/2$. As an example the molecule Na_2 for which the relevant atomic orbital is the $3s$ for each atom, has $d = 3.08\,\text{A}$. Then we find: $V_2 = -1.32\,\hbar^2/m_e d^2 = 1.06\,\text{eV}$ and, by choosing $c_U \approx 1$, $D = |V_2| \approx 1\,\text{eV}$, while the experimental value is $D = 0.78\,\text{eV}$. For the molecule O_2, for which d = 1.208 A, there are two *independent* pairs of relevant atomic orbitals giving rise to a double bond. The p_{x1}/p_{x2} (with V_{2x} =$2.22\,\hbar^2/m_e d^2 = 11.58\,\text{eV}$) produces a strong bond and the p_{y1}/p_{y2} (with $V_{2y} = -0.63\,\hbar^2/m_e d^2 = 3.29\,\text{eV}$), a weak one. The dissociation energy in this case is given by the following formula: $D = (2 - c_U)\left(|V_{2x}| + |V_{2y}|\right) = 5.95$ eV versus 5.165 eV for the experimental value, where in the numerical results we have taken the experimental value of d = 1.208 A and we have chosen $c_U \approx 1.6$. For the nitrogen molecule N_2 (see Fig. 11.3) there are three independent pairs making the triple bond: p_{x1}/p_{x2} (with $V_{2x} = 2.22\,\hbar^2/m_e d^2 = 14\,\text{eV}$), p_{y1}/p_{y2} (with $V_{2y} = -0.63\,\hbar^2/m_e d^2 = -4\,\text{eV}$), and p_{z1}/p_{z2} (with $V_{2z} = V_{2y}$). By choosing $c_U = 1.6$ and using the actual value of d = 1.098 A, we find D = 8.8 eV compared to D = 9.79 eV measured experimentally.

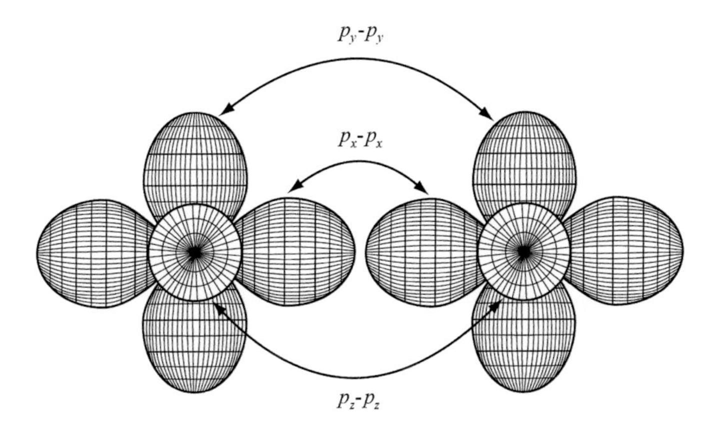

Fig. 11.3 The nitrogen molecule. The half occupied atomic orbitals p_{x1}, p_{y1}, p_{z1} of atom 1 of nitrogen and the p_{x2}, p_{y2}, p_{z2} of atom 2 of nitrogen (the p_z orbitals are perpendicular to the page and are drawn as *circles*) are combined as shown to create in the N_2 molecule three bonds (a situation usually called triple bond), one strong between the p_x s and two weak ones between the p_y s and the p_z s

11.5 Hybridization of Atomic Orbitals

In the applications of the LCAO method we have presented up to now, we considered cases of a single atomic orbital per atom. Even for the O_2, N_2 molecules the four $(p_{xi}, p_{yi}, i = 1, 2)$ or six $(p_{xi}, p_{yi}, p_{zi} \; i = 1, 2$, see Fig. 11.3) relevant atomic orbitals were naturally grouped into two or three *independent* pairs respectively. Thus the calculation was reduced to simply solving systems of two linear homogeneous algebraic equations with two unknowns. In general, nature is more complex. To see why, let us consider the important molecule CO_2. If we employ only singly occupied atomic orbitals, this molecule will look as shown in Fig. 11.4. In other words, according to this choice of bonding, the stereochemistry of the molecule would be non-colinear as shown Fig. 11.4; this is so because the bonding molecular orbital (01, perpendicular shading) between the carbon atom (no. 0) and the oxygen atom (no. 1) will involve the pair of atomic orbitals p_{x0} and p_{x1}, while the (02, horizontal shading) bond will involve the pair $p_{y0}, \; p_{y2}$ The problem is that the oxygen orbitals p_{y1} and p_{x2} (unshaded) are left dangling and, hence, their possibility to form bonds was not exploited. Such a configuration of CO_2 would be unstable. The only way out of this problem is to use the 2*s* atomic orbital of carbon which in combination with its three 2*p* orbitals will make four atomic orbitals with four electrons to pair with the four orbitals of the two oxygens $(p_{x1}, p_{y1}, p_{x2}, p_{y2})$ with their four electrons. The calculation now is considerably more complicated than before: We have to find, by minimizing the energy, which linear combinations of the four atomic orbitals $(s, \; p_x, \; p_y, \; p_z)$ of carbon will combine with the four orbitals of the two oxygens to produce the ground state of the CO_2. To avoid the calculation of the total energy, setting its derivatives equal to zero, and solving the resulting system of eight equations with eight unknowns (four for the carbon orbitals and four for the orbitals of the two oxygens), a drastic approximation has been devised. This

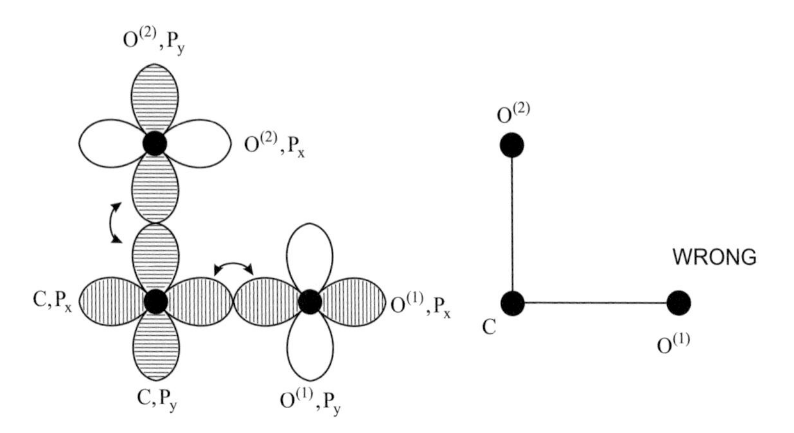

Fig. 11.4 The hypothetical structure of the molecule CO_2. One would end up with this wrong structure, if he/she were forced to employ only singly occupied atomic orbitals

approximation consists of having three standard preselected combinations, called hybridized atomic orbitals, of the four atomic orbitals (one s and three p's of the same atom, in the present case of carbon) and for each case to check which one of the three preselected combinations seems to fit best.

The three standard preselected hybridizations of the orbitals of the same atom are the following (see the first few pages of [1]):

(1) $sp^{(1)}$ **hybridization**: There are two hybrid orbitals: $\chi_1 = \frac{1}{\sqrt{2}}(s + p_x)$, $\chi_2 = \frac{1}{\sqrt{2}}(s - p_x)$ and two unhybridized ones, p_y, p_z; χ_1 is oriented along the positive x direction, while χ_2 along the negative x direction (see Fig. 11.5).

The energy associated with either χ_1 or χ_2 is $\varepsilon_h = (\varepsilon_s + \varepsilon_p)/2$. There are two drawbacks associated with the hybridization: The first one is that there is a non-zero matrix element of the Hamiltonian between χ_1 and χ_2 equal to $V_1 = -(\varepsilon_p - \varepsilon_s)/2$, although $\langle \chi_1 | \chi_2 \rangle = 0$; this implies that the hybrids are not true eigenstates of the atom and consequently every molecular orbital based on them is not fully localized in the corresponding bond, but tends to spread from bond to bond. The second is that there is an energy cost in the hybridization process equal to $(\varepsilon_s + \varepsilon_p) + 2\varepsilon_p - 2\varepsilon_s - 2\varepsilon_p = \varepsilon_p - \varepsilon_s$. The latter must be viewed as an energy investment to be more than recovered with the molecule formation. The $sp^{(1)}$ hybridization is the appropriate one for the CO_2: One C–O bonding molecular orbital is made from the pair χ_1, p_{x1} with $V_2 = 2.57\hbar^2/m_e d^2$ and another from the pair p_{y0}, p_{y1} with $V_2 = -0.63\hbar^2/m_e d^2$ both to the right of the carbon atom. Two similar bonds, one strong and one weak, are formed to the left of carbon between the atomic orbitals χ_2, p_{x2} and p_{z0}, p_{z2} (see also the caption of Fig. 11.6). Thus CO_2 is a linear molecule: C=O=C. See Fig. 11.6; see also Fig. 11.7 presenting the molecule C_2H_2 employing also the $sp^{(1)}$ hybridization

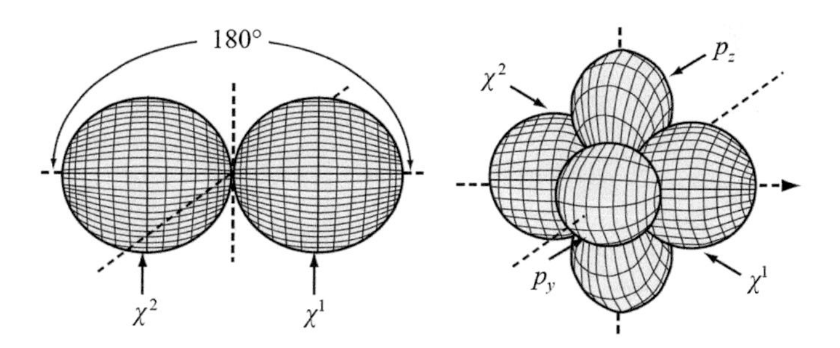

Fig. 11.5 The two orthogonal to each other hybridized atomic orbitals sp^1, $\chi_1 = (s + p_x)/\sqrt{2}$ pointing mostly along the positive x-direction and $\chi_2 = (s - p_x)/\sqrt{2}$ pointing mostly along the negative x-direction, are obtained as an equal weight linear combination of s and p atomic orbitals of the same atom (in carbon the relevant main quantum number is 2). In this figure the p_x orbital has been chosen for the sp^1 hybridization, but in general the choice could have been $c_x p_x + c_y p_y + c_z p_z$ with $c_x^2 + c_y^2 + c_z^2 = 1$ instead of p_x. The requirements of orthogonality of the hybrids and that of equal weight between s- and p-character uniquely determines the sp^1 hybrids apart from rigid rotations. In the *right figure* besides the two hybridized sp^1 atomic orbitals the two unhybridized p_y and p_z atomic orbitals are also shown. The average energy of the sp^1 hybrids is $(\varepsilon_s + \varepsilon_p)/2$. These sp^1 atomic hybrids are usually shown with only one lobe (*the largest one*)

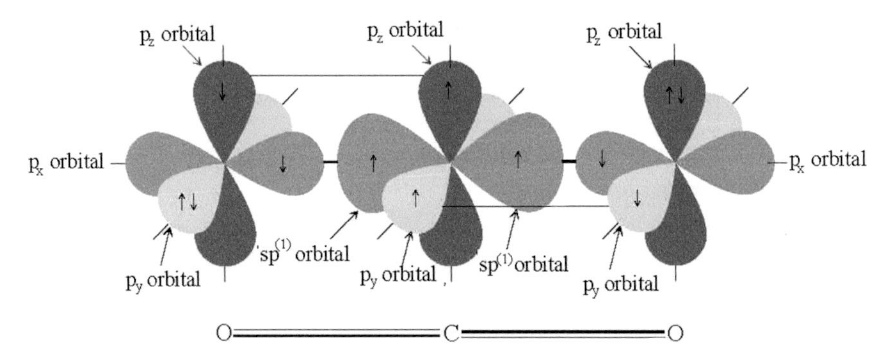

Fig. 11.6 The CO_2 molecule is employing the two sp^1 hybrid atomic orbitals of carbon each of opposite direction to form a strong bond with each of the two p_x' s of oxygens, one to its *left* and the other to its *right* and a weak bond (between the p_z s to the *left*) and another weak bond (between the p_y s to the *right*) as shown in the figure. So the molecule is linear as in reality. The p_y of the oxygen to the *left* and the p_z of the oxygen to the *right* are doubly occupied and so do not participate in the bonding. Actually, according to the rules of QM, the real ground state is an equal weight symmetric combination of the state just described and its mirror image with respect to a plane passing through the carbon atom and being perpendicular to the axis of the molecule. The bond length is 1.163 A ($1\ \mathrm{A} = 10^{-10}$ m)

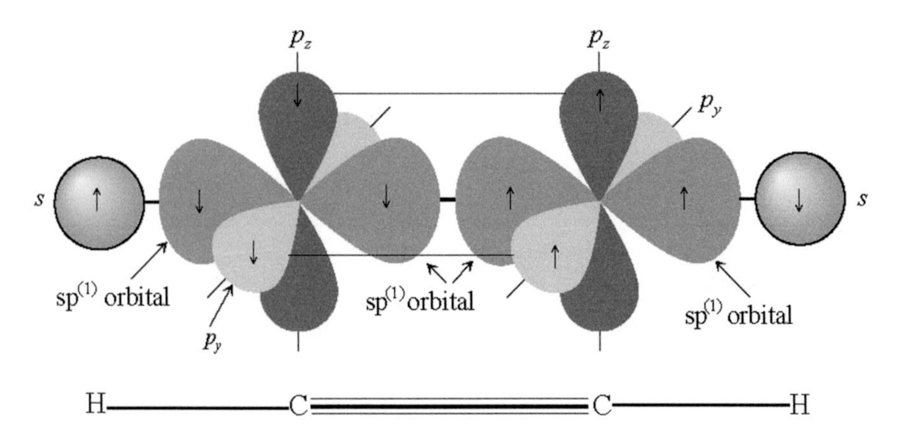

Fig. 11.7 The bonding orbitals of the molecule C_2H_2 are shown. Each carbon atom employs two sp^1 atomic hybrids, two of which are used for a strong bond between the two carbon atoms (*thick line*) and the other two to form bonds with the two hydrogens (*thick lines*). The singly occupied p_ys and p_zs of the two carbon atoms are used for the creation of two weak bonds (*thin lines*) between the carbons. So the molecule is linear with a triple C–C bond. The bond lengths are: 1.203 A for the triple C–C bond and 1.06 A for the C–H bonds ($1\ \mathrm{A} = 10^{-10}$ m)

(2) ***sp*$^{(2)}$hybridization**: There are three hybrid atomic orbitals, $\chi_1 = \frac{1}{\sqrt{3}}(s + \sqrt{2}p_x)$, and χ_2, $\chi_3 = \frac{1}{\sqrt{3}}(s - \frac{1}{\sqrt{2}}p_x \pm \frac{\sqrt{3}}{\sqrt{2}}p_y)$ each with energy $\varepsilon_h = (\varepsilon_s + 2\varepsilon_p)/3$. These three atomic hybrids are in the same plane and the angle between two successive ones is 120° (see Figs. 11.8 and 11.9).

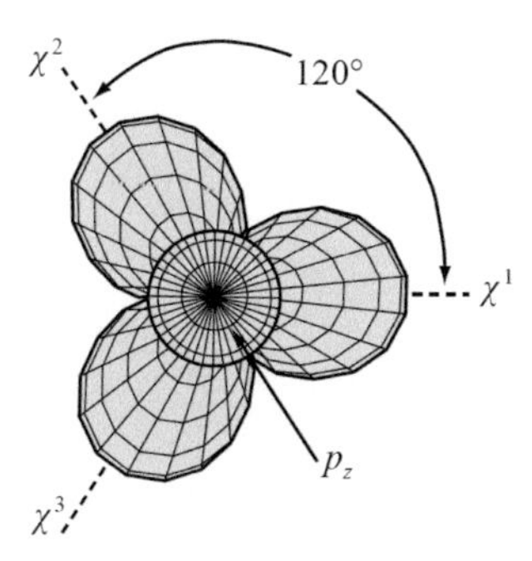

Fig. 11.8 The three sp^2 atomic hybrids are orthogonal to each other, lie on the same plane (taken as the x, y plane) and the directions of their main lobes form 120° angles as shown in the figure, in which the x-axis coincides with one of these hybrids. The weight of the orbital(s) p (p's) is twice that of the s. This condition together with their mutual orthogonality uniquely determines the three sp^2 hybrids apart from a rigid rotation. The average energy of each sp^2 hybrid is $\varepsilon_h = (\varepsilon_s + 2\varepsilon_p)/3$.

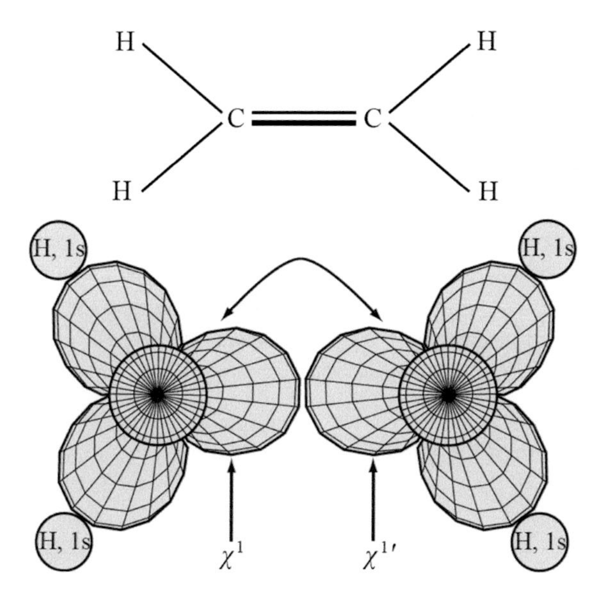

Fig. 11.9 The 12 atomic orbitals participating in the formation of the molecule C_2H_4 are separated (with the help of the six in total sp^2 hybrids of the carbon atoms) to six almost independent pairs each one forming a bonding (and an antibonding) molecular orbital as shown in the figure. The sixth one consists of the p_z s of the two carbons

Figure 11.9 shows how the sp^2 orbitals are employed to account for the structure of the planar molecule C_2H_4. The off-diagonal matrix element of the Hamiltonian between any pair of the three hybrids of the same atom is $V_1 = -(\varepsilon_p - \varepsilon_s)/3$. The fourth orbital, p_z, is unhybridized. The energy cost of the $sp^{(2)}$ hybridization is again $\varepsilon_p - \varepsilon_s$. An important example of $sp^{(2)}$ hybridization is the benzene molecule, C_6H_6, (see problem 11.3, and its solution, where the stereochemistry C_6H_6 is presented). Benzene is a planar ring molecule forming a regular hexagon; the six unhybridized p_z atomic orbitals of the carbon atoms in C_6H_6 combine to give rise to six *delocalized* molecular orbitals shown in (11.16) and having the form of standing waves going around the ring:

$$\phi_\nu = \sum_{n=1}^{n=6} c_n^{(\nu)} \psi_n, \; c_n^{(\nu)} = c_o \exp(i\varphi_\nu n), \text{ and } \varphi_\nu = \frac{2\pi}{6}\nu, \; \nu = 0, \pm 1, \pm 2, 3 \quad (11.16)$$

The reader may prove that the energy corresponding to each molecular orbital ϕ_ν is $\varepsilon_\nu = \varepsilon_p + 2V_{2,zz}\cos\varphi_\nu$. The formation of these delocalized molecular orbitals gives an additional energy reduction, which is equal to $(4|V_{2,zz}|/3)$ per p_z orbital, besides that associated with the $sp^{(2)}$ hybrids. Notice that the formation of three independent bonds, involving two p_z orbitals each, would give a smaller energy reduction equal to $|V_{2,zz}|$ per p_z orbital (the U term has been omitted). Thus delocalization pays off.

(3) $\boldsymbol{sp^{(3)}}$**hybridization**: In this case, all four atomic orbitals are hybridized, taking the form $\chi_1 = \frac{1}{2}(s + p_x + p_y + p_z)$, $\chi_2 = \frac{1}{2}(s + p_x - p_y - p_z)$, while χ_3, and χ_4 are obtained from χ_2 by cyclic permutations of the two minuses among the p's. The directions of these hybrids are the same as those in a regular tetrahedron from its center to the four vertices. It follows that the cosine of the angle between any two of these four directions is $-1/3$ and the angle is 109.47° (Fig. 11.10).

The energy of each of these four hybrids is $\varepsilon_h = (\varepsilon_s + 3\varepsilon_p)/4$ and the off-diagonal matrix element of the Hamiltonian between any two sp^3 hybrids of the same atom is non-zero and equal to $V_1 = -(\varepsilon_p - \varepsilon_s)/4$. The energy cost of this hybridization is again $\varepsilon_p - \varepsilon_s$. Important molecules, such as methane, CH_4, ethane, C_2H_6, silane, SiH_4, etc., employ the $sp^{(3)}$ hybridization of C or Si. (The reader is strongly encouraged to determine the stereochemistry of these molecules)

11.6 Summary of Important Relations

$$E = -\frac{A}{d'^6}, \quad \text{with } A > 0, \; d' \gg d; \quad \text{van der Waals interaction} \quad (11.1)$$

$$D = c_d \frac{\hbar^2}{m_e d^2} = c_d \frac{\hbar^2}{m_e a_B^2} \frac{a_B^2}{d^2} \approx c_d\, 27.2 \frac{a_B^2}{d^2}\,\text{eV} \approx 27.2 \frac{1}{d^2}, \quad c_d \approx 1 \quad (11.4)$$

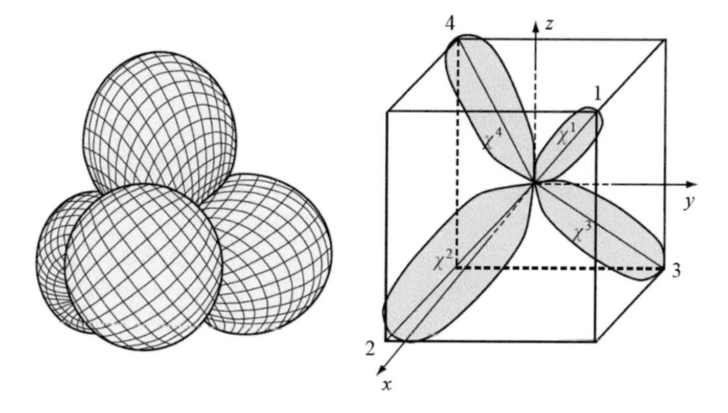

Fig. 11.10 The four sp^3 atomic hybrids are orthogonal to each other with the p-character having three times the weight of the s-character. These two properties uniquely determined the four sp^3 hybrids apart from a rigid rotation of the set. The specific (quite common) form of the sp^3 hybrids given in the text orients them along the non-coplanar opposite diagonal directions of a cube as shown in the figure to the *right*. Any two of the four sp^3 directions form an angle equal to 109.47°. Their average energy is equal to $\varepsilon_h = (\varepsilon_s + 3\varepsilon_p)/4$ and the off-diagonal matrix element of the Hamiltonian between any two of the sp^3 hybrids of the same atom is $V_1 = -(\varepsilon_p - \varepsilon_s)/4$

$$\hbar\omega = c_v' \frac{\hbar^2}{m_e d^2}\sqrt{\frac{m_e}{m_r}} \approx \frac{3026}{d^2\sqrt{A_{Wr}}}\,\text{meV} \tag{11.5b}$$

$$\boxed{\varepsilon_{b,a} = \varepsilon \mp \sqrt{V_2^2 + V_3^2}, \quad \varepsilon \equiv \frac{\varepsilon_1 + \varepsilon_2}{2}, \quad \text{and} \quad V_3 \equiv \frac{\varepsilon_1 - \varepsilon_2}{2} > 0} \tag{11.17}$$

$$V_2 = c_2 \frac{\hbar^2}{m_e d^2}, \tag{11.15}$$

$$\boxed{a_p \equiv \frac{V_3}{\sqrt{V_2^2 + V_3^2}} > 0}, \quad \text{polarity index} \tag{11.12}$$

$$sp^1: \quad \chi_1 = \frac{1}{\sqrt{2}}(s + p_x), \quad \chi_2 = \frac{1}{\sqrt{2}}(s - p_x) \tag{11.18}$$

$$sp^2: \quad \chi_1 = \frac{1}{\sqrt{3}}\left(s + \sqrt{2}p_x\right), \quad \chi_{2,3} = \frac{1}{\sqrt{3}}\left(s - \sqrt{1/2}p_x \pm \sqrt{3/2}p_y\right) \tag{11.19}$$

$$sp^3: \quad \chi_1 = \frac{1}{2}(s + p_x + p_y + p_z), \quad \chi_2 = \frac{1}{2}(s - p_x - p_y + p_z), \quad \chi_{3,4} \tag{11.20}$$

11.7 Multiple-Choice Questions/Statements

1. The coefficient A of the van der Waals interaction is:
(a) $A \propto e^2 r_{a1}^3 r_{a2}^3 / (r_{a1} + r_{a2})$ (b) $A \propto e^2 r_{a1}^2 r_{a2}^2 / (r_{a1} + r_{a2})$
(c) $A \propto e^2 r_{a1}^3 r_{a2}^3 / (r_{a1}^2 + r_{a2}^2)$ (d) $A \propto e^4 r_{a1}^3 r_{a2}^3 / (r_{a1} + r_{a2})$

2. The bond length of H_2 is about (in units of 10^{-10} m)
(a) 3.12 (b) 2.06 (c) 1.57 (d) 0.74

3. The bond length of N_2 is about (in units of 10^{-10} m)
(a) 0.5 (b) 1.1 (c) 1.62 (d) 2.13

4. The bond length of O_2 is about (in units of 10^{-10} m)
(a) 3.22 (b) 2.45 (c) 1.82 (d) 1.21

5. The bond length of Na_2 is about (in units of 10^{-10} m)
(a) 1.02 (b) 1.52 (c) 3.08 (d) 4.93

6. The dissociation energy of Na_2 is about (in units of eV)
(a) 5.2 (b) 3.4 (c) 1.7 (d) 0.8

7. The dissociation energy of N_2 is about (in units of eV)
(a) 1 (b) 21 (c) 10 (d) 2

8. The van der Waals interaction between two neutral atoms at a distance d $\left(1\text{A} \ll d \ll 20\text{A}\right)$ is of the form
(a) $-A/d^3$ (b) $-A/d^4$ (c) $-A/d^5$ (d) $-A/d^6$

9. The vibrational quantum energy of N_2 in meV is about
(a) 20 (b) 300 (c) 800 (d) 1100

10. The vibrational quantum of Na_2 in meV is about
(a) 20 (b) 300 (c) 800 (d) 1100

11. The rotational quantum $\hbar^2/J$ of Na_2 in meV is about
(a) 2 (b) 28 (c) 0.04 (d) 0.001

12. The rotational quantum $\hbar^2/J$ of H_2 in meV is about
(a) 1.5 (b) 15 (c) 150 (d) 0.15

13. The polarity index a_p is defined as follows ($V_3 = |\varepsilon_1 - \varepsilon_2|/2$):
(a) $a_p = V_3/\sqrt{V_2^2 + V_3^2}$ (b) $a_p = V_2/\sqrt{V_2^2 + V_3^2}$
(c) $a_p = \sqrt{|V_2^2 - V_3^2|}/\sqrt{V_2^2 + V_3^2}$ (d) $\sqrt{|V_2^2 - V_3^2|}/V_3$

14. We assume that in the formation of a diatomic molecule each atom is employing only one atomic orbital ψ_i of eigenenergy ε_i, $(i = 1, 2)$. In terms of the quantities $V_2 \equiv \langle \psi_1 | \hat{H} | \psi_2 \rangle$, $\bar{\varepsilon} = (\varepsilon_1 + \varepsilon_2)/2$, and $V_3 = |\varepsilon_1 - \varepsilon_2|/2$ the ground state energy of the molecule is given by the formula $2\varepsilon_b$(omit U)
(a) $\varepsilon_b = \bar{\varepsilon} - V_2$, (b) $\varepsilon_b = \bar{\varepsilon} - V_3$,
(c) $\varepsilon_b = \bar{\varepsilon} - |V_2| - V_3$ (d) $\varepsilon_b = \bar{\varepsilon} - \sqrt{V_2^2 + V_3^2}$

15. We assume that in the formation of a diatomic molecule each atom is employing only one atomic orbital ψ_i of eigenenergy ε_i, $(i = 1, 2)$. In terms of the quantities $a_p = V_3/\sqrt{V_2^2 + V_3^2}$, $V_2 \equiv \langle \psi_1 | \hat{H} | \psi_2 \rangle$, and $V_3 = \{(\varepsilon_1 - \varepsilon_2)/2\} > 0$ the ground state of the molecule is given by the formula
(a) $\phi_b = \frac{1}{\sqrt{2}}\left(\sqrt{1 - a_p}\,\psi_1 + \sqrt{1 + a_p}\,\psi_2\right)$ (b) $\phi_b = \frac{1}{\sqrt{2}}\left(\sqrt{1 + a_p}\,\psi_1 + \sqrt{1 - a_p}\,\psi_2\right)$

(c) $\phi_b = \frac{1}{\sqrt{2}}(\psi_1 + \psi_2)$ (d) $\phi_b = \frac{1}{\sqrt{2}}(\psi_1 - \psi_2)$,

16. In the figure below the two sp^1 hybrids, χ^1 and χ^2, are shown pointing along the $\pm$x-axis. Which one of the following formulae is the right one?

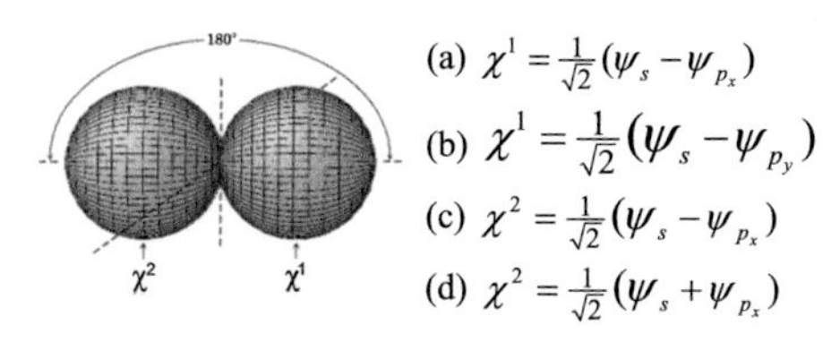

(a) $\chi^1 = \frac{1}{\sqrt{2}}(\psi_s - \psi_{p_x})$

(b) $\chi^1 = \frac{1}{\sqrt{2}}(\psi_s - \psi_{p_y})$

(c) $\chi^2 = \frac{1}{\sqrt{2}}(\psi_s - \psi_{p_x})$

(d) $\chi^2 = \frac{1}{\sqrt{2}}(\psi_s + \psi_{p_x})$

17. In the figure below the three sp^2 hybrids are shown. Which one of the following formulae is the right one?

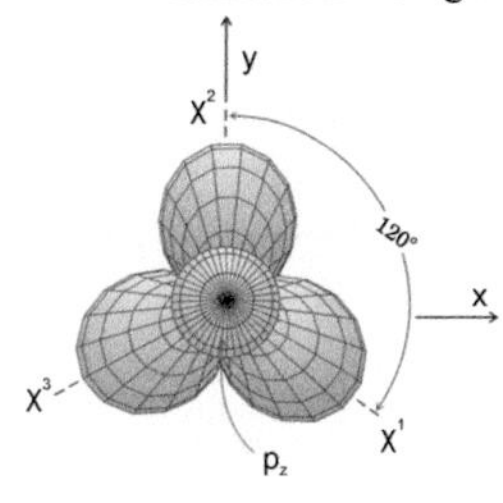

(a) $\chi^1 = \frac{1}{\sqrt{3}}(\psi_s + \sqrt{2}\psi_{p_x})$,

(b) $\chi^2 = \frac{1}{\sqrt{3}}(\psi_s - \frac{1}{\sqrt{2}}\psi_{p_x} + \frac{\sqrt{3}}{\sqrt{2}}\psi_{p_y})$

(c) $\chi^3 = \frac{1}{\sqrt{3}}(\psi_s - \frac{1}{\sqrt{2}}\psi_{p_x} - \frac{\sqrt{3}}{\sqrt{2}}\psi_{p_y})$,

(d) $\chi^2 = \frac{1}{\sqrt{3}}(\psi_s + \sqrt{2}\psi_{p_y})$

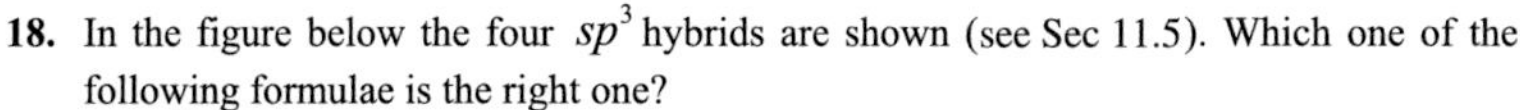

18. In the figure below the four sp^3 hybrids are shown (see Sec 11.5). Which one of the following formulae is the right one?

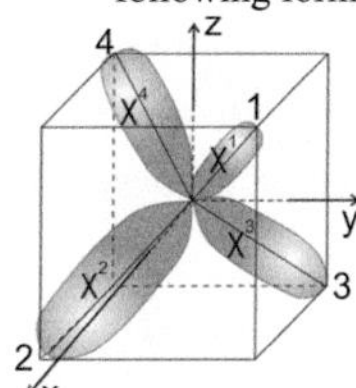

(a) $\chi^4 = \frac{1}{2}(\psi_s + \psi_{p_z} - \psi_{p_x} - \psi_{p_y})$,

(b) $\chi^3 = \frac{1}{2}(\psi_s + \psi_{p_x} - \psi_{p_y} - \psi_{p_z})$

(c) $\chi^2 = \frac{1}{2}(\psi_s + \psi_{p_z} - \psi_{p_x} - \psi_{p_y})$,

(d) $\chi^1 = \frac{1}{2}(\psi_s + \psi_{p_z} - \psi_{p_x} - \psi_{p_y})$

19. The matrix element V_{2h} of the Hamiltonian between two sp^1 hybrids belonging to neighboring atoms, lying on the same line, and pointing in opposite directions is given by $\frac{1}{2}\left\langle \psi_s^1 + \psi_{p_x}^1 \middle| \hat{H} \middle| \psi_s^2 - \psi_{p_x}^2 \right\rangle$. Taking into account that $\left\langle \psi_i^1 \middle| \hat{H} \middle| \psi_j^2 \right\rangle =$ $= \eta_{ij}(\hbar^2 / m_e d^2)$, $\eta_{ij} = -1.32 \;\; 1.42 \;\; 2.22$ for $ij = ss \;\; sp_x \;\; p_x p_x$ we find that

(a) $V_{2h} = -1.77\frac{\hbar^2}{m_e d^2}$ (b) $V_{2h} = -1.42\frac{\hbar^2}{m_e d^2}$

(c) $V_{2h} = -3.19\frac{\hbar^2}{m_e d^2}$ (d) $V_{2h} = 2.22\frac{\hbar^2}{m_e d^2}$

20. The matrix element V_{2h} of the Hamiltonian between two sp^2 hybrids belonging to neighboring atoms, lying on the same line, and pointing in opposite directions is given by $\frac{1}{3}\left\langle \psi_s^1 + \sqrt{2}\psi_{p_x}^1 \middle| \hat{H} \middle| \psi_s^2 - \sqrt{2}\psi_{p_x}^2 \right\rangle$. Taking into account that $\left\langle \psi_i^1 \middle| \hat{H} \middle| \psi_j^2 \right\rangle =$ $= \eta_{ij}(\hbar^2 / m_e d^2)$, $\eta_{ij} = -1.32 \;\; 1.42 \;\; 2.22$ for $ij = ss \;\; sp_x \;\; p_x p_x$ we find that

(a) $V_{2h} = -3.26 \frac{\hbar^2}{m_e d^2}$ (b) $V_{2h} = -1.92 \frac{\hbar^2}{m_e d^2}$

(c) $V_{2h} = -1.48 \frac{\hbar^2}{m_e d^2}$ (d) $V_{2h} = 4.44 \frac{\hbar^2}{m_e d^2}$

21. The bond of the diatomic molecule O_2 is

(a) single between p_x, p_x,

(b) double, a strong one between p_x, p_x and a weak one between $((p_y, p_y) + (p_z, p_z))/\sqrt{2}$

(c) double, a strong one between p_x, p_x and a weak one between (p_y, p_z)

(d) triple, a strong one between p_x, p_x and two weak ones between p_y, p_y and p_z, p_z

22. In the figure below the four normal vibrational modes of the CO_2 molecule are shown (mode (c) is doubly degenerate). Which one(s) are capable of absorbing or emitting electromagnetic (EM) radiation?

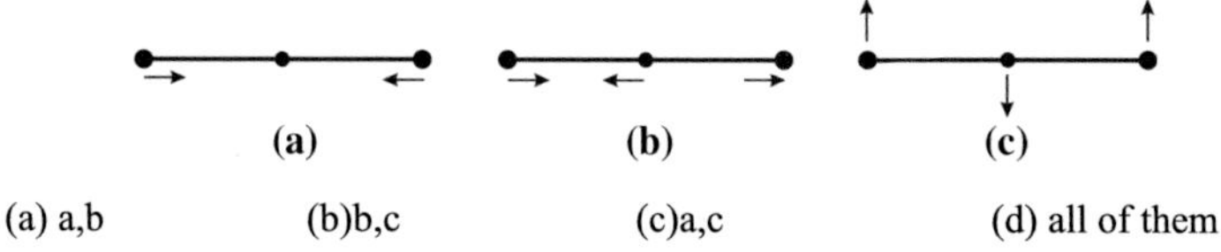

(a) a,b (b)b,c (c)a,c (d) all of them

23. In the figure below the four normal vibrational modes of the CO_2 molecule are shown. (Mode (c) is doubly degenerate and is responsible for the greenhouse effect). The eigenfrequency of this vibration is approximately (in meV):

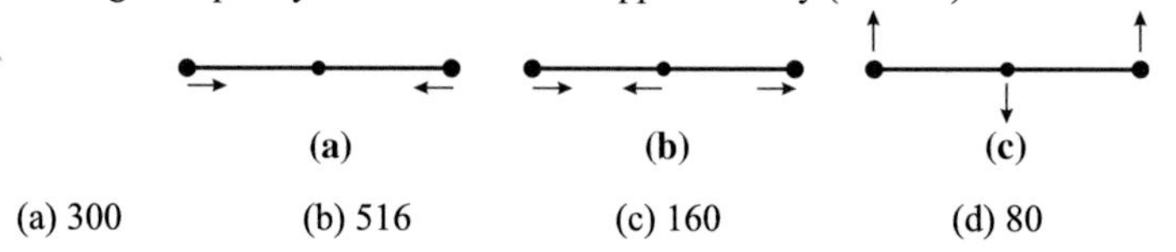

(a) 300 (b) 516 (c) 160 (d) 80

24. The energy of the system of two neutral hydrogen atoms as a function of the distance d' between their protons is given approximately by the relation, $E = (0.343/d'^2) - (0.49/d')$ in atomic units and for d' around the bond length *d*. The bond length (in Angstroms) of the molecule H_2 is according to this relation:

(a) 0.74 (b) 0.53 (c) 1.06 (d) 1.41

25. The energy of the system of two neutral hydrogen atoms as a function of the distance d' between their protons is given approximately by the relation, $E = (0.343/d'^2) - (0.49/d')$ in atomic units and for d' around the bond length *d*. The vibrational frequency of the molecule H_2 in meV is according to this relation:

(a) 516 meV (b) 258 meV (c) 380 meV (d) 760 meV

11.8 Solved Problems

(1) The energy of the system of two neutral hydrogen atoms as a function of the distance d' between their protons is given approximately by the relation, $E = (0.343/d'^2) - (0.49/d')$ in atomic units and for d' around the bond length d. Obtain the dissociation energy of the molecule H_2.

Solution The dissociation energy D is equal to $|E_0| - \frac{1}{2}\hbar\omega$, where E_0 is the minimum of E, $\omega = \sqrt{\kappa/m_r}$, $\kappa = \partial^2 E/(\partial d')^2_{d'=d}$, and $m_r = m_p/2 = 918m_e$. Thus the distance at minimum E is $d = 2 \times 0.343/0.49 = 1.4$ and $E_0 = -0.343/d^2 = -0.175$; $\kappa = (1/d)(0.49/d^2) = 0.1786$; $\omega = \sqrt{0.1786/918} = 0.0139$. Thus, finally, $D = 0.175 - \frac{1}{2} \times 0.0139 = 0.168 = 0.168 \times 27.2 = 4.57\,\text{eV}$

(2) *Show that the matrix element between two sp^3 hybrids belonging to two neighboring atoms, lying on the same line, and having opposite directions is given by the formula $V_{2h} = -3,22(\hbar^2/m_e d^2)$*

Solution We shall choose the line connecting the two atoms as the x-axis. Then the sp^3 hybrid lying along the x-axis and belonging to the atom on the left has the form $\chi_l = \frac{1}{2}(s + \sqrt{3}p_x)$, while the sp^3 hybrid lying along the x-axis and belonging to the atom on the right has the form $\chi_r = \frac{1}{2}(s - \sqrt{3}p_x)$. Thus

$$\langle\chi_l|H|\chi_r\rangle = \frac{1}{4}\left\langle s + \sqrt{3}p_x\middle|H\middle|s - \sqrt{3}p_x\right\rangle = \frac{1}{4}\left(-1.32 - 1.42\sqrt{3} - 1.42\sqrt{3} - 2.22 \times 3\right) \times \frac{\hbar^2}{m_e d^2} = -3.22\frac{\hbar^2}{m_e d^2}$$

(3) *Consider the benzene molecule shown below and its six unhybridized p_z atomic orbitals. Show that the energy gain would be equal to $6|V_{2zz}|$, if three bonds were formed between three pairs of neighboring p_z orbitals, while the gain is $8|V_{2zz}|$, if delocalized molecular orbitals according to (11.16) are formed. (Coulomb repulsion effects have been omitted)*

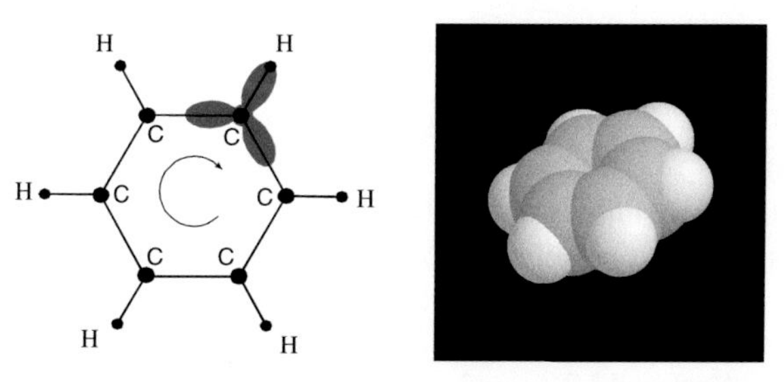

Solution According to (11.10), by placing two electrons at the bonding level for each of the three bonds between neighboring p_z atomic orbitals, the lowering of the energy is equal to $2|V_{2zz}|$, since in the present case $V_3 \equiv 0$. Thus the total lowering of the energy for the three bond case is indeed $3 \times 2|V_{2zz}| = 6|V_{2zz}|$. On the other hand for the actual case where the delocalized molecular orbitals are given by (11.16) we shall occupy by two electrons the lowest energy given by $\varepsilon_z + 2V_{2zz} = \varepsilon_z - 2|V_{2zz}|$ so that the contribution to the lowering of the energy will be $4|V_{2zz}|$; the next doubly degenerate level at $\varepsilon_z + 2V_{2zz}\cos(\pm\pi/3) = \varepsilon_z - |V_{2zz}|$ will be occupied by the remaining four electrons contributing thus to the lowering of the energy by another $4|V_{2zz}|$ to a total reduction by $8|V_{2zz}|$. We see that the delocalized configuration according to (11.16) produces lower total energy compared to the localized three bond configuration which is definitely not the ground state of the six p_z orbitals in benzene.

(4) *In the figure below the four normal vibrational modes of the* CO_2 *molecule are shown (the mode (c) is doubly degenerate). Prove that the four vibrational eigenfrequencies of this molecule satisfy the following double inequality:* $\omega_c < \omega_a < \omega_b$ *and estimate their values*

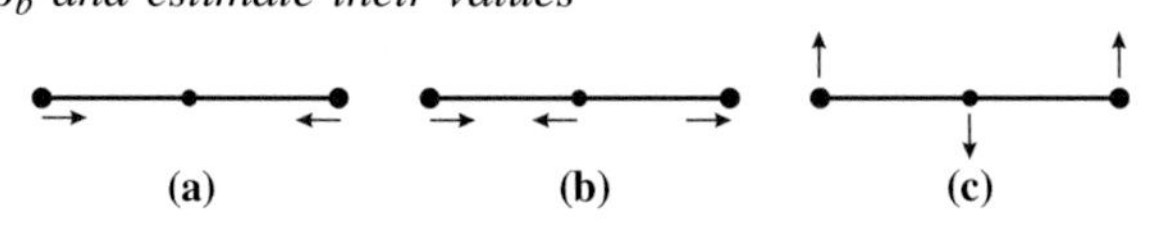

Solution We assume that the streching/compression spring constant of the C/O bond is κ, while the bending one is κ'. Since bending is much easier than streching/compressing we expect that $\kappa' \ll \kappa$. For the mode (a) the carbon atom remains unmoved so that $\omega_a = \sqrt{\kappa/m_O} = \sqrt{\kappa/(16u)}$. For the mode (b), as for every pure vibrational mode, the center of mass of the system is not moving so that $2m_O x_O + m_C x_C = 0 \;\Rightarrow\; x_C = -(2m_O/m_C)x_O$; the equation of motion is $-\omega_b^2 m_O x_O = k(x_C - x_O) \;\Rightarrow\; -\omega_b^2 m_O x_O = -k\{(2m_O/m_C) + 1\}x_O$. Thus we have $\omega_b = \sqrt{\kappa(2m_C^{-1} + m_O^{-1})} = \sqrt{\kappa/(4.36u)}$, hence $\omega_a < \omega_b$. For the mode (c) we have as the result of the immobile center of mass $2m_O y_O + m_C y_C = 0 \;\Rightarrow\; y_C = -(2m_O/m_C)y_O$ and $-\omega_c^2 m_O y_O = k'(y_C - y_O) \;\Rightarrow\; -\omega_c^2 m_O y_O = -k'\{(2m_O/m_C) + 1\}y_O$ from the equation of motion. Thus $\omega_c = \sqrt{\kappa'(2m_C^{-1} + m_O^{-1})} = \sqrt{\kappa'/(4.36u)}$ and the ordering of the eigenfrequencies is $\omega_c < \omega_a < \omega_b$.

To estimate the values of these eigenfrequencies we shall start with ω_a and we shall use (11.5b) with $\bar{d}$ the experimental value of 2.19.

$$\hbar\omega_a = c'_v \frac{\hbar^2}{m_e d^2}\sqrt{\frac{m_e}{m_r}} \approx \frac{3026}{\bar{d}^2\sqrt{A_{Wr}}} = \frac{3026}{2.19^2\sqrt{16}} = 158\,\text{meV}$$

while the experimental value is 165 meV. From the above analysis we have that $\omega_b/\omega_a = \sqrt{16/4.36} = 1.91$ so that $\hbar\omega_b = 1.91 \times 158 = 302$ meV versus 291 meV for the experimental value. Finally, to estimate ω_c we shall use the information that mode c is responsible for the greenhouse effect; this implies that the eigenfrequency of this mode is close to the frequency ω_m of maximum emission from the ground of Earth. Assuming an average temperature of the ground equal to 290 K (see Sect. 13.3) and a black body emission we have $\hbar\omega_m = 2.822\,k_B T = 70.5$ meV. Actually the eigenfrequency $\hbar\omega_c$ is equal to 82. 7 meV which means that $\kappa' = 0.08\kappa$.

11.9 Unsolved Problems

1. The interaction energy between two identical noble atoms as a function of the distance d between their nuclei is given by the formula

$$\mathcal{V}_{ij}(d) = 4\varepsilon\left\{\left(\frac{\sigma}{d}\right)^{12} - \left(\frac{\sigma}{d}\right)^{6}\right\}, \quad d \geq (\sigma/2)$$

   For the Ar atoms $\varepsilon = 10$ meV and $\sigma = 3.4$ A ($1\,\text{A} = 10^{-10}$ m). Does the diatomic argon molecule exist at absolute zero temperature? If yes, what is the bond length? What is its vibration frequency? Will this molecule survive at room temperature?
2. Consider the CO molecule. Which atomic orbitals are involved in its bond? Is the latter single, double, or triple? Any similarity with the nitrogen molecule?
3. Find the energy gap (i.e. the energy difference between the lowest unoccupied molecular orbital (LUMO) and the highest occupied molecular orbital (HOMO)) of the molecule C_6H_6

References

1. J. McMurry, *Organic Chemistry* (Brooks/Cole, KY, 1996)
2. W. Harrison, *Electronic Structure and the Properties of Solids* (Dover Publications, New York, 1989)

Chapter 12
From Atoms (or Molecules) to Solids (or Liquids)

Abstract Solids are made from a huge number of close-packed atoms, quite often periodically arranged, and undergoing small vibrations around their equilibrium positions. The same information as in molecules are sought in solids in order to deduce their various macroscopic properties. Similar theoretical and calculational tools as in molecules are employed to obtain quantities such as density, compressibility, velocity of sound, specific heats, electrical and thermal conductivity, etc.

12.1 Introduction

Atoms (or molecules) come close together in huge numbers (A human body of M kg contains $N_a = 0.96 \times 10^{26} \times M$ atoms) to form solids, liquids, and states intermediate between solids and liquids, such as polymers; all these formations for obvious reason are collectively called *condensed matter*. In solids, atoms have fixed positions in space and undergo small oscillations around them. In contrast, each atom in liquids could move over the whole extent of the liquid, while in touch with whichever atoms happen to be around it. In crystalline solids a microscopic fixed cluster of atoms (or even a single atom) located within the so-called primitive cell is repeated periodically in space. This periodic order, although not perfect, due to the unavoidable presence of defects, greatly facilitates the study of crystalline solids because it reduces it to the study of the cluster of atoms of the primitive cell, thanks to the so-called *Bloch theorem*. (It is this theorem which led to the form shown in (11.16) for the coefficients $c_n^{(v)}$ of the p_z orbitals in benzene).

Among the most important properties of condensed matter are its *density*, which is determined by the average atomic weight and the volume per atom, and its compressibility, the inverse of which is called bulk modulus, is denoted by B, and has dimensions of pressure. Solids, in contrast to liquids, put up resistance to stresses which tend to change their shape but not their volume. This resistance property is quantified by the so-called *shear modulus* denoted by μ_s (or G) and having dimensions of pressure. Other mechanical properties of solids, such as failure by breaking or plastic deformation, are of obvious interest. Moreover,

E.N. Economou, *From Quarks to the Universe*,
DOI 10.1007/978-3-319-20654-7_12

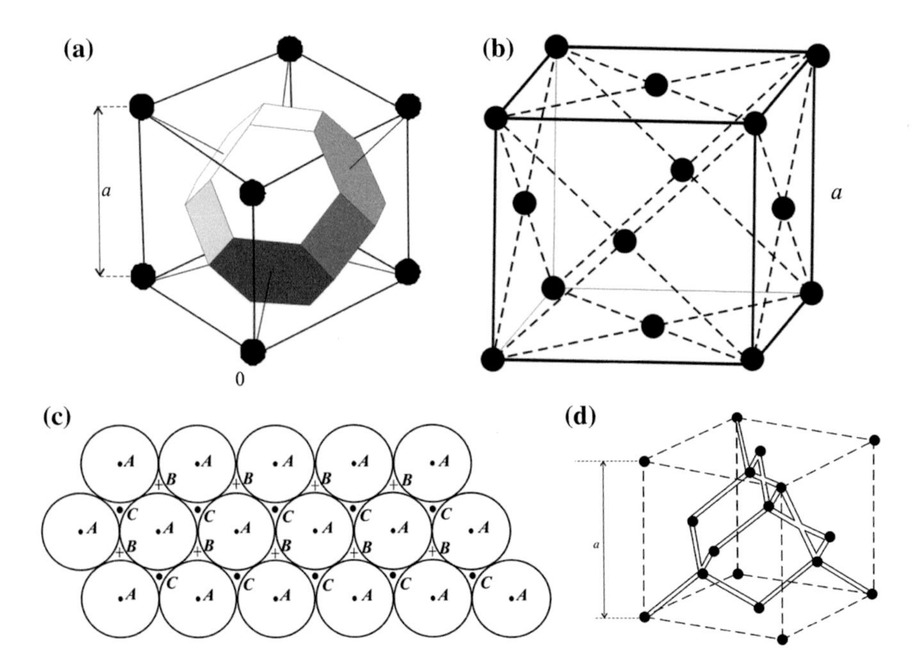

Fig. 12.1 **a** The cubic unit cell of a *body centered cubic* (bcc) lattice. The polyhedron covering the central atom is a primitive cell, the volume of which is half the volume of the unit cell. The length a is called the *lattice constant*. **b** The cubic unit cell of the *face centered cubic* (fcc) lattice. The volume of the primitive cell, which coincides with the volume per Bravais point, is 1/4 of that of the unit cell. **c** The fcc lattice can be constructed by starting with a two-dimensional hexagonal layer of touching equal spheres; then a second hexagonal layer of equal spheres is placed at the points B. A third layer is placed at the points C. After that the structure is repeated periodically. If the structure is repeated after the second layer the *hexagonal close-packed* (hcp) lattice is obtained. **d** The *diamond lattice* results from the fcc lattice by placing at every one of its points a two atom cluster of length $\sqrt{3}\,a/4$ oriented along the (111) direction

electrical properties such as *resistivity* (or its inverse called *conductivity*), magnetic properties such as *ferromagnetism*, optical properties such as *index of refraction*, thermal properties such as *specific heats*, *thermal conductivity*, etc., are of great interest both from the point of view of basic science and from that of technology.

In this chapter we extract a few basic properties of solids first by dimensional analysis, then by employing the so-called *jellium model* [1], and finally by the *LCAO method* [2]. Before implementing this program it is not pointless to present first (in Sects. 12.2 and 12.3) some elementary information regarding the structures of solids (Fig. 12.1).

12.2 Some Common Crystal Lattices

To describe crystal lattices we introduce some concepts such as *the Bravais lattice* (the set of points coinciding with the tips of the vectors $R_n = \sum_i n_i \mathbf{a}_i,\ \ \{n_i\}$ all sets of integers, and $\mathbf{a}_i$, $i = 1, 2, 3$ are three non-coplanar vectors), *the primitive cell* (the volume that, if translated by all Bravais vectors fills up all space without any overlap), and *the unit cell* (the smallest volume displaying the full symmetry of the lattice and being equal to a multiple of the volume of the primitive cell). In the figures below the positions of the Bravais points in the unit cell of some common lattices are shown. At every Bravais point the same kind of atom (or cluster of atoms) is placed. The common types of bonding in solids are shown in Fig. 12.2.

12.3 Types of Bonding in Condensed Matter

See Fig. 12.2.

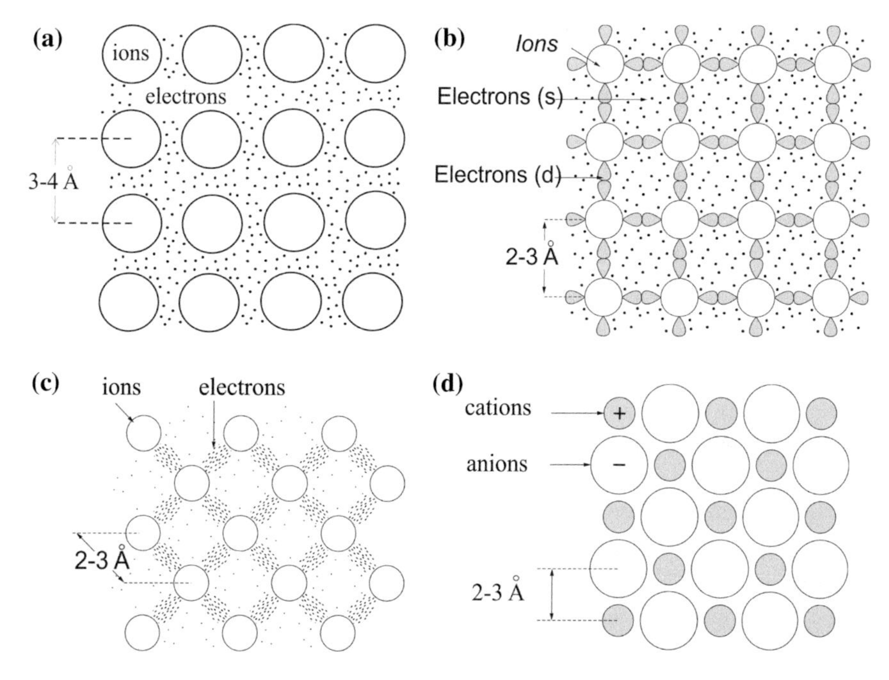

Fig. 12.2 **a** In simple metals the bonding is achieved by the valence electrons which are spread almost uniformly over the whole metal. **b** In transition (and rare earth) metals the *d* (and *f*) valence electrons, in contrast to the *s* ones, remain partly localized in the vicinity of the parent atom. **c** In semiconductors the bonding is achieved by the valence electrons being spread over the whole material but mostly along the bond-network. **d** In ionic solids the bonding is achieved by Coulomb attraction among the anions and cations, the creation of which had an energy cost. **e** In *van der Waals* bonding the attraction among the atoms (or molecules) is due to mutually induced dipole moments without any electronic detachment. **f** The hydrogen atom, as a result of its unique combination *of valence one and high ionization potential*, acts as a special bridge based on dipole/dipole interaction between the molecule to which it belongs and a nearby molecule. The *hydrogen bonding* of water molecules in ice is shown schematically

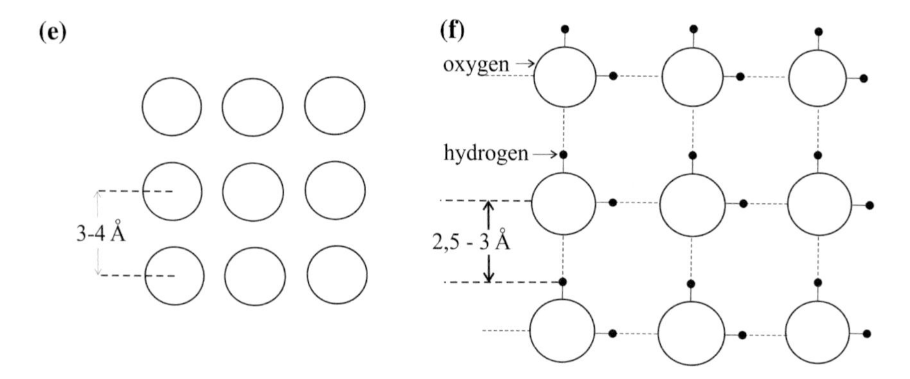

Fig. 12.2 (continued)

12.4 Dimensional Analysis Applied to Solids

As it was argued in the introduction of Part IV the properties of condensed matter depend necessarily on $e,\ \hbar,\ m_e$, or, equivalently, on $a_{B,}\ \hbar,\ m_e$. They depend also on the type of atom(s) participating in its formation, i.e. on their atomic number Z. They may depend also explicitly on the masses of the atoms, m_{ai}, on the temperature, T, on the pressure, P, on the velocity of light, c, etc. Following the general recipe for dimensional analysis presented in Sect. 5.1 and in the introduction of Part IV, we obtain the following general formula for any quantity X pertaining to the condensed matter:

$$\boxed{X \equiv X_o \bar{X} = X_o f_x\left(Z_i, \frac{m_{\alpha i}}{m_e}(Z_i), \frac{T}{T_o}, \frac{P}{P_o}, \frac{c}{\upsilon_o}, \dots\right)} \tag{12.1}$$

where X_o is the combination $\mathrm{X}_o = \hbar^{\nu_1} m_e^{\nu_2} a_B^{\nu_3}$ which has the same dimensions as X (see Table I.1 in Appendix I), and $\bar{X}$ is the value[1] of X in atomic units which may depend, through the function f_x, on the *dimensionless* quantities $Z_i,\ m_{\alpha i}/m_e, \dots$; $T_o = \hbar^2/m_e a_B^2 k_B = 27,2\,\mathrm{eV} \times 11,600\,\mathrm{K/eV} = 315,775\,\mathrm{K}$, is the atomic unit of temperature, $P_o = \hbar^2/m_e a_\mathrm{B}^5 = 2.9421 \times 10^{13}\,\mathrm{N/m}^2 = 2.9421 \times 10^8\,\mathrm{bar}$ is the atomic unit of pressure, and $\upsilon_o = \hbar/m_e a_\mathrm{B} = c/137 = 2187.69\,\mathrm{km/s}$ is the atomic unit of velocity.

The first quantity of condensed matter to be considered is the volume per atom, V/N_a, or its inverse, the *atomic concentration,* $n_a \equiv N_a/V$, or, equivalently, the *radius per atom,* $r \equiv (3V/4\pi N_a)^{1/3}$. This radius is expected to be comparable to the atomic radius r_a but somewhat larger, because many touching spherical atoms leave always empty spaces among them, which the above definition of r incorporates; how much larger $\bar{r} \equiv r/a_B$ is over $\bar{r}_a$ depends mainly on the average number

[1] A bar over any physical quantity X denotes its value at the atomic system of units: $\bar{X} \equiv X/X_o$.

Table 12.1 Comparison of the results of (12.4), (12.5), (12.7), (12.8a) and (12.8b, lowest line) with the experimental data [1, 3] for four solids of special importance (both technologically and historically)

	Fe		Al		Cu		Si	
	Estimate	Exp	Estimate	Exp	Estimate	Exp	Estimate	Exp
A_w		55.85		26.98		63.55		28.09
$\bar{r}$ (exp)		2.70		2.99		2.67		3.18
u_c (eV/atom)	3.73	4.28	3.04	3.39	3.82	3.49	2.69	4.63
B (10^{11} N/m^2)	1.29	1.68	0.73	0.72	1.29	1.37	0.54	0.99
c_o (km/s)	4.06	4.63	5.28	5.68	3.85	3.93	4.87	6.48
Θ_D (K)	366 380	464	476 446	426	347 364	344	438 386	645

of nearest neighbors (which is 12 for fcc (e.g. copper), 8 for bcc (e.g. iron), and only four for diamond lattices). Thus

$$r = \bar{r}\, a_B \tag{12.2}$$

where the average value of the dimensionless quantity $\bar{r}$ over all the elemental solids is 3, while the corresponding value for the atomic radius is $<\bar{r}_a> = 2.6$. The value of $\bar{r}$ for each solid allows us to determine its *density* and vice versa

$$\rho_M = \frac{m_a}{\frac{4\pi}{3} r^3} = \frac{A_W u}{\frac{4\pi}{3} a_B^3 \bar{r}^3} = 2.675 \frac{A_W}{\bar{r}^3}\ \mathrm{g/cm^3} \tag{12.3}$$

where A_W is the average dimensionless atomic weight of the condensed matter. As in the case of atoms and molecules, the replacement $a_B \rightarrow r$ everywhere in (12.1) is expected to improve the estimated values of various quantities X.

One such quantity is the *cohesive energy* u_c defined as the minimum energy per atom required to separate the condensed matter into (neutral) atoms under conditions of $T \rightarrow 0$ and $P = 1$ atm. The latter is too small to appreciably influence u_c. Similarly, the atomic mass is expected to play only a minor role, since it enters through the ionic kinetic energy which is smaller than the electronic one by a factor of $\sqrt{m_e/m_a}$, as we have seen in Chap. 11. Hence, from (12.1) we have

$$u_c = f_u(Z) \frac{\hbar^2}{m_e r^2} = f_u \frac{\hbar^2}{m_e a_B^2 \bar{r}^2} = f_u \frac{27.2}{\bar{r}^2}\ \mathrm{eV}; \quad f_u \approx 1 \tag{12.4}$$

The *bulk modulus* B at $T = 0$ is defined as $B \equiv -V(\partial P/\partial V) = V\partial^2 U/\partial V^2$. Taking into account that B has dimensions of pressure, i.e. energy over volume, and using similar arguments as in the case of u_c, we obtain from (12.1) and Table I.1 in Appendix I.

$$B = f_B(Z)\frac{\hbar^2}{m_e r^5} = f_B\frac{\hbar^2}{m_e a_B^5 \bar{r}^5} = f_B\frac{2.94 \times 10^{13}}{\bar{r}^5}\mathrm{N/m^2};\quad f_B \approx 0.6 \tag{12.5}$$

The *shear modulus* μ_s (also denoted by the symbol G) is of the same order of magnitude as the bulk modulus but usually smaller. For most metals, $x \equiv \mu_s/B$ is between 0.3 and 0.6, while for liquids is zero. The shear stress τ_s under which a solid fails depends mainly on linear defects called *dislocations* and, as a result, it varies in a very wide range of values with a gross average about equal to $<\tau_s> \approx 0.005\mu_s \approx B/500$.

The *velocity of sound*, c_o, in a liquid[2] is obtained from the unit of velocity, $\upsilon_o = \hbar/m_e a_B$, by implementing the replacement $a_B \to r$ and multiplying by the dimensionless factor $\sqrt{m_e/m_a}$. The latter is necessarily present whenever atomic oscillation is involved, as in sound propagation in which atomic (or ionic) oscillation migrates from atom to atom (or from ion to ion):

$$c_o = f_c\frac{\hbar}{m_e r}\sqrt{\frac{m_e}{m_a}} \approx \frac{82}{\bar{r}\sqrt{A_w}}\frac{\mathrm{km}}{\mathrm{s}};\quad f_c \approx 1.6 \tag{12.6}$$

Notice that (12.3), (12.5), and (12.6) together with the choices $f_B \approx 0.6$ and $f_c \approx 1.6$ lead to the simple relation

$$\boxed{c_o = \sqrt{\frac{B}{\rho_M}},} \tag{12.7}$$

which happens to be exact for liquids (see also (5.13)).

As was mentioned in Chap. 11, the number of independent ionic oscillations (natural modes) is huge, equal to $3N_a - 6$, and the corresponding natural frequencies cover a range from 0 to a maximum value denoted by ω_D. Connected with ω_D is the so-called *Debye temperature*, $\Theta_D \equiv \hbar\,\omega_D/k_B$, which determines the contribution of ionic oscillations to the thermodynamic properties. The quantities ω_D and Θ_D, in analogy with (11.5), are given by

$$\omega_D = f_\omega\frac{e^2}{\hbar r}\sqrt{\frac{m_e}{m_a}} = f_\omega\frac{96.8}{\bar{r}\sqrt{A_w}} \times 10^{13}\ \mathrm{rad/s};\quad \Theta_D = f_\omega\frac{7390}{\bar{r}\sqrt{A_w}}\mathrm{K};\quad f_\omega \approx 1 \tag{12.8a}$$

or, if we start from $\omega_0 = \hbar/m_e a_B^2$,

$$\boxed{\omega_D = f'_\omega\frac{\hbar}{m_e r^2}\sqrt{\frac{m_e}{m_a}} = f'_\omega\frac{96.8}{\bar{r}^2\sqrt{A_w}} \times 10^{13}\ \mathrm{rad/s};\quad \Theta_D \approx \frac{20{,}700}{\bar{r}^2\sqrt{A_w}}\mathrm{K};\quad f'_\omega \approx 2.8} \tag{12.8b}$$

[2]In isotropic solids there are two sound velocities, the transverse one, $c_t = c_o\sqrt{x}$, (in which the direction of motion of each atom is perpendicular to the direction of sound propagation), and the longitudinal one, $c_\ell = c_o\sqrt{1+\frac{4}{3}x}$ (in which the direction of motion of each atom is along the propagation direction); $x \equiv \mu_s/B$; $c_o = \sqrt{B/\rho_M}$.

Equation (12.8b) can be obtained also by multiplying an average speed of sound $\bar{c}$ by the maximum wavenumber k_D. The resulting value of f'_ω is in the range between about 2.25 and 3.2 depending on the material; the value $f'_\omega \approx 2.8$ is a kind of average among various materials.

In Table 12.1 we test the results of our simple dimensional approach against experimental data.

The largest discrepancy between experimental values and our estimates appears for Si; this is due mainly to the fact that Si has only 4 nearest neighbors and, as a result, the ratio $\bar{r}/\bar{r}_a$ in Si is overestimated by a factor $(12/4)^{1/3} \approx 1.44$ relative to that in close-packed solids where the number of nearest neighbors is 12. If this factor of 1.44 is included, our estimates for Si will become 5.58, 3.34, 7.02, 631 (12.8a), and 800 (12.8b) respectively. A corresponding correction for Fe involves the factor $(12/8)^{1/3} \approx 1.14$ and makes our estimates 4.88, 2.52, 4.64, 419 (12.8a), and 494 (12.8b) respectively. The above *partial* improvement of the estimates indicates that probably there is not a scaling length more representative than $\bar{r}$. The results for the bulk modulus will be improved by replacing $\bar{r}^5$ by $\tilde{r}^5 = \bar{r}^5(z/12)^{2/3}$.

12.5 Resistivity According to Dimensional Analysis: An Instructive Failure

Next, we shall attempt to estimate the electrical resistivity, ρ, which is connected to the resistance, R, of a wire of length ℓ and cross-section s by the well known formula, $R = \rho\,\ell/s$. It is easy to show that the dimensions of resistance is energy $\times$ time over charge squared and, hence, its unit in the atomic system is

$$R_0 = \frac{\hbar}{e^2} = 4108.236\,\text{ohm}. \tag{12.9}$$

It follows that the unit of resistivity in the atomic system is

$$\rho_0 = \frac{\hbar\, a_B}{e^2} = 21.73985\,\mu\text{ohm}\cdot\text{cm} \tag{12.10}$$

Therefore, if we make the assumptions that T and P do not play an important role, that the atomic mass is not so important,[3] and that a_B must be replaced everywhere by $\bar{r}a_B$ (or, more accurately, by $\bar{r}_e a_B$, where $\bar{r}_e = \bar{r}/\zeta^{1/3}$ is defined through the volume per free electron as $V/N_{\text{ef}} = V/N_a\zeta = 4\pi r_e^3/3$ and ζ is the valence), we estimate that the resistivity is $\rho \approx 21.74\,\bar{r}_e$ μohm · cm, i.e., of the order of a few tens of μohm · cm.

[3]These assumptions seem justified if we adopt the picture that the electrical resistivity is due to the collisions of the current-carrying electrons (behaving as classical particles) with the more or less immobile ions. This picture, as we shall see, leads to an electronic mean free path of the order of 10 A.

By equating this estimate with the general formula for ρ, $\rho = m_e/e^2 n_f \tau = m_e \upsilon_F / e^2 n_f \tau \upsilon_F = p_F / e^2 n_f \ell$ valid for metals (see (6.24), $\rho = 1/\sigma$, and set $\omega = 0$) and taking into account that $p_F = 1.919\hbar/r_e$ (see Chap. 3, (3.22a) and (3.22b) and that $E_F = p_F^2/2m_e$) we obtain for the electronic mean free path ℓ the relation

$$\ell \approx 8r_e, \qquad \text{if} \quad \rho = 21.74\bar{r}_e \; \mu\text{ohm} \cdot \text{cm}$$

where $(4\pi r_e^3/3)^{-1} = n_{ef} = \zeta n$ is the free electron concentration and ζ is the valence. This means that the electronic mean free path ℓ according to our dimensional estimate is four times or less the bond length. This sounds like a reasonable estimate, given the strong Coulomb interaction of the negatively charged electrons with the positively charged ions. However, although there are metals with resistivities at room temperature close to this estimate (e.g., Ti with $\rho = 43.1$ μohm · cm, Pr with $\rho = 67$ μohm · cm, Mn with $\rho = 139$ μohm · cm, etc.), in general our estimate for ρ and consequently ℓ fails very badly: The resistivity of copper at room temperature is 1.7 μohm · cm which implies a mean free path of the order of 400 A $(1 \text{ A} = 10^{-10} \text{ m})$. Moreover, clean single crystals of copper at helium temperatures have reached resistivities as low as 10^{-5} μohm · cm implying an electronic mean free path 25 millions the bond length! In other words, each electron is passing next to 25 million ions in a row and still seems to continue undisturbed its initial path! To the other extreme there are substances, such as yellow sulfur, exhibiting resistivities at room temperatures as high as 10^{23} μohm · cm. What did go wrong in deriving an estimate of ρ or ℓ which failed so dismally?

12.6 Resistivity as a Wave Phenomenon

As we mentioned before, our assumptions, such as weak temperature dependence, etc., rely on the picture of current-carrying electrons colliding with ions as classical particles. Hence, classical mechanics seems to be the culprit for the failure to account for the observed resistivity. We should have used quantum instead of classical mechanics for the motion of electrons. More specifically we should have taken into account that the detached electrons propagate in the solid under the action of an electric field as waves, and consequently *they may exhibit constructive interference.* If the scattered electronic waves interfere constructively in *a systematic way,* the effects of scattering could be compensated and the electrons could propagate as if they were free of any force; in other words their mean free path would be infinite. However, such a systematic constructive interference requires placement of the scattering centers (i.e., the ions) in an ordered way to make sure that all scattered waves arrive in phase. The perfect crystalline structure, due to the periodic positions of the ions, satisfy this requirement and thus provides the mechanism for essentially free-like propagation, infinite mean free path, and zero electrical resistivity. According to this argument, the observed non-zero small metallic resistivity is due to deviations from periodicity; these deviations, as a result of their randomness, produce scattered waves unable to undergo constructive interference and hence to compensate the scattering. Thus the non-zero resistivity is due to structural imperfections, the presence of foreign atoms and other defects, as well as inelastic scattering by the inevitable ionic oscillations (at $T \neq 0$ K). For a pure good conductor, such as noble metals and at room temperatures, the non-compensated scattering by the ionic oscillations is the dominant one. This scattering is expected to be proportional to the amplitude-square of the ionic oscillations, which in turn is proportional to the energy of these oscillations. The latter at room and higher temperatures is given approximately by the classical expression of $3k_BT$ per atom. We conclude that *for good conductors and for not so low temperatures the metallic resistivity is proportional to the absolute temperature.* Hence, we have the following expression for the resistivity of good conductors at room and higher temperatures, according to (12.1):

$$\rho ='' \text{const.}'' \frac{\hbar a_B}{e^2} \frac{3k_BT}{k_BT_o} \to'' \text{const.}'' \frac{\hbar a_B \bar{r}_e}{e^2} \frac{3k_BT}{\left(\hbar^2/m_e a_B^2 \bar{r}_e^2\right)}, \; a_B \to r_e = \bar{r}_e a_B \quad (12.11)$$

or $\rho ='' \text{const.}''2.07 \times 10^{-4} \bar{r}_\alpha^3 T$ μohm · cm, for good metals (T in K and not very low).

A periodic placement of the scatterers (i.e., the ions) could lead not only to perfectly systematic constructive interference but to perfectly *systematic destructive interference* as well. So it is possible in a periodic medium to have energy regions (called *bands*) where the interference is systematically constructive and *the propagation is almost free-like,* and regions of energy called *gaps* where it is systematically destructive and the electronic waves cannot even exist.[4] This wave-based, alternating band/gap structure of the energy spectrum in periodic materials explains the huge differences in electrical resistivities among various solids. If in a solid there is a partially filled band at $T = 0$ K, the electrons can easily be excited by an electric field to nearby empty levels and produce thus large electric currents. Such materials behave as good conductors and their resistivity is as in (12.11). On the other hand, if every band is either fully occupied by electrons or completely empty, the resistivity (at $T = 0$ K) is infinite and the material is insulating. This is so, because the electrons in a *fully occupied band, even if they are infinitely mobile and free-like*, cannot respond to the electric field and produce a current, because any such response requires electrons to be excited from occupied to empty states. But in a *fully occupied band* there are no empty states.[5] In reality, the resistivity in insulators and semiconductors is huge but finite, because of the thermal excitation (at $T > 0$ K) of a few electrons from the highest fully occupied band, called *valence band* (VB), to the

---

[4]It should be pointed out that gaps may appear even in non-periodic systems, if a large fraction of space is classically inaccessible to the mobile electrons. On the other hand, destructive interference in two and three dimensional periodic media is not always capable of opening gaps.

[5]There are empty states in the nearest empty band, called conduction band. However, the excitations of electrons there, requires usually a huge electric field; this is the phenomenon of electric breakdown of insulators.

lowest empty band, called *conduction band* (CB).[6] The resistivity is inversely proportional to the number of carriers, i.e., the number of electrons excited in the conduction band from the valence band plus the missing electrons (*holes*) in the valence band. In other words the electronic conductivity (i.e. the inverse of the resistivity) in semiconductors has in general two contributions (see (6.24) for $\omega = 0$)

$$\boxed{\sigma = \sigma_c + \sigma_v = \frac{e^2 n_c \tau_c}{m_c^*} + \frac{e^2 p_v \tau_v}{m_v^*}} \tag{12.12}$$

where n_c is the concentration of electrons in the conduction band (CB), τ_c is the relaxation time for electrons in the CB, m_c^* is the effective electronic mass in the CB (the cancellation of scattering by constructive interference has as a "side effect" the replacement of m_e by m_c^*, see Appendix D), p_v is the concentration of missing electrons (holes) in the valence band (VB), and τ_v, m_v^* are the relaxation time and the effective mass of holes in the VB. Notice that in pure semiconductors the concentrations n_c, p_v depend very strongly on the temperature, in contrast to metals where $n_e = n_a \zeta$ is temperature independent, while the relaxation times in both metals and semiconductors depend relatively weakly on temperature (usually proportionally to $1/T$ as deduced from (12.11)). In pure semiconductors the concentrations depend on thermal excitation of electrons from the VB to the CB and, as a result, the dependence is exponential on the dimensionless ratio $E_g/2k_BT$, where E_g is the so-called *energy gap*, i.e., the energy separation between the bottom of the conduction band and the top of the valence band. Thus, for very pure semiconductors and insulators the resistivity is proportional to $\exp\left[E_g/2k_BT\right]$

$$\boxed{\rho_e \propto \exp\left[\frac{E_g}{2k_BT}\right], \text{ pure semiconductors, insulators}} \tag{12.13}$$

It must be stressed that the resistivity in semiconductors may be reduced by several orders of magnitudes if appropriate substitutional impurities are incorporated in the lattice, because then the gap is *effectively* reduced substantially (see Appendix D).

12.7 The Jellium Model and Metals

The Jellium Model (JM) adopts the general picture that each atom participating in the formation of solids splits into ζ detached electrons and a cation of positive charge ζe. Moreover, the JM assumes that each cation is mashed as to cover uniformly the volume per atom, associated with it. Thus, the JM replaces the

[6] In insulators may exist non-negligible electric currents due to ionic migration, especially in the presence of impurities and imperfections.

discrete lattice of the cations by a homogeneous and isotropic continuous background of ionic mass and charge. This positive uniform ionic charge density is fully compensated by the negative charge density of the detached electrons, which, being free of any force, spread themselves uniformly over the whole volume, V, of the solid. The input for any calculation based on the JM is the valence ζ and the atomic weight A_W of the atoms making up the solid. The drastic approximation of ignoring any force on every (detached) electron sounds less unreasonable in view of the free-like propagation occurring in a periodic potential. On the other hand, the JM misses completely the presence of gaps which a periodic potential may produce. As a result, the JM is a model for metals and it is absolutely inappropriate for electronic transport properties of semiconductors and insulators.

The starting point of any calculation within the JM is the expression of the total energy per atom, U/N_a, in terms of the radius per atom $r' = \bar{r}' a_B$:

$$\boxed{\frac{U}{N_a \mathrm{E}_o} = \frac{\alpha}{\bar{r}'^2} - \frac{\gamma}{\bar{r}'}, \qquad E_o \equiv \frac{\hbar^2}{m_e a_B^2}} \tag{12.14}$$

The first term in the rhs of (12.14) is the kinetic energy of the detached electrons divided by N_a (see (3.19) and take into account that $r'^2 = R'^2/N_a^{2/3}$), and the second is the Coulomb energy among all the charged particles (excluding the self-interaction of each cation). The values of α and γ become more realistic than those resulting from the JM, if we reintroduce the cations as spherical particles of radius $r_c = \bar{r}_c a_B$, while still considering the detached electrons as uniformly spread in the remaining volume $V' = V - N_a(4\pi/3)r_c^3$. We have then $\alpha = 1.3\zeta^{5/3} + \eta \bar{r}_c \zeta^2$ and $\gamma = 0.56\zeta^{4/3} + 0.9\zeta^2$ where η varies between 0.4 and 0.9 (the lower value is for ions of half-filled subshells and the highest for ions of completed subshells) (see [1]). The presence of the valence ζ in α and γ is due to the relation $N_{ef} = \zeta N_a$.

By minimizing U/N_a, with respect to r' we obtain its equilibrium value

$$\bar{r} = \frac{2\,\alpha}{\gamma} \tag{12.15}$$

For the solids Fe, Al, Cu, and Si the results for $\bar{r}$, according to (12.15), are 2.54, 3.01, 2.74, 2.733 respectively to be compared with the experimental values given in Table 12.1 (2.70, 2.99, 2.67, 3.18 respectively). Using the definition of B (see line below (12.4)), and (12.14) we find that,

$$B = \frac{a}{6\pi\,\bar{r}^5} P_o, \qquad P_o \equiv \frac{\hbar^2}{m_e a_B^5}, \tag{12.16}$$

under zero pressure and temperature. The results for the four solids of Table 12.1 are 1.97, 0.85, 1.04, 2.51 $10^{11}\,\mathrm{N/m^2}$ respectively to be compared with the corresponding experimental results 1.62, 0.72, 1.37, 0.99. Results for the sound velocity,

$$c_o = 24.15\sqrt{\alpha}/\left(\bar{r}\sqrt{A_W}\right)$$

and the Debye temperature,

$$\Theta_D = 8426(\alpha/A_W)^{1/2}(f/\bar{r}^2)\,\mathrm{K},$$

can be obtained as well. The value of f in the expression for Θ_D depends on the ratio $x = \mu_s/B$ and is found to be 0.78, 0.64, 0.67, 0.913 for the Fe, Al, Cu, and Si respectively. Within the JM the normal modes of ionic oscillations are plane acoustic waves of frequency $\omega = 2\pi c_o/\lambda$ (for liquids, while for solids we have transverse and longitudinal acoustic waves (see Footnote 2).

The JM supplemented by some scattering processes allows us to calculate the response of a metal to external stimuli, such as an electric field, and/or a magnetic field and/or a temperature gradient [1]. This response is characterized in the linear regime by quantities such as the electrical conductivity, the dielectric function, the magnetic permeability, the thermal conductivity, etc. (Figs. 12.3 and 12.4).

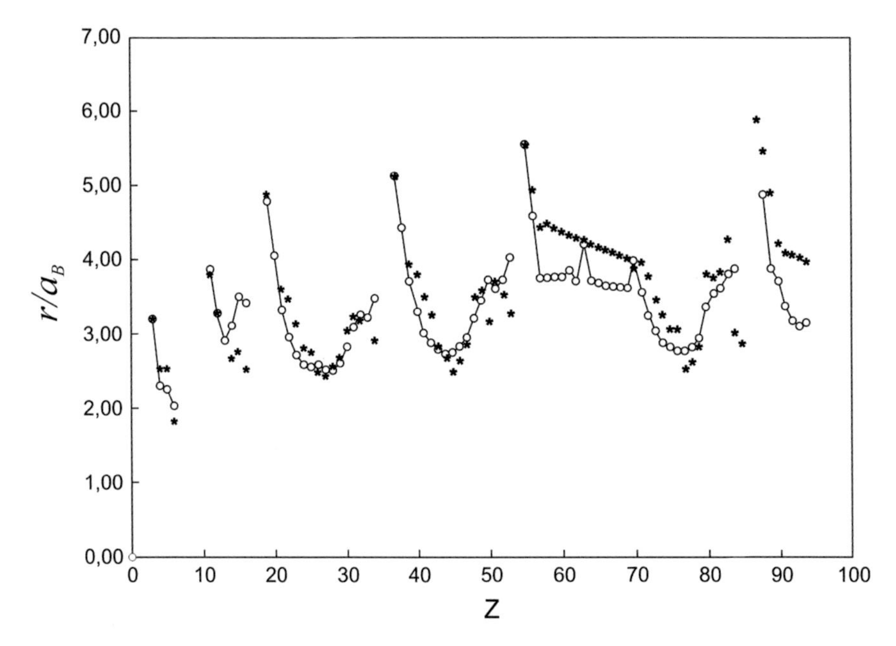

Fig. 12.3 Comparison of the estimated values of $\bar{r}$ (*asterisks*, *) based on $\bar{r} = 2\,\alpha/\gamma$ with the corresponding experimental data (*open circles*, o) for all metallic elemental solids. The *solid lines* is a guide to the eye through the experimental data of $\bar{r}$. Notice the non-monotonic variations with Z and connect it with the data of Fig. 10.1

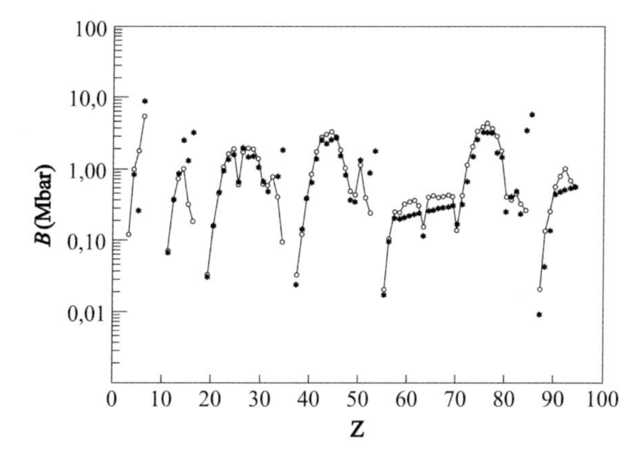

Fig. 12.4 Estimated values (*) for B based on $B = (\alpha/6\pi\bar{r}^5)P_o,\ P_o \equiv \hbar^2/m_e a_B^5$, compared with the corresponding experimental data (o) for all the elemental solids. The *solid lines* through the data are a guide to the eye

12.8 The LCAO Method

The LCAO method lets us study, besides metals, solids such as semiconductors, ionic solids, and molecular solids where the JM is not appropriate. In Fig. 12.5 we plot schematically one-dimensional[7] models for the basic kinds of solids: In cases (a), (b), and (c) each atom A or B has one electron in an atomic orbital ψ_A or ψ_B with corresponding energies ε_A or ε_B respectively; In case (d) there is one electron in each of the two relevant atomic orbitals s and p_x. For all cases there are also non-zero off-diagonal matrix elements V_2 which can transfer the electron from one atom to each one of the two nearest neighbour atoms. The electronic waves for the periodic systems of Fig. 12.5, called Bloch-states (B-states), and the corresponding energies are as follows (see [1]):

Case (a) The B-states are as in (11.16) with 6 replaced by the number of atoms N_a. The energy of each B-state is

$$\boxed{E_\nu = \varepsilon_A + 2V_2 \cos\varphi_\nu, \quad \varphi_\nu \equiv k\,a, \quad -\pi < \varphi_\nu \leq \pi} \tag{12.17}$$

Two successive φ_νs differ by $2\pi/N_a$; hence, as $N_a \to \infty$, the energy spectrum becomes dense giving rise to a band of total energy width equal to $4|V_2|$. We define the so-called crystal wavenumber $k \equiv \varphi_\nu/a$. The number of B-states in this band is N_a. Placing two electrons in each of the lower B-states we end up with a half filled band and, hence, with a conducting, metallic behavior (see Fig. 12.6).

[7] One-dimensional models of solids, instead of 3-D real solids, allow us to present basic concepts in the simplest possible way.

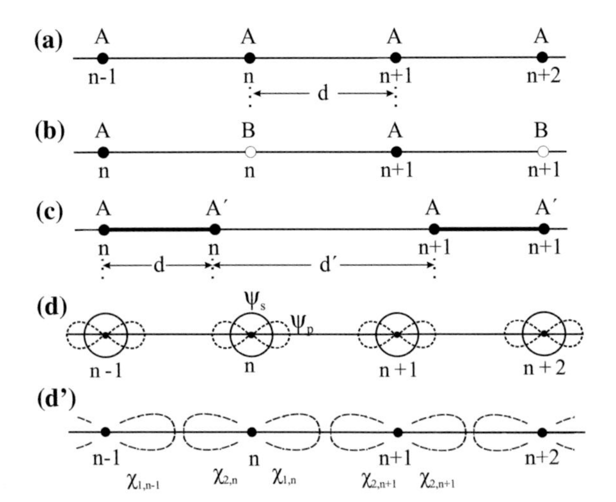

Fig. 12.5 **a** An elemental 1-D periodic metallic "solid"; the period is $a = d$. **b** An ionic 1-D "solid" consisting of two kinds of atoms; the period is $a = 2d$. **c** A molecular 1-D "solid"; the period is $a = d + d'$. **d** An elemental periodic 1-D "semiconductor" with two relevant atomic orbitals and two electrons per atom. **d'** As in (*d*) but with the χ_1, χ_2 hybrids instead of the atomic orbitals ψ_s, ψ_p

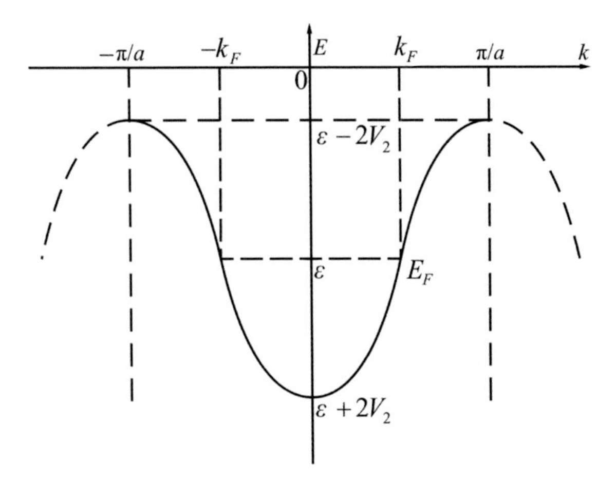

Fig. 12.6 The *band structure*, i.e., the energy $E(k)$ versus k relation, for a single particle state in the one-dimensional model for a metal shown in Fig. 12.5a with $V_2 < 0$. For k outside the first Brillouin zone (BZ) defined as the region $[-\pi/a,\ \pi/a]$ in k space, the eigenfunction is the same as $k + G_p$, where $G_p \equiv (2\pi/a)n,\ n = 0,\ 1,\ 2,\ \ldots$ is a vector such that $k + G_p$ is in the first BZ. The Fermi energy E_F and the Fermi wavenumber k_F are also shown assuming one electron per atom

Case (b) The B-states now are of the form, $\phi = \sum_n c_{n,A}\psi_{n,A} + \sum_n c_{n,B}\psi_{n,B}$. The coefficients, $c_{n,A}$, $c_{n,B}$ satisfy the relation $c_{n,j} = c_j \exp(in\varphi), j = A, B$, as a result of Bloch's theorem stating that a translation by the period a multiplies the state by a phase factor $\exp(i\varphi)$. Instead of φ we use the so-called crystal wavenumber $k \equiv \varphi/a$. Its values range from $-\pi/a$ to π/a in steps of $2\pi/[(N_a/2)a] = 2\pi/L$. The energy of the B-state can be found as in (11.8). By minimizing this energy with respect to c_A, c_B, we find a system of two equations for these unknown coefficients,

$$(\varepsilon_A - E_k)c_A + V_2\left(e^{-ika} + 1\right)c_B = 0, \quad (\varepsilon_B - E_k)c_B + V_2\left(1 + e^{ika}\right)c_A = 0 \quad (12.18)$$

This system has non-zero solutions only if the energy E_k satisfies the relations

$$E_k = E_\pm(k) = \varepsilon \pm \sqrt{V_3^2 + 4V_2^2 \cos^2(\tfrac{ka}{2})} \quad (12.19)$$

where ε and V_3 are defined as in (11.10). Obviously (12.19) implies that there are two bands, the lower one extending from $\varepsilon - \sqrt{V_3^2 + 4V_2^2}$ to $\varepsilon - V_3 = \varepsilon_B$ and the upper one from $\varepsilon + V_3 = \varepsilon_A$ until $\varepsilon + \sqrt{V_3^2 + 4V_2^2}$ with a gap equal to $E_g = \varepsilon_A - \varepsilon_B$. Each band has $N_a/2$ B-states. Hence, at $T = 0$ the lower band is fully occupied and the upper one is completely empty. Therefore, we have the case of an ionic insulator, e.g. NaCl.

Case (c) Following a similar procedure as in case (b) we find for the energies

$$E_\pm(k) = \varepsilon_A \pm \sqrt{V_2^2 + V_2'^2 + 2V_2V_2' \cos ka} \quad (12.20)$$

where $V_2 \propto d^{-2}$, $V_2' \propto d'^{-2}$ are the large and the small matrix elements respectively. Again we have two bands, the lower fully occupied and the upper completely empty; the gap is $E_g = 2\left|V_2 - V_2'\right|$. This is the case of a molecular insulator, such as solid N_2.

12.9 LCAO and Semiconductors

The last case of Fig. 12.5 deserves special attention because of its relevance to the technologically critical group of semiconductors. This case can be analysed by employing hybrids as shown in Fig. 12.5d′. Two neighbouring hybrids, e.g., $\chi_{1\cdot n}$ and $\chi_{2\cdot n+1}$, form a molecular bonding and a molecular antibonding level at each bond as in diatomic molecules. Because of the non-zero matrix element V_1 connecting $\chi_{1.n}$ and $\chi_{2.n}$, each electron in the bonding molecular orbital propagates from bond to neighbouring bond. This propagation is equivalent to that in case 12.5a with ε_b playing the role of ε_A and $V_1/2 = (\varepsilon_s - \varepsilon_p)/4$ playing approximately the role of V_2. Hence, a band is created around the bonding level ε_b of total width approximately equal to $(\varepsilon_p - \varepsilon_s)$ and a band of approximate width $(\varepsilon_p - \varepsilon_s)$ is formed around the antibonding molecular level ε_a. It follows that the gap is (see also [1], Chap. 6):

$$E_g \approx (\varepsilon_a - \varepsilon_b) - (\varepsilon_p - \varepsilon_s) = 2|V_{2h}| - (\varepsilon_p - \varepsilon_s) \approx 6.4\frac{\hbar^2}{m_e d^2} - (\varepsilon_p - \varepsilon_s) \quad (12.21)$$

Each of the bonding and antiboning bands has N_a states. Hence, the bonding band can accommodate exactly all the electrons, becoming thus the valence band, while the antibonding band is empty at $T = 0$, being thus the conduction band. The gap for the elemental solids of the carbon column and for diamond, Si, Ge, Sn, and Pb is, according to (12.21), 12.28, 1.66, 0.42, −0.02, and −1.95 eV respectively, while the corresponding experimental values are 5.5, 1.17, 0.74, 0, no gap, i.e. metal. An appreciable negative value of the gap indicates substantial overlap of the two bands and, hence, a metallic behaviour (Fig. 12.7). The tendency for the gap to be reduced as we go down the carbon column is mainly due to the systematic increase of the size of the corresponding atoms and the resulting increase of the bond length, which implies a smaller $|V_{2h}|$ and, according to (12.21), a smaller gap. In Figs. 12.8, 12.9 and 12.10 the above steps are also shown schematically. Semiconductors in 3-D employ the $sp^{(3)}$ hybridization and, hence, are arranged in space in a tetrahedral way with four nearest neighbours.

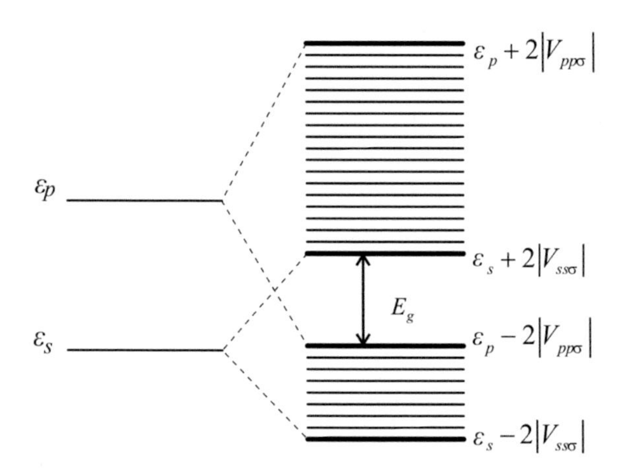

Fig. 12.7 An alternative determination of the band edges is obtained from the splitting of the ε_p and ε_s atomic levels by $\pm 2|V_{pp\sigma}|$ and $\pm 2|V_{ss\sigma}|$ respectively. The size of the gap is given then by $E_g = 2(|V_{ss\sigma}| + |V_{pp\sigma}|) - (\varepsilon_p - \varepsilon_s) = (7.08\hbar^2/m_e d^2) - (\varepsilon_p - \varepsilon_s)$, which almost coincides with (12.21). The interior of the bands are of mixed s and p character, in contrast to the band edges which are of pure either p or s character. For $|V_{ss\sigma}| + |V_{pp\sigma}| > (\varepsilon_p - \varepsilon_s)/2$, the tops of both bands are of p character, while the bottoms of both bands are of s character. For $|V_{ss\sigma}| + |V_{pp\sigma}| < (\varepsilon_p - \varepsilon_s)/2$, $\varepsilon_p - 2|V_{pp\sigma}|$ is higher than $\varepsilon_p + 2|V_{ss\sigma}|$ and the gap becomes negative, meaning no gap

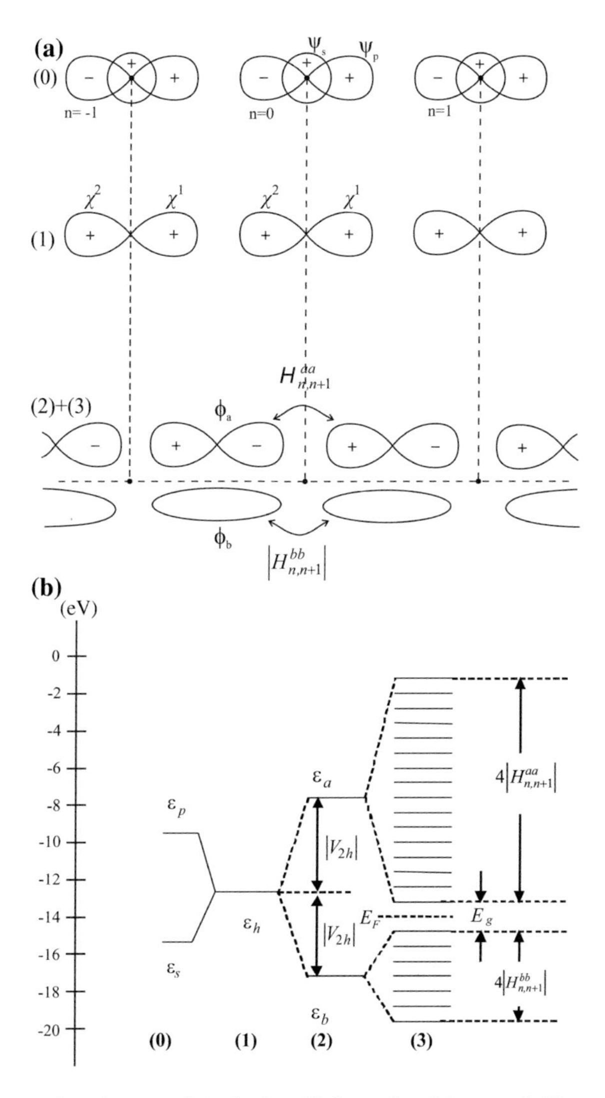

Fig. 12.8 **a** Successive changes of the basis orbitals produced by step 1 (from atomic orbitals ψ_s, ψ_p to hybrid atomic orbitals χ^1 , χ^2), step 2 (from χ^1 , χ^2 to molecular orbitals ϕ_b, ϕ_a); the matrix elements responsible for the creation of the VB and the CB are also shown. **b** The corresponding energy levels for Si (associated with the changes in **a**) for the 1-D model shown in Fig. 12.5d. The VB and the CB are formed around the bonding, ε_b, and the antibonding, ε_a, molecular levels respectively. The band edges can be obtained by the simpler approach shown in Fig. 12.7, which implies that the top of both bands is of pure p-character, while the bottom of both bands is of pure s-character

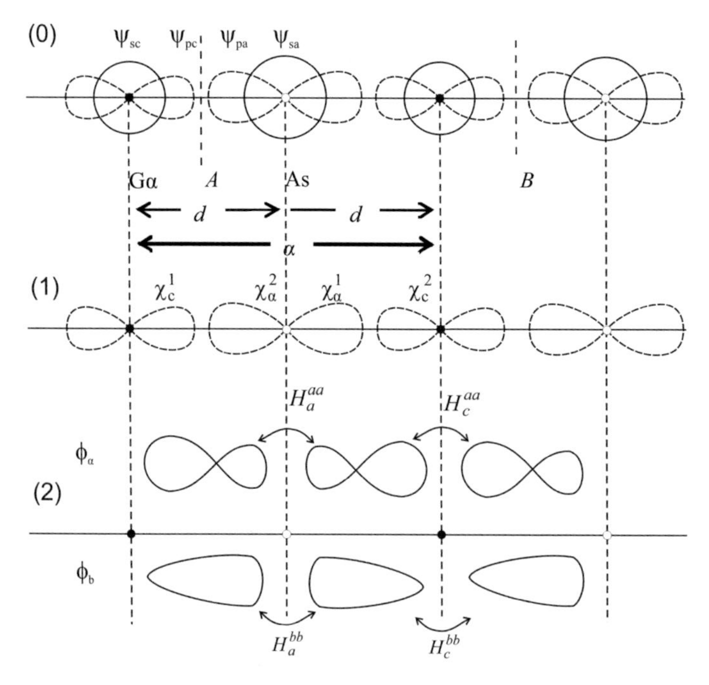

Fig. 12.9 *(0)* 1-D model of a compound semiconductor with two different atoms per primitive cell (the latter extends from A to B) and two orbitals (ψ_s and ψ_p) per atom. The type c (cation) atom carries two electrons; so does the type a (anion) atom. In *(1)* the sp^1 hybridized atomic orbitals for each atom are shown and in (2) the bonding ϕ_b and antibonding ϕ_a molecular orbitals, together with the corresponding off-diagonal nearest neighbor matrix elements H_i^{bb} and H_i^{aa} ($i = a$ or c) of the Hamiltonian are displayed

From Fig. 12.10 follows that the gap in the case of compound semiconductors is given by:

$$\boxed{E_g \approx 2\sqrt{V_{2h}^2 + V_{3h}^2} - \left(\bar{\varepsilon}_p - \bar{\varepsilon}_s\right)} \tag{12.22}$$

where $\bar{\varepsilon}_p = \frac{1}{2}\left(\varepsilon_{pc} + \varepsilon_{pa}\right), \bar{\varepsilon}_s = \frac{1}{2}\left(\varepsilon_{sc} + \varepsilon_{sa}\right)$.

12.10 Summary of Important Relations

Among the many formulae of this chapter, the reader should pay special attention to the following equations: (12.1), (12.7), (12.8a, b), (12.10), (12.11), (12.12), (12.13), (12.14), (12.16), (12.17), (12.21), (12.22).

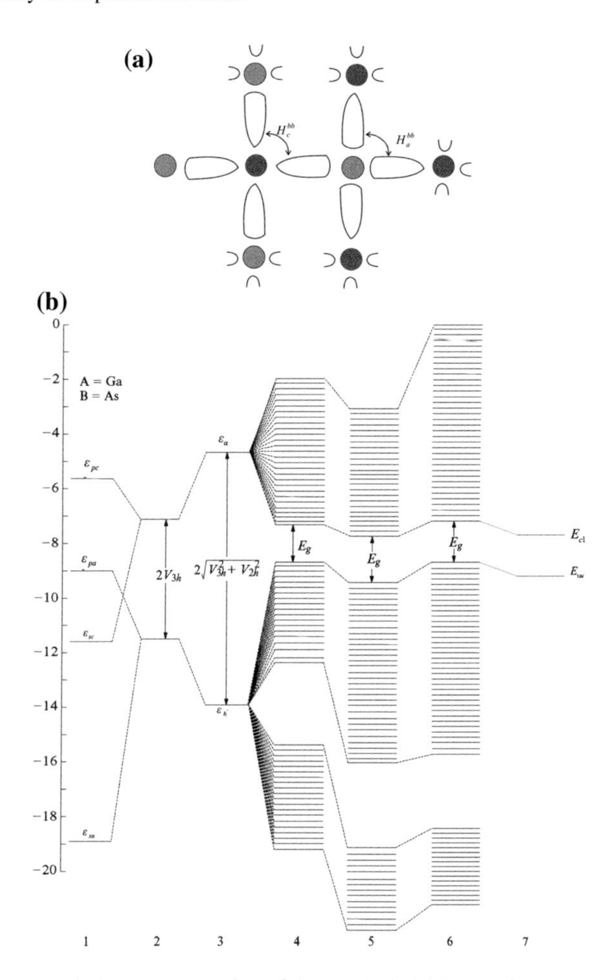

Fig. 12.10 **a** Schematic 2-D representation of the tetrahedral 3-D actual structure of a compound III-V semiconductor such as GaAs and the (asymmetric) molecular bonding orbitals. **b** The atomic *p* and *s* energy levels of each atom in eV (starting step 1); the sp^3 hybrid level of each atom (step 2); the bonding and antibonding molecular levels (step *3*); the resulting valence and conduction bands around the bonding and the antibonding molecular levels respectively (step *4*). Notice that the valence band is split to two subbands as a result of H_a^{bb} and H_c^{bb} being different (in analogy with the case of Fig. 12.5c). The (*5*), (*6*), and (*7*) columns show how successively more sophisticated approximations modify the bands and the gap

12.11 Multiple-Choice Questions/Statements (See also Appendix D)

1. All quantities characterizing the properties of any solid or liquid depend necessarily but not exclusively on the universal physical constants appearing in one of the following triads:
   (a) $\hbar$, c, m_e (b) $\hbar$, c, m_p (c) $\hbar$, e, m_p (d) $\hbar$, e, m_e
2. In an fcc lattice with a being the lattice constant the distance between nearest neighbors is:
   (a) $a/\sqrt{3}$ (b) $a/\sqrt{2}$ (c) $a/2$ (d) $a/3$
3. In a diamond lattice with a being the lattice constant of the unit cubic cell the distance between nearest neighbors is:
   (a) $\sqrt{3}a/4$ (b) $a/\sqrt{6}$ (c) $2a/\sqrt{6}$ (d) $a/4$
4. In a solid we define the length r by the relation: $(4\pi/3)r^3 \equiv V/N_a$. In terms of r the density is given by one of the following equation ($\bar{r} \equiv r/a_B$):
   (a) $\rho = (2.675A_B/\bar{r}^2)\,\text{g/cm}^3$ (b) $(2.675A_B/\bar{r}^3)\,\text{g/cm}^3$
   (c) $(2.675A_B^2/r^2)\,\text{g/cm}^3$ (d) $(2.675A_B/\bar{r}^4)\,\text{g/cm}^3$
5. In a solid we define the length r by the relation: $(4\pi/3)r^3 \equiv V/N_a$. In terms of r the cohesive energy is given by one of the following equation ($\bar{r} \equiv r/a_B$):
   (a) $\varepsilon_c \approx 27.2/\bar{r}^2\,\text{eV}$ (b) $\varepsilon_c \approx 27.2/\bar{r}\,\text{eV}$
   (c) $\varepsilon_c \approx 27.2/\bar{r}^3\,\text{eV}$ (d) $\varepsilon_c \approx 27.2/\bar{r}^5\,\text{eV}$
6. In a solid we define the length r by the relation: $(4\pi/3)r^3 \equiv V/N_a$. In terms of r the bulk modulus B is given by one of the following equation ($\bar{r} \equiv r/a_B$):
   (a) $B \approx 175\times10^6/\bar{r}^2\,\text{bar}$ (b) $B \approx 175\times10^6/\bar{r}^3\,\text{bar}$
   (c) $B \approx 175\times10^6/\bar{r}^4\,\text{bar}$ (d) $B \approx 175\times10^6/\bar{r}^5\,\text{bar}$

7. In a liquid the longitudinal speed of sound can be estimated by the formula : $(\bar{r} \equiv r / a_B)$

(a) $\upsilon \approx \frac{e^2}{\hbar}$ (b) $\upsilon \approx \frac{\hbar}{m_e a_B \bar{r}}$ (c) $\upsilon \approx \frac{\hbar}{m_p a_B \bar{r}}$ (d) $\upsilon \approx \frac{\hbar}{\sqrt{m_e m_a}\, a_B \bar{r}}$

8. In a liquid the longitudinal speed of sound is given by the formula:

(a) $\upsilon = B / \rho$ (b) $\upsilon = (B / \rho)^2$ (c) $\upsilon = (B / \rho)^{1/2}$ (d) $\upsilon = (B / \rho)^{1/3}$

9. In a solid a kind of average between the longitudinal and the transverse velocity of sound given by the formula $\upsilon = (B / \rho)^{1/2}$ can be estimated as follows:

(a) $\upsilon \approx 80 / (\bar{r}^2 \sqrt{A_B})\,\mathrm{km/s}$ (b) $\upsilon \approx 80 / (\bar{r}^2 A_B)\,\mathrm{km/s}$

(c) $\upsilon \approx 80 / (\bar{r} \sqrt{A_B})\,\mathrm{km/s}$ (d) $\upsilon \approx 80 / (\bar{r}\, A_B)\,\mathrm{km/s}$

10. The Debye temperature Θ_D, which is involved in the thermodynamic quantities due to lattice vibrations, can be estimated as follows:

(a) $\Theta_D \approx \frac{7390}{\bar{r}^3 \sqrt{A_B}}\mathrm{K}$ (b) $\Theta_D \approx \frac{21000}{\bar{r}^2 \sqrt{A_B}}\mathrm{K}$

(c) $\Theta_D \approx \frac{7390}{\bar{r}^2 A_B}\mathrm{K}$ (d) $\Theta_D \approx \frac{21000}{\bar{r}^2 A_B}\mathrm{K}$

11. The unit of resistivity in the atomic system of units is:

(a) $= \hbar / e^2$ (b) $= \alpha \hbar / e^2$ (c) $= \hbar a_B / e^2$ (d) $= \hbar a_B \alpha / e^2$

12. The unit of resistivity in the atomic system of units has the following value:

(a) 4108 Ω (b) 188.4 Ω (c) 0.1586 μΩ·cm (d) 21.74 μΩ·cm

13. The resistivity ρ of crystalline metals is due to the existence of deviations from the perfect periodicity such as defects, foreign atoms and, mainly, thermal oscillations of the ions. The resistivity is given by the approximate formula:

(a) $\rho = 3c_1(\hbar a_B / e^2)(T / T_0) \approx c_1\,(0.0002T)\,\mu\Omega \cdot \mathrm{cm}$

(b) $\rho = 3c_1(\hbar a_B / e^2)\bar{r}^3 (T / T_0) \approx c_1\,(0.006T)\,\mu\Omega \cdot \mathrm{cm}$

(d) $\rho = 3c_1(\hbar a_B / e^2) \approx c_1\, 65\,\mu\Omega \cdot \mathrm{cm}$

(d) $\rho = 3c_1(\hbar a_B / e^2)(T / T_0)^2 \approx c_1\, 6{,}5 \times 10^{-10}\, T^2\,\mu\Omega \cdot \mathrm{cm}$

14. The resistivity ρ depends on the concentration of carriers, $n_{e,h}$, (electrons or holes) on their electric charge, $\mp e$, and, finally, on the scatterings which macroscopically appear as a friction force proportional to the average velocity; the proportionality constant for dimensional reasons has the form $m^*_{e,h} / \tau_{e,h}$, where $m^*_{e,h}$ is the effective mass and $\tau_{e,h}$ is the so-called relaxation time for electrons or holes respectively. Dimensional analysis suggests that the formula for the conductivity has the following form in the G-CGS system.

(a) $\sigma_{e,h} = \frac{e\, n_{e,h}}{m^*_{e,h} / \tau_{e,h}}$ (b) $\sigma_{e,h} = \frac{e^2\, n_{e,h}}{m^*_{e,h} / \tau_{e,h}}$

(c) $\sigma_{e,h} = \dfrac{n_{e,h}}{e m^*_{e,h} / \tau_{e,h}}$ (d) $\sigma_{e,h} = \dfrac{1}{e n_{e,h} m^*_{e,h} / \tau_{e,h}}$

15. Dimensional analysis leads to the following formula for the conductivity of metals in terms of the mean free path $\ell = \upsilon_F \tau$ and in the G-CGS system:

(a) $\sigma = \dfrac{m_e \upsilon_F}{e^2 n \ell}$ (b) $\sigma = \dfrac{p_F \ell}{e^2 n}$ (c) $\sigma = \dfrac{e^2 p_F \ell}{n}$ (d) $\sigma = \dfrac{e^2 n \ell}{p_F}$

16. From the formula for the metallic conductivity and the dimensional result for the size of the conductivity (using the replacement $a_B \to \overline{r}\, a_B$), the magnitude of the electronic mean free ℓ path is estimated as:

(a) $\ell = 1$ cm (b) $\ell = 0.1$ mm (c) $\ell = 1$ μm (d) 1 nm

17. From the formula $\sigma_{e,h} = e^2 n_{e,h} / (m^*_{e,h} / \tau_{e,h})$ for the electric conductivity of electrons or holes we deduce that the general expression for the conductivity of a semiconductor is the following (Usually we define the mobility $\mu_{e,h}$ by the relation $|\mu_{e,h}| \equiv |e| \tau_{e,h} / m^*_{e,h}$):

(a) $\sigma = (e^2 n_e \tau_e / m^*_e) + (e^2 n_h \tau_h / m^*_h)$

(b) $\rho = \rho_e + \rho_h = (e^2 n_e \tau_e / m^*_e)^{-1} + (e^2 n_h \tau_h / m^*_h)^{-1}$

(c) $\sigma^{-2} = (e^2 n_e \tau_e / m^*_e)^{-2} + (e^2 n_h \tau_h / m^*_h)^{-2}$

(d) $\sigma^2 = (e^2 n_e \tau_e / m^*_e)^2 + (e^2 n_h \tau_h / m^*_h)^2$

18. The graphs below present the typical temperature dependence of the resistivity, ρ, of metals and semiconductors. What is the correct explanation for this opposite behavior?

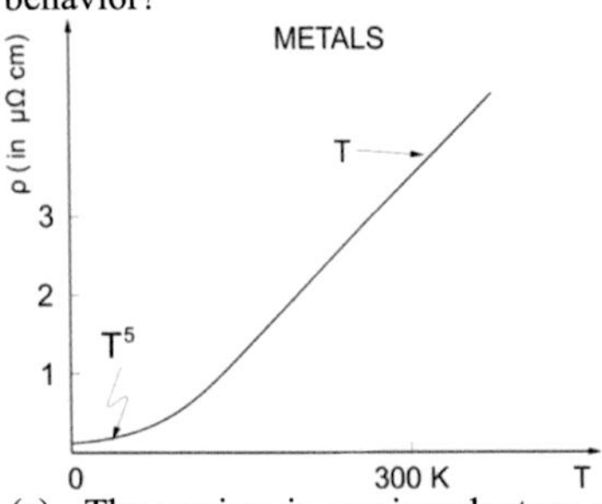

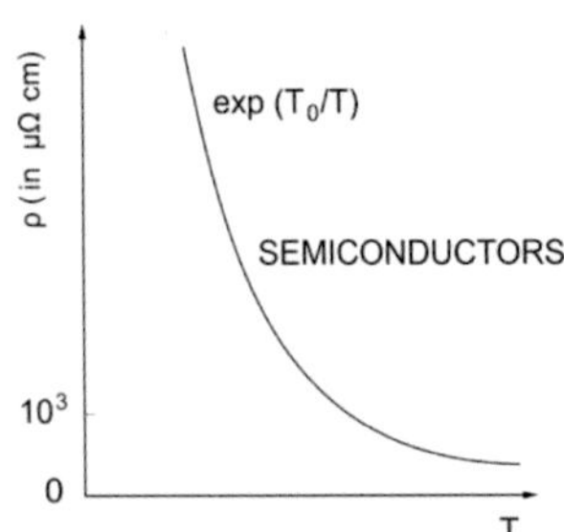

(a) The carriers in semiconductors are in general both electrons and holes. Thus the electric current due to electrons partially cancels that due to holes; this cancellation tends to become complete as the temperature tends to the absolute zero.

(b) From the relation $\rho_\eta = m_e / (e^2 n \tau)$ we conclude that the relaxation time τ is the only one which depends on the temperature. Thus τ must be a decreasing function of temperature for metals and an increasing one for semiconductors.

(c) In the formula $\rho_\eta = m_e / (e^2 n \tau)$ the concentration n is an exponentially increasing function of temperature for semiconductors and a decreasing function in metals.
(d) For both metals and semiconductors τ is decreasing as the temperature is raised. The concentration of carriers n in metals does not depend on the temperature, while is increasing exponentially for semiconductors as the temperature is raised.

19. In the jellium model the cohesive energy U of a metal is given by:
(a) $U / N(\hbar^2 / m_e a_B^2) = (a / \bar{r}'^{12}) - (\gamma / \bar{r}'^6)$
(b) $U / N(\hbar^2 / m_e a_B^2) = (a / \bar{r}'^6) - (\gamma / \bar{r}')$, $\bar{r} \equiv r / a_B$
(c) $U / N(\hbar^2 / m_e a_B^2) = (a / \bar{r}'^2) - (\gamma / \bar{r}')$
(d) $U / N(\hbar^2 / m_e a_B^2) = (a / \bar{r}'^4) - (\gamma / \bar{r}'^2)$

20. In the jellium model the bulk modulus B of a metal is given by the formula (in Mbar):
(a) $B = 15.6 a / \bar{r}'^2$ (b) $B = 15.6 a / \bar{r}'^3$
(c) $B = 15.6 a / \bar{r}'^4$ (d) $B = 15.6 a / \bar{r}'^5$, $\bar{r} \equiv r / a_B$

21. According to the jellium model the parameters a and γ for Al are equal to 13.6 and 9.03 respectively. The resulting value of $\bar{r}$ is:
(a) $\bar{r} = 3.01$ (b) $\bar{r} = 1.51$ (c) $\bar{r} = 6.02$ (d) $\bar{r} = 9.03$

22. According to the jellium model the parameters a and γ for Al are equal to 13.6 and 9.03 respectively. The resulting value for the bulk modulus B is in Mbar:
(a) 13.6 (b) 15.6 (c) 9.03 (d) 0.86

23. For a compound semiconductor we define the polarity index by $a_p \equiv V_{3h} / \sqrt{V_{2h}^2 + V_{3h}^2}$ and the metallicity index by $a_m \equiv 2\bar{V}_1 / \sqrt{V_{2h}^2 + V_{3h}^2}$, where $\bar{V}_1 = (\varepsilon_{pc} - \varepsilon_{sc} + \varepsilon_{pa} - \varepsilon_{sa})/8$. The approximate formula for the size of the gap E_g is:
(a) $E_g = 2\sqrt{V_{2h}^2 + V_{3h}^2} - \bar{V}_1$ (b) $E_g = 2\sqrt{V_{2h}^2 + V_{3h}^2} - 2\bar{V}_1$
(c) $E_g = 2\sqrt{V_{2h}^2 + V_{3h}^2} - 3\bar{V}_1$ (d) $E_g = 2\sqrt{V_{2h}^2 + V_{3h}^2} - 4\bar{V}_1$

24. For the compound III/V semiconductor GaAs, the relevant atomic levels are (in eV) -5.67, and -11.55 for Ga and -8.98 and -18.91 for As. Find the hybrids atomic levels, the bonding and antibonding molecular levels, and the energy bands. The bond length in GaAs is 2.448 A. Estimate the magnitude of the gap E_g (in eV) in GaAs;
(a) 0.7 (b) 1.1 (c) 1.4 (d) 2.9

25. Estimate the size of the gap E_g (in eV) in Si, given the following data: $\varepsilon_s = -14.79\,\text{eV}$, $\varepsilon_p = -7.58\,\text{eV}$ and $d = 2.3517\,\text{A}$
(a) 0.5 (b) 1.66 (c) 2.7 (d) 4.1

26. For a 1-D LCAO model with one relevant atomic orbital per unit cell (of length *d*), the diagonal matrix element is ε_0 and the one between nearest neigbhors is V_2. The dispersion relation between the eigenenergy ε and the wavenumber k is:

(a) $\varepsilon = \varepsilon_0 + V_2 d^2 k^2$ (b) $\varepsilon = \varepsilon_0 - V_2 d^2 k^2$

(c) $\varepsilon = \varepsilon_0 + 2V_2 \cos(k\,d)$ (d) $\varepsilon = \varepsilon_0 + V_2 d\,k$

The following multiple questions/statements require studying the Appendix IV

27. For a 1-D LCAO model with one relevant atomic orbital 1s per unit cell (of length *d*), the diagonal matrix element is ε_0 and the one between nearest neigbhors is V_2. The effective mass in the limit $k \to 0$ is given by:

(a) $m^* = -\hbar^2 / 2V_2 d^2$ (b) $m^* = -\hbar^2 / V_2 d^2$

(c) $m^* = \hbar^2 / 2V_2 d^2$ (d) $m^* = \hbar^2 / V_2 d^2$

28. The definition of the effective mass for a particle moving in a 3-D periodic potential such that its eigenenergy $\varepsilon_n(\boldsymbol{k})$ is a function of the wavevector $\boldsymbol{k}$ and the band index n, is as follows:

(a) $1/m^*_{i,n} \equiv \hbar^{-1}\nabla_k \varepsilon_n(\boldsymbol{k})$ (b) $1/m^*_{i,n} \equiv \hbar^{-2}\Delta_k \varepsilon_n(\boldsymbol{k})$

(c) $1/m^*_{i,n} \equiv \hbar^{-2}\nabla_k \times (\nabla_k \varepsilon_n(\boldsymbol{k}))$ (d) $1/m^*_i \equiv \hbar^{-2}\partial^2 \varepsilon_n(\boldsymbol{k}) / (\partial k_i)^2$

29. For a particle of effective mass m^* moving in a 3D volume V its energy is given by $\varepsilon(\boldsymbol{k}) = E_a + \hbar^2 k^2 / 2m^*$. The density of states $\rho(E)$ is given by the following relation:

(a) $\rho(E) \approx \left[V \left| m^* \right|^{3/2} / \sqrt{2}\pi^2 \hbar^3) \right] (E - E_a)$

(b) $\rho(E) \approx \left[V \left| m^* \right|^{3/2} / \sqrt{2}\pi^2 \hbar^3) \right] \left| E - E_a \right|$

(c) $\rho(E) \approx \left[V \left| m^* \right|^{3/2} / \sqrt{2}\pi^2 \hbar^3) \right] \left| E - E_a \right|^{3/2}$

(d) $\rho(E) \approx \left[V \left| m^* \right|^{3/2} / \sqrt{2}\pi^2 \hbar^3) \right] \left| E - E_a \right|^{1/2}$

30. The dielectric constant of Si is 12.1 and the average effective mass in the CB is $m^*_e = 0.329\, m_e$. Replacing a Si atom by a P atom will produce an impurity level at ε_i in the gap. If E_g is the lower edge of the CB, the difference $E_g - \varepsilon_i$ is:

(a) 1 eV (b) 0.362 eV (c) 0.091 eV (d) 0.030 eV

31. The dielectric constant of GaAs is 12.85 and the average effective mass in its VB is $m^*_e = 0.51 m_e$. Replacing an As atom by a Ge atom will produce a bound state of "radius" a^*_B which is approximately equal to:

(a) 0.529 A (b) 6.8 A (c) 13.3 A (d) 171 A

32. The electronic concentration $n \equiv N_{e,ZA}/V$ in the CB of a semiconductor is given by the relation $n = (2/V)\int_{E_g}^{\infty} \rho_{ZA}(E) f(E) dE$. Taking into account that the DOS is $\rho_{ZA}(E) \approx (A_c/\pi^{1/2})(E - E_g)^{1/2}$ and approximating the Fermi distribution by the Boltzmann one we find that:

(a) $n \approx A_c (k_B T)^{3/2} \exp[-(E_g / k_B T)]$

(b) $n \approx A_c (k_B T)^{3/2} \exp[(-\mu / k_B T)]$

(c) $n \approx A_c (k_B T)^{3/2} \exp[(E_g / k_B T)]$

(d) $n \approx A_c (k_B T)^{3/2} \exp[-(E_g - \mu) / k_B T)]$

33. The concentration of holes $p \equiv N_{h,\mathrm{VB}}/V$ in the VB of a semiconductor is given by the relation $p = (2/V)\int_{-\infty}^{0} \rho_{VB}(E)[1 - f(E)] dE$. Taking into account that the DOS is $\rho_{VB}(E) \approx (A_h/\pi^{1/2})(-E)^{1/2}$ and approximating the Fermi distribution by the Boltzmann one we find that:

(a) $p \approx A_h (k_B T)^{3/2} \exp[(-\mu / k_B T)]$

(b) $p \approx A_h (k_B T)^{3/2} \exp[(E_g - \mu) / k_B T)]$

(c) $p \approx A_h (k_B T)^{3/2} \exp[(\mu / k_B T)]$

(d) $p \approx A_h (k_B T)^{3/2} \exp[-(E_g - \mu) / k_B T)]$

34. The dielectric constant ε of Si is 12.1 and the average effective mass m_e^* in the CB is $m_e^* = 0.329\, m_e$. Replacing a Si atom by a P atom will produce an impurity level at ε_i in the gap. If E_g is the lower edge of the CB, the difference $E_1 = E_g - \varepsilon_i$ is:

(a) $E_1 = e^2 / 2a_B$, (b) $E_1 = e^2 m_e^* / 2a_B m_e \varepsilon^2$

(c) $E_1 = e^2 \varepsilon\, m_e^* / 2a_B m_e$ (d) $E_1 = e^2 \varepsilon^2 / 2a_B$

35. In a semiconductor with N_a concentration of acceptors and N_d $(N_a > N_d)$ concentration of donors, which is the relation connecting the electronic concentrations n, n_d in the CB and in donors respectively with the hole concentrations p, p_a in the VB and in acceptors respectively? (Hint: Consider first the case for $T = 0\,\mathrm{K}$ before you give the answer for $T \neq 0\,\mathrm{K}$)

(a) $n + n_d = p + p_a$, (b) $n + n_d = p + p_a + N_a + N_d$,

(c) $n + n_d = p + p_a - N_a + N_d$ (d) $n + n_d = p + p_a + N_a - N_d$

36. In a semiconductor the concentrations of electrons n and holes p in the CB and in the VB respectively are given by the formulae:

$n \approx A_c (k_B T)^{3/2} \exp[-(E_g - \mu) / k_B T)]$, $p \approx A_h (k_B T)^{3/2} \exp[(-\mu / k_B T)]$,

$A_i \propto m_i^*$, $i = c, h$. Then in the absence of donors and acceptors (intrinsic case) the chemical potential is:

(a) $\mu = (E_g / 2) - (3k_B T / 4)\ln(m_h^* / m_c^*)$,

(b) $\mu = (E_g / 2) - (3k_B T / 4)\ln(m_c^* / m_h^*)$

(c) $\mu = E_g - (3k_B T / 4)\ln(m_h^* / m_c^*)$

(d) $\mu = E_g + (3k_B T / 4)\ln(m_h^* / m_c^*)$

37. In a semiconductor the concentrations of electrons n and holes p in the CB and in the VB respectively are given by the formulae: $n \approx A_c (k_B T)^{3/2} \exp[-(E_g - \mu) / k_B T)]$, $p \approx A_h (k_B T)^{3/2} \exp[(-\mu / k_B T)]$, $A_i \propto m_i^*$, $i = c, h$. Then in the absence of donors and acceptors (intrinsic case) the concentration of holes in the VB is:

(a) $p \approx (A_e A_h)^{1/2} (k_B T)^{3/2} \exp(-E_g / k_B T)$

(b) $p \approx (A_e A_h)^{1/2} (k_B T)^{1/2} \exp(-E_g / k_B T)$

(c) $p \approx (A_e A_h)^{1/2} (k_B T)^{3/2} \exp(-E_g / 2k_B T)$

(d) $p \approx (A_e A_h)^{1/2} (k_B T)^{1/2} \exp(-E_g / 2k_B T)$

38. In a semiconductor of N_a and N_d concentrations of acceptors and donors respectively ($N_d - N_a \equiv \delta N \gg n_i$, where n_i is the electron concentration in the CB in the intrinsic case), the concentration of electrons n in the CB is the following: (Take into account that $n + n_d = p + p_a - N_a + N_d$)

(a) $n \approx \frac{1}{2}(\sqrt{\delta N^2 + 4n_e^2} - \delta N)$, (b) $n \approx \frac{1}{2}(\sqrt{\delta N^2 + 4n_e^2} + \delta N)$,

(c) $n \approx \delta N$, (d) $n \approx n_i$

39. In a pure crystalline Si one atom per million has been replaced by a P atom. As a result of this replacement the conductivity will increase by a factor x, where:

(a) x=1.000001, (b) $x = 5.3 \times 10^6$, (c) x=15.4, (d) x=2

40. In a copper wire the current density is equal to $16\,\text{A}/\text{mm}^2$. For copper the valence $\zeta = 1$ and $\bar{r} = 2.67$. The average velocity (in m/s) of electrons along the direction of the applied electric field is:

(a) 1.18×10^{-3}, (b) 1.18 (c) 1.18×10^3 (d) 1.18×10^6

12.12 Solved Problems

1. *For water determine the radius per molecule $\bar{r}$, the velocity of sound, and the bulk modulus.*

Solution From the relations $\rho_M = 2.675(A_W/\bar{r}^3)\,\text{g/cm}^3 = 1$ g and $A_W = 18$ it is obtained that $\bar{r} = 3.64$. The velocity of sound for dimensional reasons is given by the following formula $\upsilon = \eta(\hbar/m_e a_B \bar{r}) \times \sqrt{m_e/(u A_B)}$. The numerical factor η (in the present case of hydrogen bonding of the molecules) is not around 1.6 as in the presence of strong bonds; it is expected to be about 3.5 smaller. Taking into account that $(\hbar/m_e a_B) = c/137$, $(m_e/u) = 1/1823$ we find $\upsilon = 1.52$ km/s. From the relation $\upsilon = \sqrt{B/\rho_M}$ we obtain $B = 2.29 \times 10^9$ N/m ^{2} = 22,900 bar

2. *Estimate the coefficient of surface tension for a liquid (e.g. water)*

Solution The surface tension is due to the difference appearing in the cohesive energy per atom (or molecule, if we are considering water) depending on whether the atom is in the bulk or at the surface. In the first case the cohesive energy is equal to $\frac{1}{2}A_{nn,b}\varepsilon$, where $A_{nn,b} \approx 8$ is the number of nearest neighbors and ε is the interaction energy per pair of atoms (or molecules), while in the second case is equal to $\frac{1}{2}A_{nn,s}\varepsilon$, where $A_{nn,s} \approx 5$ is the number of nearest neighbors for an atom (or molecule) located at the surface. The surface tension coefficient σ is the difference $\frac{1}{2}A_{nn,b}\varepsilon - \frac{1}{2}A_{nn,s}\varepsilon$ divided by the surface area s per atom (or molecule). For water, where the molecules are held together by the weak hydrogen bond, the quantity $\frac{1}{2}A_{nn,b}\varepsilon$ is about 0.3 eV so that $\varepsilon \approx 0.075$ eV; the area s is expected to be about twice the quantity $\pi a_B^2 \bar{r}^2 = \pi(0.529 \times 3.64)^2 \times 10^{-20}$ m^2 = 11.6 $\times 10^{-20}$ m^2. Thus $\sigma = \frac{3}{8}(0.3/23.2) \times 10^{20}$ eV/m^2 = 0.075 J/m^2 versus 0.073 J/m^2 for the experimental value at $T = 20\,^\circ$C. For mercury $\bar{r} = 3.36$ and $\frac{1}{2}A_{nn,b}\varepsilon$ can be estimated at about 1 eV; the resulting value for σ is about 0.3 J/m^2 versus 0.486 J/m^2 for the experimental value at $T = 20\,^\circ$C. The discrepancy may be due to the factor of 2 we included in the surface area per atom which is probably too big for mercury.

12.13 Unsolved Problems

1. Does sound propagate faster in aluminum or in lead? Justify your answer.
2. In n-type Si($N_d = 5 \times 10^{16}\,\text{cm}^{-3}$) the current density is 0.1 A/cm^2 at $T = 300$ K. What is the average velocity of the electrons producing this current density?

References

1. E. Economou, *The Physics of Solids, Essentials and Beyond* (Springer, Heidelberg, 2010)
2. W. Harrison, *Electronic Structure and the Properties of Solids* (Dover Publications, NY, 1989)
3. E. Cohen, D. Lide, G. Trigg (eds.), *AIP Physics Desk Reference* (Springer, NY, 2003)

Part V
Gravity at the Front Stage

General Remarks

When is the Gravitational Self-energy Comparable with the Coulomb Self-energy in a Neutral Condensed System?

The gravitational self-energy U_G of an electrically neutral system of mass M (such as a planet or an active star) depends on M as $U_G = aM^{5/3}$ (assuming that the density does not depend on M), while the Coulomb self-energy grows linearly with M as $U_C = bM$ with the quantities a and b to be examined below. This difference in the dependence on the mass is due to the long range and always attractive character of the gravitational interaction vs. the *effectively* short-range character of the Coulomb interaction in an electrically neutral system, in which positively and negatively charged particles combine to form *locally* neutral configurations. The linear dependence of U_C on M follows from the general thermodynamic classification of intensive and extensive quantities; the latter quantities are proportional to the number of particles in the system, which is proportional to the mass of the system. See Sect. 4.6, where it is explained that the only requirement for this thermodynamic classification is the short range of the effective interactions among the particles of the system. The gravitational energy will become equal to the Coulomb energy when the total mass will be equal to

$$M = (b/a)^{3/2} \tag{V.1}$$

This characteristic exponent of 3/2 will appear very often in calculations involving gravity.

The gravitational self-energy of a spherical body of mass M and radius R, according to Newton's law, equals to

$$U_G = -a_G \frac{GM^2}{R}, a_G = 3x/5 \tag{V.2}$$

where x equals 1, if the mass density is homogeneous and isotropic; if the density is decreasing as the distance from the center is increasing, x is larger than 1. (For Earth, $x \approx 1.025$; see (13.11)). Taking into account that $M = (4\pi/3)R^2\rho_M$ and (12.3) we obtain that

$$a = -\frac{3x}{5}\frac{GA_W^{1/3}u^{1/3}}{a_B\bar{r}} \tag{V.3}$$

where AW is the average atomic weight and u is 1/12 of the mass of carbon-12.

The Coulomb energy was obtained before (see (12.14) and text below it). Substituting in this formula, N_a from the relation $M = N_aA_Wu = N_\nu u$, where N_ν is the total number of nucleons, we find for b

$$b = -y\gamma\frac{e^2}{\bar{r}a_B}\frac{1}{A_Wu} \tag{V.4}$$

where y is a correction numerical factor to take into account a non-uniform concentration of particles; y is expected to be a little larger than 1, but probably smaller than x; γ, according to the text below (12.14), is equal to $0.56\zeta^{4/3} + 0.9\zeta^2$ (see Sect. 12.7). Combining (V.1), (V.3), (V.4), and $M = N_\nu u$,we find that, in an electrically neutral system, the number of nucleons N_ν, which makes the gravitational self-energy equal to the Coulomb self-energy is:

$$N_V = a_\nu\frac{\gamma^{3/2}}{A_W^2}\left(\frac{e^2}{Gu^2}\right),\quad a_\nu = \left(\frac{5y}{3x}\right)^{3/2} \tag{V.5}$$

Table V.1 Total number of nucleons as well as their mass in the following celestial bodies

Earth	$3.5978 \times 10^{51} \Rightarrow 5.9743 \times 10^{24}$ kg
Sun	$1.1979 \times 10^{57} \Rightarrow 1.9891 \times 10^{30}$ kg
Moon	$4.4263 \times 10^{49} \Rightarrow 7.35 \times 10^{22}$ kg
Jupiter	$1.1437 \times 10^{54} \Rightarrow 1.899 \times 10^{27}$ kg
Pluto	$8.9945 \times 10^{48} \Rightarrow 1.49 \times 10^{22}$ kg
Smallest active star	$1056 \Rightarrow 1.66 \times 10^{29}$ kg
Larger active star	$1059 \Rightarrow 1.66 \times 10^{32}$ kg

Table V.2 Astronomical units of length

Light year	1 ly = 9.4607305×10^{15} m
Parsec	1 pc = 3.0856776×10^{16} m = 3.2616 ly
Astronomical unit	1 AU = 1.4959787×10^{11} m

The dimensionless ratio $e^2/Gu^2 \approx 1.2536 \times 10^{36}$ is the ratio of the strengths of the electromagnetic to the gravitational interactions (see Table 2.2). The radius R corresponding to N_v is given by

$$R = N_a^{1/3} \bar{r} a_B = \frac{N_V^{1/3} \bar{r} a_B}{A_W^{1/3}} \tag{V.6}$$

If we choose $AW \approx 40$ (as in Earth),$\varsigma = 1$, $(y/x) = 1$, and $\bar{r} = 2.75$, we find $N_\nu = 3.3 \times 10^{51}$ and $R = 6.35 \times 10^6$ m, while the corresponding measured values for Earth are $N_\nu = 3.596 \times 10^{51}$ and $R = 6.371 \times 10^6$ m. If we choose $A_W \approx 2, \zeta \approx 1, x \approx 1, \bar{r} \approx 1.6$, values close to those of Jupiter, we obtain $N_\nu \approx 1.33 \times 10^{54}$ and $R \approx 7.36 \times 10^7$m versus $N_\nu \approx 1.143 \times 10^{54}$ and $R \approx 7.15 \times 10^7$ m for Jupiter.

We conclude that for Earth and for Jupiter the Coulomb self-energy is comparable to the gravitational self-energy. For the Sun the ratio U_G/U_C (assuming that its A_W and ζ are about those of Jupiter) is equal to

$$\frac{U_G}{U_C} \approx \frac{a}{b} M_S^{2/3} \approx \left(\frac{M_s}{M_J}\right)^{2/3} \approx 100 \tag{V.7}$$

where M_S, $M_J$are the masses of the Sun and the Jupiter, respectively.

Chapter 13
Planets

Abstract In planets the gravitational self-energy is comparable to the electrostatic self-energy. The central star sends to each of its planets high temperature power which is reemitted at low temperature leading thus to a net flow of entropy out of the planet and thus to an information inflow. This exchange determines not only the temperature, but establishes also another necessary condition for rocky planets to be candidate cradles of life.

13.1 Summary

In this chapter we shall examine a few properties pertaining to planets. First, why have planets spherical shape? Yet, why are there mountains in rocky planets? We find that the maximum possible height of a mountain in a rocky planet is determined by the competition between the weight of the mountain and the mechanical failure of the rocky material at the base of the mountain. Second, what determines the average temperature of a planet? The answer is: The balance between the E/M power absorbed by a planet and the E/M power emitted by the planet; the source of the former is the central star. We show that the temperature difference between the planet and the central star leads to a net emission of entropy by the planet, a fact that makes the phenomenon of life possible: Rocky planets are candidate cradles of life. Finally, we briefly examine the question of why winds appear in the atmospheres of planets. Other information and data regarding planets can be found in [1, 2].

13.2 How Tall Can a Mountain Be?

Planets and other celestial large bodies are of spherical shape, because this shape is the one which minimizes the total gravitational self-energy (assuming isotropy in the chemical composition). This seems rather obvious, since any other shape can

E.N. Economou, *From Quarks to the Universe*,
DOI 10.1007/978-3-319-20654-7_13

come from the spherical one by removing appropriate pieces of matter and placing them at a larger distance from the center, increasing thus the gravitational energy. Hence, the mountains or ridges in a rocky planet, since they represent deviations from the spherical symmetry, increase its total energy. Mountains (and ridges) were created during the formation of a planet and/or during major "geological" activities over the ages. What keeps them existing is an energy barrier that does not allow the spontaneous flow of matter, which would lead to a locally flattened surface. The overcome of this barrier involves the deformation and eventual breaking of chemical bonds, which macroscopically lead to failure of the solid state either by plastic flow or by propagation of cracks. This failure will occur when the *shear* stress, τ, at the base of a mountain reaches the critical value τ_s at which the solid material at the base of the mountain either breaks or undergoes plastic deformation. The quantity τ_s has been estimated very crudely in Sect. 12.4 to be of the order of $B/500$. The shear stress at the base of a mountain is of the order of the pressure $P = W/S$, where W is the weight of the mountain and S is its cross-section area at its base. Assuming a conical shape of the mountain, we get $W = \frac{1}{3}SH\rho_M g$, where H is the height of the mountain. Hence, by employing the relation $\tau = \tau_s$ and using (12.5), we obtain the criterion determining the highest possible mountain in a rocky planet:

$$\frac{1}{3}xH\rho_M g = y\frac{0.6}{500}\frac{\hbar^2}{m_e a_B^5 \bar{r}^5} \tag{13.1}$$

where x, y are numerical factors of the order of unity. We shall take as rough estimate $x/y = 1$. The density has been calculated in Chap. 12: $\rho_M = A_W m_u / (\frac{4\pi}{3} a_B^3 \bar{r}^3)$. The acceleration of gravity g is given by $g = GM/R^2 = G(\frac{4\pi}{3}R^3)\rho'_M / R^2 = \frac{4\pi}{3} G R \rho'_M$, where the prime indicates that we refer to the whole planet, while the unprimed quantities refer to the crust. Substituting in (13.1) we obtain

$$\frac{HR}{a_B^2} = B\frac{1}{A_W A'_W}\frac{e^2}{Gu^2}, \qquad B \equiv 0.015\frac{y}{x}\frac{\bar{r}'^3}{\bar{r}^2} \tag{13.2}$$

Choosing the Earth's values, $\bar{r} = 2.89$, $\bar{r}' = 2.69$, $A_W = 24.25$, $A'_W = 40$, we find $H R \approx 10^{11}\,\mathrm{m}^2$ or $H \approx 16\,\mathrm{km}$. This means that it is not possible for a mountain on Earth to be taller than about 16 km. For Mars with $R = 3396$ km and the product $H R \approx 1.97 \times 10^{11}\,\mathrm{m}^2$ (assuming that $\bar{r}$, $\bar{r}'$ are the same as in Earth and that $(A_W A'_W)_E / (A_W A'_W)_M \approx (\rho_{M,E}/\rho_{M,M})^2$) the largest possible height of a mountain cannot exceed 59 km. The highest mountain in Mars, Mt. Olympus, is about 24 km [3].

A rocky asteroid may have a potato-like shape, which implies that $H \approx R$. Combining this last relation with $H R \approx 10^{11}\,\mathrm{m}^2$ (as in Earth), we obtain a rough estimate of the biggest asteroid having such a potato-like shape: its linear extent is about 300 km.

13.3 Temperature of a Planet

The surface[1] temperature of a planet is determined by equating the absorbed E/M power, I_a, with the emitted one, I_e; the latter, according to (6.3), is given by $I_e = 4\pi R_P^2 \sigma T_P^4$; $\sigma \equiv \pi^2 k_B^4 / 60\hbar^3 c^2$ is the Stefan-Boltzmann constant. The power, I, emitted by the central star, equals $I = 4\pi R_S^2 \sigma T_S^4$; all this power passes through the surface of a sphere of radius R equal to the mean distance between the star and the planet. The planet intercepts a small fraction of the energy flux I equal to $\pi R_P^2 / 4\pi R^2$. Hence, the power it absorbs is $I_a = (1 - A) I \, \pi R_P^2 / 4\pi R^2 = (1 - A)(4\pi\sigma R_S^2 T_S^4)\pi R_P^2 / 4\pi R^2$, where A is the reflection coefficient of the planet's surface, the so-called albedo (for Earth $A \approx 0.3$). Substituting the above relations in the equality $I_a = I_e$ we obtain the planet's surface temperature T_P:

$$\boxed{T_P = (1 - A)^{1/4} T_S \sqrt{\frac{R_S}{2R}} = (1 - A)^{1/4} T_S \sqrt{\frac{\theta}{4}}} \tag{13.3}$$

The ratio R_S/R is equal to $\tan(\theta/2) \approx \theta/2$, if θ is measured in radian, where θ is the viewing angle of the star from the planet. For the Earth/Sun system, $\theta = 32'$ $= (32/60)(2\pi/360) = 0.0093$ rad and the Sun's surface temperature is 5800 K. Hence, the surface temperature of Earth is, according to (13.3), 256 K, very close to the accepted value of 254 K.

Actually, for Earth, about 25 % of I_e is emitted from the ground at a temperature T_g, another 25 % is emitted by the clouds at about 3 km with a temperature of T_c, and the rest 50 % is emitted from the top of the troposphere at about 10 km and with a temperature T_t. Thus

$$I_e = 4\pi R_P^2 \sigma T_P^4 = 4\pi R_P^2 \sigma \left(\frac{1}{4} T_g^4 + \frac{1}{4} T_c^4 + \frac{1}{2} T_t^4 \right) \tag{13.4}$$

By assuming mechanical equilibrium in the atmosphere and constant entropy as a function of height we find that the temperature goes down by 7–8° for each km above the ground. Thus the temperatures T_c, T_t can be expressed in terms of T_g as follows: $T_c = T_g - 25\,\text{K}$, $T_t = T_g - 65\,\text{K}$. Substituting these relations in (13.4) we find that $T_g \approx 290\,\text{K}$, $T_c \approx 265\,\text{K}$, $T_t \approx 225\,\text{K}$. The result $T_g \approx 290\,\text{K}$ is very important for the appearance and continued existence of life on Earth, since it implies that water is kept mostly in a liquid form on Earth's ground.

It is worth to point out that the flow of E/M radiation in and out of a planet implies a rate of change of the external contribution to its entropy given by $đ S_e/dt = (I_a/T_S) - (I_e/T_P)$, which is clearly negative due to the facts that $T_P \ll T_S$ and that long term steady state implies $I_a = I_e$. Consider now an initial state of the planet such that the rate of increase of the internal entropy $đ S_i/dt$ is smaller than $|đ S_e/dt|$;

[1]The surface includes also the main volume of the atmosphere, if any, of the planet.

then, the total entropy of the planet will be decreasing which implies increasing organization possibly through new structures. The latter ought to be of dynamical rather than of static nature for the following reason: These new structures have to *continuously* produce entropy at such a rate as to approach its external net outflow and to reach eventually a steady state such that $d\,S_{total}/dt = 0$. In conclusion, the in and out flow of E/M radiation in a planet leads to a net outflow of entropy; this may produce a decrease of the planet's total entropy which implies more ordering by the setting up of additional structures; if eventually a steady state is to be achieved, these additional ordered structures must be of dynamical nature as to continuously produce an internal increase of entropy up to a rate of fully compensating the external one. Such a dynamical structure of extreme complexity, extent, and rarity is possibly the living matter on Earth with its continuous evolution. We conclude that life may exist in open systems, such as planets, not in spite of, but because of the Second Law of Thermodynamics. Of course, the appearance of living matter requires many other conditions to be satisfied; very few of them are known with certainty.

13.4 Winds in the Atmospheres of Planets

Winds in the atmospheres of planets appear mainly because of the non-uniform absorption of the fraction of E/M radiation emitted by the central star. This non-uniformity is due to the rotation of the planet around its axis and the resulting day/night temperature difference, as well as to the fact that usually the equatorial region absorbs more radiation than the poles. The average of the square of the velocity of winds, υ^2, is directly connected with the average macroscopic kinetic energy, E_K, of the atmosphere. This kinetic energy depends obviously on the E/M power absorbed, I_a, since this is the only appreciable source of energy. For dimensional reasons we need a characteristic time, τ, by which to multiply I_a to get E_K:

$$E_K = \eta\, I_a \tau \tag{13.5}$$

where $E_K = \frac{1}{2} M_a \upsilon^2$, M_a is the total mass of the atmosphere, and η is a numerical factor of the order of one. What is the characteristic time τ? It turns out that it is the time required for a sound wave to go around half the circumference of the planet, i.e. the shortest time required for communication between the night region of minimum absorption to the day region of maximum absorption: $\tau = \pi R_P / c_o$; the velocity of sound, c_o, according to (12.7) and the text in Sect. 12.4, is given by $c_o = \sqrt{B/\rho_M}$, where $B \equiv -V(\partial P/\partial V)$. The derivative in the definition of B is to be taken under constant entropy, since during the sound oscillations there is no time either for exchange of heat or for appreciable entropy production. If the atmosphere is assumed to be a perfect gas obeying the equation of state $PV = N k_B T$, we have that $V(\partial P/\partial V)_T = -P$ which combined with the general relation (to be proved by the reader) $(\partial P/\partial V)_S = (C_P/C_V)(\partial P/\partial V)_T$ gives $B_S \equiv -V(\partial P/\partial V)_S = (C_P/C_V)P$. The density is $\rho_M = N A_W u/V$, where N is the total number of atoms (or molelules) within the volume V, A_W is the average atomic (or molecular) weight ($A_W = 28.96$ for Earth's atmosphere). Substituting in the expression for c_o, we have

$$c_o = \sqrt{\frac{C_P}{C_V}\frac{RT}{A_W \times 1g}} \tag{13.6}$$

where $R \equiv N_A k_B = 8.31$ J/mol K is the gas constant. Setting $\tau = \pi R_P/c_o$ in (13.5) we end up with the following rather complicated formula, which was first obtained empirically:

$$\sqrt{\upsilon^2} = a\,\sigma^{1/16} q^{7/16} c_P^{-1/4} R_P^{1/2} \mu^{-1/2} \tag{13.7}$$

where $a = \sqrt{\eta/2}$ ($\approx$0.6 for Earth), $\sigma \equiv \pi^2 k_B^4/60\hbar^3 c^2$, $q = (1 - \frac{A}{4})(I_S/4\pi R^2)$, R is the distance planet/star, $c_P = C_P/M$ is the specific heat per unit of mass of the atmosphere under constant pressure, and $\mu = M_a/4\pi R_P^2$.

13.5 Pressure and Other Quantities Within a Planet

The dependence of the mass density ρ_M of a planet as a function of the distance r from its center will be approximated by a linear relation of the form of (13.8) (see also Fig. 13.1):

$$\rho_M = \rho_o - a\,r, \quad a = (\rho_o - \rho)/R_P, \quad 0 \leq r \leq R_P \tag{13.8}$$

where ρ_o is the mass density at the center of the planet, $r = 0$, and ρ is that at its surface, $r = R_P$. This dependence on the distance r is due to the long range of the gravitational interaction which is responsible for the increasing pressure as we move from the surface towards the center of the planet (see Fig. 13.1). Then the mean density is $\bar{\rho} = \rho_o - 3\,a R_P/4 = \frac{1}{4}\rho_o + \frac{3}{4}\rho$. The acceleration of gravity at r is equal to $g(r) = -G M(r)/r^2$, where $M(r)$ is the mass within a sphere of radius r. The final result is

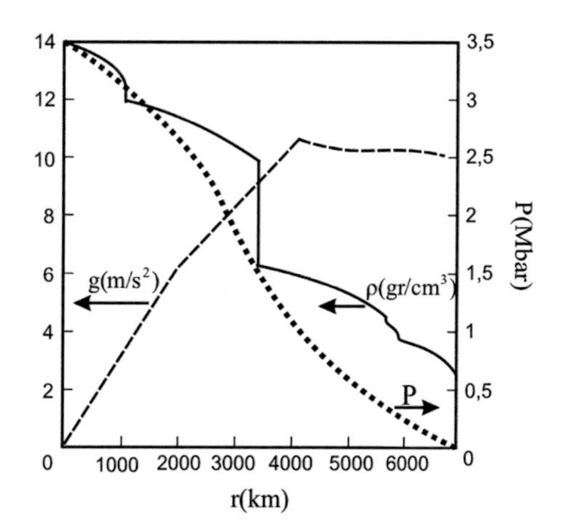

Fig. 13.1 The dependence on the distance r from the center of the Earth of the mass density ρ and the acceleration of gravity g (*left vertical axis*) and the pressure P (*right vertical axis*)

$$g(r) = -\frac{4\pi}{3} G r \left(\rho_o - \frac{3}{4} (\rho_o - \rho) \frac{r}{R_P} \right), \quad 0 \le r \le R_P \tag{13.9}$$

By integrating $g(r)$ over r and determining the integration constant from the value at $r = R_P$ we obtain the potential $\mathcal{V}(r)$:

$$\mathcal{V}(r) = \tfrac{2\pi}{3} G(\rho_o r^2 - \tfrac{1}{2} a\, r^3) + C, \quad C = -\frac{3}{2}\frac{GM}{R_P} + \tfrac{\pi}{3} G\, a\, R_P^3 \tag{13.10}$$

where $M \equiv M(R_P)$. By integrating $\frac{1}{2}\mathcal{V}(r)\rho(r)d^3r$ we can calculate the gravitational self-energy U_G (the factor $\frac{1}{2}$ is included to avoid double counting, since the potential $\mathcal{V}(r)$ is not an external one, but one produced by the mass of the planet):

$$U_G = -\frac{3}{5}\frac{GM^2}{R_P}\left(1 + \frac{5}{48}\frac{\rho_o - \rho}{\rho_o} - \frac{5}{56}\left(\frac{\rho_o - \rho}{\rho_o} \right)^2 \right) \tag{13.11}$$

Finally, the pressure at r can be obtained as the weight per unit cross-section of a *cylindrical* column along the radial direction extending from r to R_P:

$$P(r) = \int_r^{R_P} g(r')\rho_M(r')dr' \tag{13.12}$$

The pressure at the center is

$$P(0) = \frac{F}{2}\frac{GM}{R_P^2}\bar{\rho}\, R_P = \frac{F}{2} g(R_P)\,\bar{\rho}\, R_P \tag{13.13}$$

where $F = (\rho_o/\bar{\rho})^2 - \frac{7}{6}[\rho_o(\rho_o - \rho)/\bar{\rho}^2] + \frac{3}{8}[(\rho_o - \rho)^2/\bar{\rho}^2]$.

For Earth we have that $\bar{\rho} = 5.515\,\mathrm{g/cm^3}$, $R_P = 6371\,\mathrm{km}$; we choose approximately $\rho_o = 14\,\mathrm{g/cm^3}$ and $\rho = 2.69\,\mathrm{g/cm^3}$ in order to reproduce correctly the average density and to reasonably represent the actual complicated $\rho(r)$. We find for the pressure at the center of the Earth $P(0) = 3.44$ Mbar versus 3.51 Mbar for the actual value.

13.6 Summary of Important Relations

It is (13.3) which gives the average temperature on the surface of a planet (or a moon) including its atmosphere, if any. Equally important is the temperature on the ground of a rocky planet (or moon) because it determines the phase of water (if water is present). Finally, because of the need to compensate the net outgoing entropy flow from a planet, dynamical structures ought to appear of high entropy production (a possible extreme example is living matter).

13.7 Multiple-Choice Questions/Statements

1. The total gravitational self-energy $E_G \equiv U_G$ of a planet or of an active star of mass M and density equal to that of ordinary solid state is of the form:
(a) $E_G = b_G M^{5/3}$ (b) $E_G = b_G M$ (c) $E_G = b_G M^{1/2}$ (d) $E_G = b_G M^{1/3}$

2. The total electrostatic self-energy $E_E \equiv U_C$ of a planet or an active star of mass M is of the form:
(a) $E_E = b_E M^{5/3}$, (b) $E_E = b_E M$, (c) $E_E = b_E M^{1/2}$, (d) $E_E = b_E M^{1/3}$

3. The total gravitational self-energy E_G of a planet or of an active star of mass M and density equal to that of ordinary solid state is of the form $E_G = b M^{5/3}$, where:
(a) $b = -\gamma G m_a^{1/3} / \bar{r}$, (b) $b = -\gamma G m_a^{1/3} / \bar{r} a_B$,
(c) $b = -\gamma G m_a^{1/2} / \bar{r} a_B$, (d) $b = -\gamma G m_a / \bar{r} a_B$
where γ is larger than 0,6 και $\bar{r} a_B$ is the radius per atom.

4. The total electrostatic self-energy E_E of a planet or an active star of mass M is of the form $E_E = b M$, where:
(a) $b = -\gamma''(e^2 / r_a)$, (b) $b = -\gamma''(e^2 / r_a u)$
(c) $b = -\gamma''(e^2 / r_a m_a)$, (d) $b = -\gamma''(e^2 / r_a A_B)$

5. The total electrostatic self-energy E_E of a planet of mass M is about equal to the total gravitational self-energy E_G when (omitting numerical factors)
(a) $M = (e^2 / G m_a^2)^{3/2}$, (b) $M = (e^2 / G m_a)^{3/2} m_a$
(c) $M = (e^2 / G m_a^2)^{3/2} m_a$, (d) $M = (e^2 / G m_a^2) m_a$

6. The approximate formula which gives the mass M of a planet such that the total gravitational and the total electrostatic self-energies are equal is the following:
(a) $M = (3.8 / A_B^2)(e^2 / G u^2)^{3/2} u$ (b) $M = 3.8(e^2 / G u^2)^{3/2} u$
(c) $M = (3.8 / A_B^2)(e^2 / G u)^{3/2} u$ (d) $M = (3.8 / A_B^2)(e^2 / G u^2) u$

7. The approximate formula which gives the maximum possible height H of a mountain in a rocky planet is:
(a) $(H R / a_B^2) \approx (0.02\bar{r} / A_B^2)(\hbar c / G u^2)$ (b) $(H R / a_B^2) \approx 0.02\bar{r}(\hbar c / G u^2)$
(c) $(H R / a_B^2) \approx (0.02\bar{r} / A_B^2)(e^2 / G u^2)$ (d) $(H R / a_B^2) \approx 0.02\bar{r}(e^2 / G u^2)$

8. For the same reasons that there is an upper limit in the height of a mountain in a rocky planet, there is a limit in how deep can one drill a well:
(a) 5 km (b) 15 km (c) 50 km (d) 150 km
In principle the deepest drilling will occur on the land or underwater?
(a) on the land (b) underwater

9. Equating the energy absorbed with this emitted by a planet we determine the average surface temperature T_p of the planet in terms of the surface temperature of the central star T_s and the angle θ (in rad) that the star is viewed from the planet. The results is:
(a) $T_p = (1 - A)^{1/4} \theta T_s / 4$ (b) $T_p = (1 - A)^{1/4} \sqrt{\theta T_s / 4}$
(c) $T_p = (1 - A)^{1/4} \sqrt{\theta / 4}\, T_s$ (d) $T_p = (1 - A)^{1/4} \theta T_s$

10. The energy emitted by Earth is attributed to the ground (25%), the clouds at average height of 3 km (25%), and from the top of the troposphere at height of about 10 km (50%). The average temperature of the ground is 290 K and keeps decreasing at a rate of 7 K per km. The average temperatures of the surface of the Earth including the atmosphere is:
(a) 249.75 (b) 255.3 (c) 220 (d) 290

11. The average speed of winds in a planet of radius R and total atmospheric mass M_a is given by the formula:

$$\sqrt{<v^2>} = a\sigma^{1/16} q^{1/16} c_p^{-1/4} R^{1/2} \mu^{-1/2}$$

σ is the Stefan-Boltzmann constant, q is the power per unit area absorbed by the planet, $c_p = C_p / M_a$, $\mu = M_a / 4\pi R^2$. The above formula can be derived by equating the average macroscopic kinetic energy of the atmosphere with:
(a) the kinetic energy due to the rotation of the planet around its axis,
(b) the internal energy of the atmosphere of the planet
(c) the product of the solar power absorbed by the planet times the time needed for sound to go around the half circumference of the planet
(d) the product of the solar power absorbed by the planet times the half period of the rotation of the planet around its axis

12. The acceleration of gravity at a distance r from the center of a planet of radius R and total mass M and constant density ρ is given by (for $r < R$):
(a) $g(r) = GM\,r / R^3$ (b) $g(r) = GM / r$
(c) $g(r) = GM\,R / r^3$ (d) $g(r) = GM / r^2$

13. The pressure at a distance r from the center of a planet of radius R and mass M, constant density ρ and acceleration of gravity at its surface g is given by (in your calculation you may omit the atmospheric pressure):
(a) $P = 0.5g\,\rho\,r$ (b) $P = 0.5g\,\rho(R - r)$
(c) $P = 0.5g\,\rho\,R$ (d) $P = 0.5g\,\rho(R^2 - r^2)/R$

13.8 Solved Problems

1. *The period of the Moon in its motion around the Earth is* 27.32 *days, while the time between two successive full moons is* 29.5 *days. How do you explain this difference? Does your explanation survive a quantitative test?*

Solution Full moon implies that Sun, Earth, and Moon (in that order) must be on almost the same straight line. During the time of 29.5 days, Earth traces out an angle $\varphi = 2\pi(29.5/365.25)$; during the same time the moon must trace out a full circle plus φ in order to be found again on the same straight line Sun/Earth. Thus the ratio of the two angles $(2\pi + \varphi)/2\pi = 1 + (29.5/365.25) \approx 1.08$ ought to be equal to the ratio of the corresponding two times 29.5/27.32 which is actually 1.08; this is the quantitative test which was obviously passed in flying colors.

2. *The period of Mars in its orbit around the Sun lasts* 1.88 *times more than that of Earth. The average surface temperature of Earth is* 254 *K. Estimate the average surface temperature of Mars.*

Solution We assume that the main factor on which this temperature depends is the distance from the Sun. Taking into account that

$$\left(\upsilon^2/R\right) = \left(GM_S/R^2\right), \quad \upsilon = 2\pi R/t \;\Rightarrow t \propto R^{3/2}$$

where t and R are the period and the radius of the orbit of a planet. Since the surface temperature of a planet is proportional to $R^{-1/2} \propto t^{-1/3}$, we have

$$(T_M/T_E) = (t_E/t_M)^{1/3} = 0.81 \;\Rightarrow T_M = 0.81 \times 254 = 206\,\mathrm{K}$$

3. *Tidal phenomena have a period of about 12 h. Explain why. When are these phenomena more intense? At full moon, new moon, or in between?*

Solution To be specific let us consider the case of Earth, where the tidal effects are due to both the Moon and the Sun. Let us consider first the effect of Sun. If the distance from the center of Sun to the center of Earth is L, then that of the surface of Earth facing the Sun is $L - R$, while the opposite site is $L + R$, where R is the radius of Earth. Thus the force exercised by the Sun on a mass m located on the nearest site is *larger* than that on the same mass located at Earth's center by an amount equal to

$$\delta F = \frac{GM_S m}{(L-R)^2} - \frac{GM_S m}{L^2} \approx \frac{2GM_S mR}{L^3}$$

As a result of this extra force the mass on the nearest site to the Sun will tend to follow a trajectory closer to the Sun than that of the center of Earth and therefore will tend to be raised towards the Sun.

The same mass located on the most remote from the Sun site will experienced a *smaller* force than that of the center by about the same amount δF. As a result of this reduced force its trajectory will tend to be more remote from the Sun than that of Earth's center and therefore will tend to be lifted too. Since these two points will interchange positions relative to the Sun every 12 h due to Earth's rotation the period of the tidal effects will be 12 h. The ratio of the tidal forces on Earth due to the Sun and the Moon is

$$\left(M_S/L^3\right)/\left(M_m/l^3\right) = M_S l^3/M_m L^3 = \left(M_S/M_m\right) \times \left(l^3/L^3\right)$$

$$= \left[(1.9891/7.35) \times 10^8\right] \times \left[(3.83/1.496)^3 \times 10^{-9}\right] \approx 0.45$$

Thus the tidal effects on Earth due to the Moon are 2.2 times stronger than that due to the Sun. When the Moon is on the straight line defined by Earth and the Sun (i.e. when we have full or new Moon) the tidal forces due to the Moon and the Sun are parallel to each other, they reinforce each other, and the total tidal forces are maximal. When the Moon is at first or third quarter, so that the angle between the Moon, the Earth, and the Sun is 90°, the places on Earth subject to the maximal tidal forces from the Moon experiences no tidal force from the Sun.

13.9 Unsolved Problem

1. Why is the Earth slightly compressed at the poles and bulged at the equator. Estimate how much larger its radius at the equator is relative to its average value.

References

1. E. Cohen, D. Lide, G. Trigg (eds.), *AIP Physics Desk Reference* (Springer-Verlag, NY, 2003)
2. F. Shu, *The Physical Universe: An Introduction to Astronomy* (University Science Books, Mill Valley, 1982)
3. R. Eisberg, R. Resnick, *Quantum Physics*, 2nd edn. (Wiley, NY, 1985)

Chapter 14
Stars, Dead or Alive

Abstract In active stars the gravitational compressive pressure is balanced by the pressure due to the thermal motion of particles and to that of photons. This balance is maintained as long as the fusion nuclear reactions in the interior of the star take place. When the nuclear fuel is exhausted the mighty gravity may compress the star even to a black hole, unless the pressure of the non-relativistic quantum kinetic energy of either the electrons in the case of white dwarfs or the neutrons in the case of neutron stars (often referred as pulsars) is enough to balance the gravitational pressure.

14.1 Introduction

In this section we present the processes that lead to the formation of an active star, and how these depend on its mass. We also examine what physical processes determine the lower limit of the initial mass of a star (about 0.08 the present mass of the Sun) and the upper limit (about 200 times the mass of the Sun). An active star is a stellar object whose nearly steady state is maintained by hydrostatic equilibrium between the gravitational pressure on the one hand, and the radiation pressure and the pressure due to thermal kinetic energy on the other; the latter pressures are fed by the fusion nuclear reactions taking place within the central region of the star. When the nuclear fuel is exhausted, a relatively short period of instability follows before the star transforms into a stellar remnant, often called a dead star. During its active time and especially during the instability period, the star is losing mass so that its mass M_{f} at the final stage is quite lower than the initial mass M_i.

There are three kinds of dead stars depending on what their initial mass M_i was at its creation: If M_i was lower than about eight times the present mass, M_S, of the Sun, the star ends its life as a *white dwarf* of M_{f} less than $1.4M_S$ and of size comparable to that of the Earth. The white dwarf is at a state of true thermodynamic equilibrium where the gravitational squeeze is balanced almost exclusively by the non-relativistic quantum kinetic energy of the electrons. The second kind of dead

E.N. Economou, *From Quarks to the Universe*,
DOI 10.1007/978-3-319-20654-7_14

star occurs if the initial mass M_i was in the range between 8 and 20 times the mass of the Sun; then the instability phase involves the collapse of its core to a super dense state and the subsequent explosive rejection of the rest of its mass (a phenomenon known as supernova explosion). The remaining core part has a mass M_f in the range between $1.4M_S$ and about $3M_S$ and a size of about 20 km in diameter! It consists mainly of neutrons (this is the reason is called *neutron star*). Pulsars, being rotating neutron stars, present the most extreme known state of matter and apparently the most effective in resisting the gravitational squeeze by the non-relativistic quantum kinetic energy of neutrons (and to a much smaller degree by that of protons and electrons-the latter being in the extreme relativistic state). However, if the initial mass was more than about $20M_S$ and the final mass M_f exceeds a critical value of about $3M_S$, even this extreme state of matter is unable to withstand gravity. In this case the neutron star collapses to the third kind of dead star which is known as a *black hole*.

The process of star formation occurs in gaseous interstellar media, more specifically in regions of increased density called *molecular clouds*. The largest among them, called giant molecular clouds, have concentrations of about 100 particles per cm^3 (vs. 0.1–1 particles per cm^3 for the average interstellar medium), linear extent of about 100 light years (see Part V, Table V.2), and total mass of about 6×10^6 solar masses. For a number of reasons, material consisting mainly of hydrogen molecules and helium atoms starts to coalesce towards some points within the cloud, thus setting up in motion the process of the formation of several stars. The resulting higher local density around each of these points attracts more material which arrive with ever increasing kinetic energy as a result of the growing attractive gravitational force. Collisions among the coalescing particles tends to restore a temporary equilibrium which keeps evolving towards higher temperatures and concentrations as the gravity continues to bring in more particles of higher kinetic energy. Thus a gaseous spherical object of radius R, temperature T, and mass M is formed. If the mass M, reached when the surrounding material is exhausted, has a value M_i exceeding a critical value M_{i1}, the temperature at its center is high enough to ignite fusion nuclear reactions, and to turn the object into an active star. On the other hand, as we shall show in Sect. 14.6, if the mass is too high, the radiation pressure will overcome the gravitational attraction and will not allow a steady state star to be formed.

In an active star the fusion reactions taking place at its central region supply enough energy to compensate the emitted energy (mainly in the form of E/M radiation) and to maintain the kinetic and radiation pressure required to counterbalance the gravitational one. Details regarding star formation and their dynamical equilibrium maintained between the fusion nuclear reactions and the gravitational squeeze can be found in several excellent textbooks dealing with the fascinating and growing subject of astrophysics, see for example [1, 2].

When the nuclear fuel within the star is exhausted, the star enters into a relatively short period of instability during which it loses an appreciable percentage of its mass, either through the so-called stellar wind,[1] if the mass is small, as in the case of our Sun, or, mainly, through an explosive process known as supernova, if the initial mass is large (between, roughly, 8–20 solar masses). The remaining mass M_{f}, after the end of the instability period, has at its disposal only the quantum kinetic energy to counterbalance the gravitational squeezing pressure.[2] As it will become evident below, the first stage of halting the gravitational collapse is through the quantum kinetic energy of the electrons, which have been fully detached from the atoms. This state of matter resembles the metallic one examined in Sect. 12.7. Indeed, in metals, the quantum kinetic energy of the detached (valence only) electrons balance the Coulomb energy. In the present case, more detached electrons are needed (usually all of them) to balance the much stronger gravitational energy. Thus, the first resistance front against the gravitational collapse of a dead star involves a fully ionized neutral spherical body, called white dwarf, consisting of electrons and naked nuclei. This kind of structure is effective up to a mass M_{f1}. If the mass[3] M_{f} exceeds M_{f1}, the gravitational pressure will squeeze the dead star, until the second and probably the last resistance front is formed. This second front involves a structure called neutron star or *pulsar* in which all kind of particles participate: neutrons, protons, and electrons, although neutrons provide the main contribution. Pulsars are successful in halting the gravitational collapse up to a mass M_{f2}; beyond this limiting value, no structure seems to exist capable of stopping the gravitational squeeze, which continues without limit giving rise to the formation of a black hole.

14.2 White Dwarfs

Let us assume that the quantum kinetic energy of the electrons within a white dwarf has not approached the relativistic limit. Then, according to (3.22b), the total electronic kinetic energy is

$$E_K = 1.105\,\frac{\hbar^2 N_e^{5/3}}{m_e R^2} \tag{14.1}$$

(see (3.22b)), where R is the radius of the white dwarf. The numerical factor 1.105 is valid for a system of uniform density. For a white dwarf one can calculate the density as a function of the distance r from its center [3] and as a result will find that 1.105 is replaced by a larger constant denoted here by a_K. The gravitational energy E_G is given by (V.2) and (V.3) in Part V, with the constant $3x/5$ denoted by a_G.

[1] Stellar winds, powered by radiation pressure, occur during the whole life of an active star although at a low rate; this rate is greatly accelerated during the period of instability.

[2] It is assumed here that the mass M_{f} is large enough, e.g., larger than $10^{55}u$, so that the gravitational energy dominates over the Coulomb energy (see V.7 in Part V).

[3] It is not unusual for a dead star to suck up mass from a companion star so that its actual mass M_{f} to be larger than its remaining mass just after the end of its active cycle.

Minimizing the sum $E_K + E_G$ with respect to the radius R (in the present case this minimization is equivalent to $E_K = \frac{1}{2}|E_G|$ which is the virial theorem) we find

$$R = A\frac{\hbar^2}{Gu^2 m_e}\frac{1}{N_v^{1/3}}, \qquad A = \frac{2a_K}{a_G}\left(\frac{N_e}{N_v}\right)^{5/3} \tag{14.2}$$

The ratio $2a_K/a_G$ is equal to 4.51 instead of the value 2.21/0.6 = 3.68 for a uniform system. The ratio N_e/N_v is just a little lower than 0.5, if we assume that all the hydrogen has been burnt during the active life of the star; we shall choose the value 0.5 for this ratio. The universal constant $e^2/Gu^2 = \hbar^2/(m_e a_B G u^2)$ is equal to 1.2536×10^{36}. Taking for the number of nucleons in the white dwarf the typical value of about half the present number of nucleons in the Sun, $\frac{1}{2}N_{v,S} = 0.5986 \times 10^{57}$, we find that $R = 1.11 \times 10^7$ m, i.e., about 75 % larger than the radius of the Earth.

The kinetic energy per electron $\varepsilon_k \equiv E_K/N_e$, taking into account (14.1) and (14.2), is proportional to $N_e^{4/3} \propto N_v^{4/3}$. Hence, for large enough N_v the electron velocities will be relativistic and will eventually reach the extreme relativistic limit for which (3.23) is valid. We have then for the total quantum kinetic energy $E_K = \tilde{a}_K \hbar c N_e^{4/3}/R$, where the constant $\tilde{a}_K$ is 1.44 for a uniform system and higher for the actual density distribution in a relativistic white dwarf.

Notice that, in the extreme relativistic limit, the total electronic kinetic energy becomes A/R, i.e., of the same R dependence as the gravitational energy which, of course, is of the form $-B/R$. Hence, if $B < A$, the minimization implies that R will increase until we return to the non-relativistic limit, while, if $B > A$, R will decrease until the neutron star structure is reached. Therefore, the critical value M_{f1} for the collapse of the white dwarf is determined by the relation $A = B$, which gives

$$\boxed{N_{vf1} = A\left(\frac{\hbar c}{Gu^2}\right)^{3/2}, \qquad A = \left(\frac{\tilde{a}_K}{\tilde{a}_G}\right)^{3/2}\left(\frac{N_e}{N_v}\right)^2 \approx 0.77} \tag{14.3}$$

The numerical value of the ratio $\hbar c/G u^2$ of universal constants is 1.718×10^{38}. The ratio $\tilde{a}_K/\tilde{a}_G$ is actually 2.126 instead of 1.44/0.6 = 2.4 for a uniform system. Finally, for the ratio of the number of electrons over the number of nucleons we take the same number as before, 0.5. Replacing these numbers in (14.3) we find $N_{vf1} = 1.7179 \times 10^{57}$ or $N_{vf1}/N_{v,S} = 1.43$. The conclusion is that no white dwarf with mass greater than about 1.4 times the mass of the Sun can exist. This agrees with all the observational data and it is known as the *Chandrasekhar limit*. Dead stars with mass $M_f > M_{f1} = 1.4M_S$ will be either neutron stars or black holes, although the hypothetical existence of a *quark star* has been examined theoretically as a state of matter after the gravitational collapse of a neutron star and before the ultimate collapse to a black hole.

14.3 Neutron Stars or Pulsars (≡Rotating Neutron Stars)

The analysis of the previous section implies that for $M_{\mathrm{f}} > M_{\mathrm{f}1}$ the quantum kinetic energy of the electrons cannot stabilize the structure no matter how small the radius R may become. Thus the only possibility for equilibrium is through the contribution of the non-relativistic kinetic energy of either the nuclei or the nucleons, depending on the equilibrium value of R. If R becomes so small that nuclei will be squeezed partly inside each other, their individuality will be lost and the whole star will become a super huge nucleus held together by the mighty gravitational pressure. To see if this is the case we shall start with the opposite hypothesis, that the nuclei, by retaining their individual existence, produce an equilibrium value of R which does not imply squeezing the nuclei inside each other. Under this hypothesis the non-relativistic quantum kinetic energy of the nuclei must balance a fraction of the gravitational energy, for example half of it. The result for R would be as in (14.2) with m_e replaced by $m_a = A_W m_\nu$ and N_e replaced by $N_a = N_\nu / A_W$; this means that (14.2) will be multiplied by $(m_e/m_a) = (m_e/uA_W)$ and divided by $A_W^{5/3}$. This implies a reduction of (14.2) by a factor of the order of 10^{-6}, which means a radius R of about 10 meters, and a nearest neighbor distance among nuclei of the order of 10^{-2} fm! This is obviously inconsistent with the hypothesis that individual nuclei will be able to stabilize the dead star; hence, stabilization will be achieved, if at all, by the nuclei being dissolved into a closely packed "soup" of nucleons and electrons. Moreover, the electrons will react with protons producing neutrons and vice versa.[4] When this reaction reaches equilibrium, the Gibbs free energy becomes minimum which implies the following relation among the chemical potentials of neutrons, protons, and electrons: $\mu_n = \mu_p + \mu_e$. Using this relation in combination with the equality $E_K = -\frac{1}{2}E_G$, stemming from the minimization of total energy, and taking into account that $\mu_i = \partial E_{Ki}/\partial N_i,\ i = n, p, e$, we find that the numbers of neutrons, protons, and electrons at equilibrium can be expressed in terms of the number of nucleons as follows: $N_n' = xN_\nu$, $N_p' = N_e' = yN_\nu$, where $x \approx 0.934$ and $y \approx 0.066$. In obtaining these relations we assumed that the neutrons and protons are non-relativistic, in contrast to the electrons which are extreme relativistic, and we also took into account that before the reaction $p + e \rightleftarrows n$ took place almost all hydrogen has been exhausted and consequently $N_e/N_\nu \approx 0.5$. The above relations allowed us to determine also that the relative contribution of the protons to the kinetic energy of neutrons is equal to $(y/x)^{5/3} = (0.066/0.934)^{5/3} = 1.2\ \%$, The kinetic energy of the electrons (which is inversely proportional to R) is no more

[4]The neutrinos and antineutrinos produced in these reactions will not be retained within the system.

than 6.8 % of the absolute value of the gravitational energy. Thus, as long as the kinetic energy of neutrons is non-relativistic, we have the relation

$$a_K \frac{\hbar^2 (xN_\nu)^{5/3}}{m_n R^2}(1+0.012) = \tfrac{1}{2} a_G \frac{G m_n^2 N_\nu^2}{R}(1-0.068) \tag{14.4}$$

from which it follows that the radius of a neutron star is

$$R = A' \frac{\hbar^2}{G m_n^3} \frac{1}{N_\nu^{1/3}}, \qquad A' = 1.086 x^{5/3} \frac{2a_K}{a_G} = \frac{2.02 a_K}{a_G} \tag{14.5}$$

which is of the same form as that of (14.2) with the $(N_e/N_\nu)^{5/3} = 0.315$ replaced by the ratio $(m_e/m_n) = 1/1839$. This is the factor which accounts for the dramatic reduction in the radius of the neutron star in comparison with the radius of the white dwarf. For $N_\nu = 1.4 N_S$ we obtain $R = 13.7$ km.

Recently (2010) the mass, $(1.58 \pm 0.06)M_S$, and the radius, 9.11 ± 0.40 km, of a neutron star in the binary system 4U1820-30 was estimated. Equation (14.5) yields for this case 13.16 km. This rather small discrepancy can be attributed to our use for the ratio $2a_K/a_G$ the value 4.51 calculated for white dwarfs. To reproduce the observational data we must chose for the ratio $2a_K/a_G$ in a neutron star the value 3.12, so that the numerical factor A' in (14.5) to be 3.16.

As in the white dwarf case, the kinetic energy per nucleon in neutron stars is proportional to $N_\nu^{4/3}$. Thus for very large mass even the neutrons and the protons will reach the extreme relativistic limit, which means that the total kinetic energy will be of the form A/R with $A = \tilde{a}_K \hbar c N_\nu^{4/3}(x'^{4/3} + 2y'^{4/3})$, where $N'_n = x' N_\nu$ and $N'_p = N'_e = y' N_\nu$, with x' and y' being different in the extreme relativistic case from the values obtained before for the non-relativistic case. The reason is that in this extreme relativistic limit the equilibrium equation for the reaction $p + e \rightleftarrows n$, $\mu_n = \mu_p + \mu_e$, becomes $N_n'^{1/3} = N_p'^{1/3} + N_e'^{1/3} = 2N_p'^{1/3}$ which implies that the ratios of neutrons to protons and to electrons will be as follows: $N'_n{:}N'_p{:}N'_e = 8{:}1{:}1$; since $N'_n + N'_p = N_\nu$, we find that $x' = 0.8888$ and $y' = 0.1111$. Then the coefficient A becomes $A = 0.9615\, \tilde{a}_K \hbar c N_\nu^{4/3}$. Equating this with $\tilde{a}_G G N_\nu^2 m_n^2$ we obtain

$$\boxed{N_{\nu f2} = \tilde{A}\left(\frac{\hbar c}{G m_n^2}\right)^{3/2}, \qquad \tilde{A} = \left(\frac{0.9615\, \tilde{a}_K}{\tilde{a}_G}\right)^{3/2} \approx 1.84} \tag{14.6}$$

which for $\tilde{a}_K/\tilde{a}_G = 1.56$ gives $N_{\nu f2} = 3.38 N_{\nu,S}$. The actual critical value seems to be around $3N_{\nu,S}$. Another estimate of this critical value can be obtained by comparing the radius R as given by (14.5) with the Schwarzchild radius of a black hole (see next section, (14.8)).

14.4 Black Holes

As we did in the case of white dwarfs and neutron stars, we shall restrict ourselves to the simpler case of an electrically neutral, *non-rotating* black hole. Such a black hole is accompanied by a spherical surface, called the *event horizon*. Physically, the event horizon is the perfect one-way surface in the sense that, while it permits all kind of particles to cross it from its exterior, no particle, not even a photon, is allowed to escape from its interior. The spherical event horizon is determined by its radius r_S, called the *Schwarzchild* radius. The Schwarzchild radius can be determined through dimensional analysis by taking into account that it depends on the mass M of the black hole (as the only parameter characterizing the black hole), on the gravitational constant G (since we are considering a gravitational field), and on the velocity of light in vacuum c (since a black hole is an extreme relativistic phenomenon). Actually the quantities G, M must enter as the product GM, because it is exactly this product which determines the gravitational field. Thus

$$\boxed{r_S = \eta \frac{GM}{c^2}} \tag{14.7}$$

General relativity shows that $\eta = 2$; accidentally, the classical calculation of the escape velocity, $\frac{1}{2}\upsilon^2 - (GM/r_S) = 0$, from the event horizon of a particle having velocity $\upsilon = c$ gives also $\eta = 2$.

By equating the Schwarzchild radius with the radius of a neutron star as given by (14.5) with $A' = 3.16$, we obtain another estimate of the critical mass for the collapse to a black hole. The result is

$$\boxed{N_{\mathrm{vf2}} = \left(\frac{3.16}{2}\right)^{3/4} \left(\frac{\hbar c}{G m_n^2}\right)^{3/2} = 2.59\, N_S} \tag{14.8}$$

Although nothing can escape from the interior of a black hole, nevertheless, it is believed that a black hole can emit radiation, in particular E/M radiation, from a region just outside the event horizon by a quantum process proposed by Hawking. The process is based on the quantum nature of vacuum according to which particle/antiparticle pairs appear and disappear all the time existing for a very short interval of time $\Delta t \approx \hbar/\Delta\varepsilon$, where $\Delta\varepsilon$ is the total energy of the pair. Photon pairs having zero rest mass are the easiest to appear/disappear. In the region just outside the horizon of the black hole the gravitational field is so strong that one partner of the quantum fluctuation pair may be sucked in by the black hole leaving the other partner unpaired and, hence, transforming it from a virtual to a real particle. The energy during this process is conserved, since the energy of the created real particle comes at the expense of the total energy Mc^2 of the black hole. A detailed calculation of this process shows that the frequency distribution of the emitted E/M radiation by the black hole is that of a black body with an effective temperature T. Thus a temperature and, therefore, an entropy is attributed to a black hole. For these quantities, besides the product GM and the velocity of light c, the Planck constant $\hbar$ must enter in their formulae, since T and S clearly owe their existence to a quantum mechanism. Thus the temperature of a black hole, physically appearing in the combination $k_B T$, must depend on the gravitational field strength and, hence, on the product GM, on $\hbar$ (as a quantum phenomenon) and on c (as a relativistic phenomenon). The only combination of the last three quantities having dimensions of energy is $\hbar c^3/GM = 2\hbar c/r_S$. Thus

$$k_B T = \eta' \frac{\hbar c}{r_S} \tag{14.9}$$

It turns out that $\eta' = 1/4\pi$, where the factor 1/4 is coming by averaging $\cos\theta$ over half the solid angle. The $\cos\theta$ factor appears, because the emitted radiation is proportional to the *outgoing* normal component of the flux. The entropy of a black hole can be obtained from the first law of thermodynamics, $TdS = dU = c^2 dM$, by integrating the relation $dS = c^2 dM/T \propto MdM$). We find

$$\frac{S}{k_B} = 4\pi \frac{GM^2}{\hbar c} = \pi \frac{c^3 r_s^2}{\hbar G} \tag{14.10}$$

Notice that the entropy is proportional to the area of the event horizon and not to the volume enclosed by it as in ordinary thermodynamic systems (for comments on the entropy of black holes see [4]). Finally, the lifetime of a black hole of initial ($t = 0$) mass M_o can be obtained from the equation $c^2 dM/dt = -4\pi r_S^2 \sigma T^4$, where σ, is the Stefan-Boltzmann constant, (see Sect. 13.3). Substituting from (14.7) and (14.9) we find that

$$M^3(t) = M_o^3 - a\frac{\hbar c^4}{G^2} t, \qquad a \equiv \frac{1}{5120\,\pi} \tag{14.11}$$

from which the lifetime follows immediately by setting $M(t) = 0$.

Notice that, as we mentioned before, the three universal constants $\hbar$, c, and G define a system of units, the so-called Planck's system with length unit $\ell_P = \sqrt{G\hbar/c^3} = 1.6161 \times 10^{-35}$ m, time unit $t_P = \sqrt{G\hbar/c^5} = 5.3907 \times 10^{-44}$ s and mass unit given by $m_P = \sqrt{\hbar c/G} = 2.1766 \times 10^{-8}$ kg. The energy unit is $\varepsilon_P = m_P c^2 = 1.2234 \times 10^{28}$ eV$=1.9601 \times 10^9$ J and the temperature unit is $T_P = 1.4196 \times 10^{32}$. In Planck's units formulae (14.7)–(14.11) become simpler and easier to use. In particular, the entropy becomes $S/k_B = \frac{1}{4}\tilde{A},\ \tilde{A} = 4\pi\tilde{r}_S^2,\ \tilde{r}_S = r_S/\ell_P$.

14.5 The Minimum Mass of an Active Star

In Fig. 14.1 we plot the total and the kinetic energy as well as the temperature as a function of the radius R of a gaseous body in its way to become a star. Whether this will happen depends on the maximum temperature T_M being larger than the ignition temperature T_i for the $p+p \rightarrow (p,n)+e^+ + \nu_e$ fusion reaction to start taking place. Notice that the existence of the maximum temperature T_M stems from the quantum kinetic energy. Classically, the temperature will be proportional to the kinetic energy according to the formula $E_{Kc} = \frac{3}{2}N\,k_B T$ and will exhibit no maximum, while for small R and low T quantum mechanics gives $E_{Kq} = a_K\,\hbar^2 N_e^{5/3}/m_e R^2$. Interpolating between these two limits we obtain a simple approximate formula for the exact quantum expression $E_K = \mathrm{f}(R,T)$:

$$E_K \approx \left(E_{Kc}^2 + E_{Kq}^2\right)^{1/2} \tag{14.12}$$

The virial theorem implies that $E_K = -\frac{1}{2}E_G$ where $-E_G = a_G\,GM^2/R = a_G\,Gu^2 N_\nu^2/R$. Assuming full ionization of all the atoms in the star, we have $N = N_e + N_a = x\,N_\nu$ and $N_e = y\,N_\nu$, where a reasonable estimate of the composition gives $x \approx 1.5$ and $y \approx 0.8$. By combining (14.12) and $E_K = -\frac{1}{2}E_G$ we obtain $k_B T$ as a function of R. By differentiating this function with respect to R we find an approximate expression for $k_B T_M$:

$$k_B T_M \approx \gamma \left(\frac{Gu^2}{e^2}\right)^2 \frac{e^2}{a_B} N_\nu^{4/3}, \qquad \gamma = \frac{a_G^2}{12\,a_K\,x\,y^{5/3}} \tag{14.13}$$

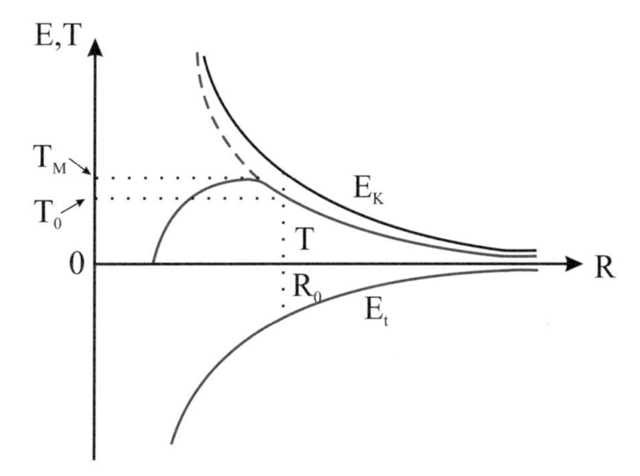

Fig. 14.1 The total energy E_t, the kinetic energy E_K, and the temperature T (*solid line*) as a function of the radius R of a star to be. The *dash-line* gives the classical result for the temperature which coincides with the quantum one for large values of R. T_o is the equilibrium average value of the temperature and R_o is that of the radius when a steady state star is formed. T_M is the maximum temperature the system would achieve, if nuclear fusion were absent

By choosing $a_K/a_G \approx 2,\ a_G \approx 0.8$, (see (14.2)) we find $\gamma \approx 0.038$. The ignition energy $k_B T_i$ can be estimated by dimensional considerations: it must depend on e^2, since it is the Coulomb repulsion which demands high kinetic energy for two protons to approach each other; it must also depend on the reduced mass of the two protons $m_p/2$ and Planck's constant $\hbar$, since the fusion is greatly facilitated by the quantum tunneling effect. Hence,

$$k_B T_i = \eta \frac{e^4 m_p}{2\hbar^2} \tag{14.14}$$

where the numerical constant η is expected to be much smaller than one, since the Boltzmann distribution allows squared velocities much larger than the average value. Actually, since the temperature 1.57×10^7 K at the center of the Sun allows fusion, it means that the ignition temperature must be clearly less than that, which implies that $\eta < 0.054$. Equating (14.13) to (14.14) we obtain for the minimum mass of a star

$$\boxed{M_{i1} = N_{vi1} u \approx \left(\frac{\eta}{2\gamma}\right)^{3/4} \left(\frac{m_p}{m_e}\right)^{3/4} \left(\frac{e^2}{Gu^2}\right)^{3/2} u} \tag{14.15}$$

By choosing $\eta = 0.04$, we find $M_{i1} \approx 2.42 \times 10^{56} u$ i.e. five times smaller than the mass of the Sun. Actually the critical value is significantly lower than this estimate, because the relevant temperature is that around the center of the star where the fusion reactions take place and not the average one estimated in (14.13); this means that the quantity γ must increase by a factor, let us say, of about 3 so that the critical value N_{i1} will became approximately

$$N_{vi1} \approx 10^{56} \tag{14.16}$$

This estimate agrees quite well with observational data.

When the mass M of the star is larger than M_{i1} as obtained by (14.16), the temperature at the center of the star will reach the ignition temperature for $R \approx R_o$, where R_o is the equilibrium value of R, as shown in Fig. 14.1. Therefore, the average equilibrium value T_o of the temperature is proportional to the ignition temperature: $T_o = z T_i$. Figure 14.1 suggests that for $R \approx R_o$ the classical result $\frac{3}{2} x N_v k_B T_o$ is a reasonable approximation to the kinetic energy E_K; the latter is equal to $-\frac{1}{2} E_G = \frac{1}{2} a_G G u^2 N_v^2 / R_o$. Hence, we end up with the equality $\frac{1}{2} a_G G u^2 N_v^2 / R_o \approx \frac{3}{2} x N_v k_B z T_i$ from which the proportionality $R_o \propto N_v$ follows. More explicitly we have taking also into account (14.14):

$$R_o = b_o \frac{Gu^2}{e^2} \frac{m_e}{m_p} N_v a_B, \qquad b_o \approx \frac{2 a_G}{3 x z \eta} \tag{14.17}$$

The proportionality factor z can be determined by comparing with the solar data: we find $z \approx 0.35$, a reasonable value, since T_o is the average temperature over the whole star and not its value at the center of the star where the highest temperatures are expected. Using the previously estimated values of a_G, x, y, and η, we obtain the following formula for the radius of an active star roughly the size of the Sun: $R_o \approx 5.837 \times 10^{-49} N_\nu$ in meters. In reality, for stars more massive than $2M_S$ but less massive than $20M_S$ the kinetic energy increases with the mass a bit faster than linearly, which in turn, implies that the radius/mass relation, becomes sublinear: $R_o \propto N_\nu^s$, $s \approx 0.75$ or even less. In the range $0.5M_S < M < 20M_S$ the surface temperature T is roughly proportional to $M^{1/2}$ so that the luminosity L which is proportional to $R_o^2 T^4$ ($L = 4\pi R_o^2 I$; I is defined in (6.3)) becomes proportional to $M^{2s} M^2 = M^{2(1+s)}$. We have seen already that $s \approx 1$ for about $0,5M_S < M < 2M_S$; thus for this range of masses $L \propto M^4$. For M larger than about 2 M_S but less than about twenty solar masses $s \approx 0.75$ and, consequently, $L \propto M^{3.5}$. The luminosity determines the life-time t of an active star. Indeed, assuming that the available energy to be emitted is proportional to the nuclear fuel which in turn is proportional to the initial mass of the star, we can conclude that $t \propto M/\bar{L} \propto M^{-1-2s}$.

14.6 The Maximum Mass of an Active Star

There are very few observational data regarding the upper mass limit of an active star for two reasons: First, such big stars need special conditions for their creation. Second, even if they manage to be formed, they die young, since $t \propto M^{-2.5}$. Thus very few of them are present at any given time. From the theoretical side, the following argument is usually advanced: High mass implies high temperature, which in turn produces high radiation pressure, P_R, as a result of its T^4 dependence versus the linear on T dependence of the thermal kinetic pressure. Thus, for high enough mass the radiation pressure P_R will exceed the thermal kinetic pressure P_K. Let us write $P_R = \Lambda P_K$, where Λ is usually taken around 2. Under these conditions we shall show that an instability will occur and, as a consequence, the star will cease to exist. Indeed, replacing in the equality $P_R = \Lambda P_K$, the expressions for $P_K = n k_B T = x n_\nu k_B T$ and $P_R = \frac{1}{3} U_R / V$, (see (6.1)), where $n_\nu \equiv N_\nu / V$ we obtain

$$(k_B T)^3 = c_1 (\hbar c)^3 x n_\nu, \quad c_1 = 45\Lambda/\pi^2 \tag{14.18}$$

The gravitational pressure P_G is determined by differentiating $E_G \propto V^{-1/3}$ with respect to $-V$ and it is found to be $E_G/3V = -(4\pi/3)^{1/3} (a_G/3) G u^2 n_\nu^{4/3} N_\nu^{2/3}$. Substituting the above expressions of the pressures in the equilibrium condition $|P_G| = P_K + P_R = P_K(1 + \Lambda)$ we have:

$$k_B T = c_2 G u^2 n_v^{1/3} N_v^{2/3}, \qquad c_2 = \frac{1}{3}\left(\frac{4\pi}{3}\right)^{1/3} \frac{a_G}{(1+\Lambda)x} \tag{14.19}$$

Eliminating $k_B T$ between (14.18) and (14.19) we obtain an equation for the maximum value of N_v to be denoted as N_{vi2}:

$$\boxed{N_{vi2} = A\left(\frac{\hbar c}{G u^2}\right)^{3/2}, \qquad A = 5.42\left(\frac{\Lambda^{1/3}(1+\Lambda)}{a_G}\right)^{3/2} x^2} \tag{14.20}$$

For $\Lambda = 2$ and the other parameters as before we find $N_{vi2} = 2.75 \times 10^{59}$ or $N_{vi2} = 230 N_{vS}$, while for $\Lambda = 1$ we find $N_{vi2} = 1.08 \times 10^{59}$ or $N_{vi2S} = 91 N_{vS}$.

14.7 Summary of Important Relations

The following (14.3), (14.6), (14.7), (14.8), (14.9), (14.10), (14.16), (14.17), and (14.20) are the ones to receive the increased attention of the reader. Notice that in all these formulae the relevant quantities referring to stars, dead or alive, are expressed in terms of universal constants.

14.8 Multiple-Choice Questions/Statements

1. In a white dwarf, as a result of the huge compression, all atoms are fully ionized. Therefore the white dwarf consists of bare nuclei and electrons which are non-relativistic. By minimizing the total energy the radius of the white dwarf is obtained:
   (a) $R = 4.51(N_e / N_v)^{5/3}(\hbar^2 / m_e G u^2) N_v^{1/3}$
   (b) $R = 4.51(N_e / N_v)^{5/3}(\hbar^2 / m_e G u^2) N_v^{-1/3}$
   (c) $R = 4.51(N_v / N_e)^{5/3}(\hbar^2 / m_e G u^2) N_v^{1/3}$
   (d) $R = 4.51(N_v / N_e)^{5/3}(\hbar^2 / m_e G u^2) N_v^{-1/3}$
2. In a white dwarf of mass $M = N_v u$, its total energy is proportional to:
   (a) M^2 (b) $M^{5/3}$ (c) $M^{7/3}$ (d) M
3. The radius of a white dwarf of mass equal to that of the Sun, $(1.98 \times 10^{30}\,\text{kg})$ is (In atomic units $G \approx 2.4 \times 10^{-43}$ and $u = 1.66 \times 10^{-27}\,\text{kg}$):
   (a) $R \approx 1.2 \times 10^6\,\text{m}$ (b) $R \approx 8.9 \times 10^6\,\text{m}$ (c) $R \approx 3.1 \times 10^7\,\text{m}$ (d) $R \approx 8.2 \times 10^7\,\text{m}$
4. In a white dwarf the average magnitude of the velocity of an electron is proportional to:
   (a) $M^{1/3}$ (b) M (c) $M^{2/3}$ (d) $M^{-2/3}$
5. Just before a white dwarf is ready to collapse to a neutron star, the total kinetic energy of its electrons is

(a) $E_K = a_K \hbar c N_\nu^{2/3} / R$ (b) $E_K = a_K \hbar c N_\nu / R$

(c) $E_K = a_K \hbar c N_\nu^{4/3} / R$ (d) $E_K = a_K \hbar c N_\nu^{5/3} / R$

6. A white dwarf collapses to a neutron star when its mass is equal to:

(a) $0.775(c\hbar / Gu^2)u$ (b) $0.775(c\hbar / Gu^2)^{2/3}u$

(c) $0.775(c\hbar / Gu^2)^{3/2}u$ (d) $0.775(e^2 / Gu^2)^{4/3}\, u$

7. The radius R of a neutron star is given by one of the following formula:

(a) $R = \eta(\hbar^2 / m_e G m_n^2) N_\nu^{-1/3}$ (b) $R = \eta(\hbar^2 / G m_n^3) N_\nu^{-1/3}$

(c) $R = \eta(\hbar^2 / G m_n^3) N_\nu^{-2/3}$ (d) $R = \eta(\hbar^2 / m_e G m_u^2) N_\nu^{-2/3}$, where η=3.2

8. The mass of a neutron star in the double system 4U1820-30 was estimated to be 1.58 times the mass of the Sun, i.e. $N_\nu = 1.885 \times 10^{57}$. What is the value of R?

(a) $R \approx 9.15$ km (b) 16.4 km (c) 28.63 km (d) 6760 km

9. In the extreme relativistic case valid when the mass of a neutron star is almost equal to the critical one for collapsing to a black hole, the total kinetic energy is given by one of the following formula:

(a) $E_K = a_K c\hbar N_\nu^{4/3} / R$ (b) $E_K = a_K c\hbar (N_n^{4/3} + N_p^{4/3} + N_e^{4/3}) / R$

(c) $E_K = a_K c\hbar N_n^{4/3} / R$ (d) $E_K = a_K c\hbar (N_n^{5/3} + N_p^{5/3} + N_e^{5/3}) / R$

10. The critical mass for the collapse of neutron star to black hole is related to that of the collapse of a white dwarf to neutron star by one of the formulae: (assume that the ratio a_K / γ remains the same)

(a) $M_{cr,\mathrm{ns}} \approx (N_\nu / N_e) M_{cr,\mathrm{wd}}$ (b) $M_{cr,\mathrm{ns}} \approx (N_\nu / N_e)^{3/2} M_{cr,\mathrm{wd}}$

(c) $M_{cr,\mathrm{ns}} \approx (N_\nu / N_e)^2 M_{cr,\mathrm{wd}}$ (d) $M_{cr,\mathrm{ns}} \approx (N_\nu / N_e)^{5/2} M_{cr,\mathrm{wd}}$

11. Consider a neutron star of mass equal to $1.6 M_S$. Its total internal kinetic energy E_K is given by the formula

(a) $1.105\hbar^2 N_n^{5/3} / m_n R^2$ (b) $1.105(\hbar^2 / R^2)\{(N_n^{5/3} / m_n) + (N_p^{5/3} / m_p) + (N_e^{5/3} / m_e)\}$

(c) $1.105(\hbar^2 / R^2)\{(N_n^{5/3} / m_n) + (N_p^{5/3} / m_p)\} + (1.44\hbar c N_e^{4/3} / R)$ (d) $1.44\hbar c N_e^{4/3} / R$

12. The radius of the event horizon of a black hole is:

(a) $r_s = GM / c^2$ (b) $r_s = GM^2 / c^2$ (c) $r_s = 2GM^2 / c^2$ (d) $r_s = 2GM / c^2$

13. Due to quantum fluctuations of the vacuum just outside the event horizon of a black hole, the latter appears to radiate as a black body of temperature T. This temperature is given by one of the following formulae:

(a) $k_B T = \eta \hbar c^2 / GM$ (b) $k_B T = \eta \hbar c^3 / GM^2$

(c) $k_B T = \eta \hbar^2 c^3 / GM^2$ (d) $k_B T = \eta \hbar c^3 / GM$

14. Since, in addition to its energy $M c^2$, a temperature is attributed to a black hole, an entropy S has to be attributed as well. It is given by the formula:

(a) $S / k_B = 4\pi G M^3 / \hbar c$ (b) $S / k_B = 4\pi G M^3 / \hbar c^2$

(c) $S / k_B = 4\pi G M^2 / \hbar c$ (d) $S / k_B = 4\pi G M^3 / \hbar^2 c$

15. In the Planck system of units $(\hbar = 1,\, c = 1,\, G = 1)$ the entropy S of a black hole is connected to its area $A \equiv 4\pi r_s^2$ as follows:

(a) $S / k_B = A$ (b) $S / k_B = A / 4$ (c) $S / k_B = A^{3/2}$ (d) $S / k_B = A^{3/2} / 4$

16. If the possibility of nuclear fusion reactions were absent, the temperature vs. the radius R of a star to be would exhibit a maximum $T_{\max}$ which is given by one of the following formulae:

(a) $k_B T_{\max} = \eta(G^2 u^4 / \hbar^2) N_v^{4/3}$ (b) $k_B T_{\max} = \eta(G^2 u^2 / \hbar^2) N_v^{4/3}$
(c) $k_B T_{\max} = \eta(G^2 u^4 / \hbar^2) N_v^{4/3} m_e$ (d) $k_B T_{\max} = \eta(G u^4 / \hbar^2) N_v^{4/3} m_e$

17. A necessary condition for an active star to be formed is that the temperature T_{ign} for the ignition of nuclear fusion reactions to be smaller than the temperature $T_{\max}$ defined in no.16 above. T_{ign} is given by one of the following formulae:
(a) $k_B T_{ign} = \eta(e^4 / m_e \hbar^2)$ (b) $k_B T_{ign} = \eta(e^4 m_e / \hbar^2)$
(c) $k_B T_{ign} = \eta(e^4 m_e / \hbar^2)\ m_e$ (d) $k_B T_{ign} = \eta(e^4 m_p / \hbar^2)$, η=0.02

18. The condition $T_{ign} \le T_{\max}$ leads to one of the following formulae for the minimum number of nucleons necessary for the existence of an active star:
(a) $N_v = 0.26(e^2 / G\sqrt{m_e u^3})$ (b) $N_v = 0.26(e^2 / G\sqrt{m_e u^3})^{3/2}$
(c) $N_v = 0.26(e^2 / G u^2)$ (d) $N_v = 0.26(e^2 / G u^2)^{3/2}$

19. The radius of an active star is given by one of the following formulae:
(a) $R = \eta(G u\, m_e / e^2) N_v a_B$ (b) $R = \eta(G u_u^2 / e^2) N_v a_B$
(c) $R = \eta(G u\, m_e / e^2) N_v$ (d) $R = \eta(G m_e^2 / e^2) N_v a_B$

20. When the mass of a star to be is very large compared to that of the Sun, the energy of the photons becomes Λ times larger than the thermal energy of its particles. Moreover, the photon pressure becomes comparable to the gravitational one. From these two relations the maximum mass of an active star follows. It is given by one of the following relations:
(a) $N_{\max} = \eta(\hbar c / G u^2)$ (b) $N_{\max} = \eta(\hbar c / G u^2)^2$
(c) $N_{\max} = \eta(\hbar c / G u^2)^{3/2}$ (d) $N_{\max} = \eta(e^2 / G u^2)^{3/2}$

21. The pressure at the center of an active star is:
(a) $P = (3/2\pi)(GM^2 / R^3)$ (b) $P = (3/2\pi)(GM^2 / R^4)$
(c) $P = (3/2\pi)(GM^2 / R^5)$ (d) $P = (3/2\pi)(GM^2 / R^6)$

22. If Earth having $N_v = 3.6 \times 10^{51}$ nucleons would become a black hole, its entropy would be: (a) $S / k_B \approx 10^{50}$ (b) $S / k_B \approx 10^{58}$ (c) $S / k_B \approx 10^{66}$ (d) $S / k_B \approx 10^{74}$

23. The entropy of the Earth with $N_v = 3.6 \times 10^{51}$ and average temperature $1000\,\text{K}$ cannot be larger than one of the following upper limits:
(a) $S / k_B = 1.1 \times 10^{51}$ (b) $S / k_B = 1.1 \times 10^{56}$
(c) $S / k_B = 1.1 \times 10^{61}$ (d) $S / k_B = 1.1 \times 10^{66}$

24. Taking into account the relation $R = 3.15(\hbar^2 / G m_n^3) N_v^{-1/3}$ connecting the radius of a neutron star with its mass $M = N_v m_n$, it follows that the neutron star would collapse to black hole when:
(a) $N_{v,cr} = 1.41(\hbar c / G m_n^2)^{1/2}$ (b) $N_{v,cr} = 1.41(\hbar c / G m_n^2)$
(c) $N_{v,cr} = 1.41(\hbar c / G m_n^2)^{3/2}$ (d) $N_{v,cr} = 1.41(\hbar c / G m_n^2)^2$

25. Taking into account the correct answer to 24 above and that the number of nucleons in the Sun is 1.1936×10^{57} we can determine that the neutron star will collapse to black hole when its mass is *x* times larger than the mass of the Sun, where *x* is equal to one of the following numbers:
(a) 1.44 (b) 2.6 (c) 5.37 (d) 7.42

14.9 Solved Problems

1. *The following approximate relations are valid for stars belonging to the main sequence. (In parenthesis are the corresponding values for the Sun. An active star belongs to the main sequence as long as its main fusion reaction is the burning of hydrogen to helium). Prove the first two formulae.*

Pressure at their center $P_c = 2\bar{\rho}GM/R$ $(6 \times 10^{14}\ \text{Pa})$
Temperature at their center $T_c = P_c \frac{N_\nu m_n}{\nu N_e k_B \rho_c}$ $(1.57 \times 10^7\ \text{K})$
Radius $R \propto M^s$, $0.5 \leq s \leq 1$, $(6.96 \times 10^8\text{m})$
Surface temperature $T_s \propto M^\beta$, $\beta \approx 0.5$ $(5778\ \text{K})$
Radiation power $L \propto R^2 T_s^4 \propto M^{2s+2}$ $(3.846 \times 10^{26}\ \text{J/s})$
Lifetime $t \approx M/L \propto M^{-(2s+1)}$ $(\approx 10\text{Gyr})$

Further information for the Sun:

$$\bar{\rho} \equiv M/(4\pi R^3/3) = 1.408\ \text{g/cm}^3, \quad \nu N_e/N_\nu \approx 1.25$$
$$M_S = 1.9891 \times 10^{30}\ \text{kg}, \quad N_{\nu S} = 1.1979 \times 10^{57}$$

Solution In (13.13) we have shown that the pressure at the center of a planet is given by the formula $P(0) = \frac{F}{2}\frac{GM}{R}\bar{\rho}$ where $F/2$ was found to be a little larger than one for a planet. For a star it is expected to be considerably larger as a result of its greater mass and the subsequent stronger dependence of ρ on r. Thus the value $F/2 = 2$ seems to be quite reasonable.

The second relation is a direct application of the perfect gas law $P_c V_c = N k_B T_c$ (applied to the core of the star) and the relations $V_c = M/\rho_c \approx N_\nu m_n/\rho_c$, $N = \nu N_e$

2. *Consider a photon gas of spherical volume V and temperature T in equilibrium. Under which conditions this gas would become a black hole?*

Solution The radius of this photon gas is $R = \{(3/4\pi)V\}^{1/3}$ and its mass $M = V\varepsilon/c^2$. Thus the radius of its event horizon is $r_S = 2GM/c^2 = 2GV\varepsilon/c^4$. The system will become a black hole if $R \leq r_S \Rightarrow R^2(k_B T)^4 \geq \frac{45}{8\pi^3}\frac{\hbar^3 c^7}{G}$

14.10 Unsolved Problems

1. Find the emitted power per kg by the Sun. How does this power per kg compares with the power per kg processed by the human body?
2. Recently (2012) a black hole was discovered at a distance of 12.4 light-Gyr and of mass equal to $2 \times 10^9 M_S$. Find the radius of its event horizon, its temperature, its entropy, and its lifetime.

3. Why does not an active star explode as a thermonuclear bomb?

Hint Do a stability analysis by considering in Fig. 14.1 fluctuations around the equilibrium value of the total energy and their consequences on the kinetic energy and on the temperature at the central region of the star.

References

1. F. Shu, *The Physical Universe: An Introduction to Astronomy* (University Science Books, Mill Valley, CA, 1982)
2. B. Caroll, D. Ostlie, *An Introduction to Modern Astrophysics* (Addison-Wesley, Reading, 2007)
3. L.D. Landau, E.M. Lifshitz, *Statistical Physics, Part 1* (Pergamon Press, Oxford, 1980)
4. S. Lloyd, Y.J. Ng, Scientific American, Nov 2004, p. 53; J.D. Bekenstein, Scientific American, **289**, 58, Aug 2003

Chapter 15
The Observable Universe

Abstract Hubble's discovery of the recession of distant galaxies, the observation of the cosmic microwave background radiation, and recent observational data coupled with Einstein's gravitational theory have established the inflationary Big Bang model as the standard scheme for determining the important milestones in the evolution of the Universe.

15.1 What Is Cosmology?

Cosmology is the study of the Universe as a whole and, in particular, its time evolution from the very beginning until the present epoch and even the future times: It is an evolution regarding its size, its temperature, its composition (including the process of creation of the elementary waveparticles), and, finally, its large structures. This "history" of the Universe is controlled mainly by the gravitational interaction. The latter is admirably described by the General Theory of Relativity (GTR), although the latter misses the quantum character necessary for understanding the events at the very early stages of the history of the Universe. Without underestimating the role of theory, it is fair to say that Cosmology became a genuine physics subject thanks to impressive observational discoveries such as the recession of distant galaxies, the cosmic microwave background (CMB) radiation, its full spectral analysis, and, more recently, its minute angular fluctuations observed mainly by the missions of WMAP and Planck.

To proceed with the theoretical study of the Universe we need to employ the simplifying assumption that the Universe is homogeneous, isotropic, and without boundaries. The assumed homogeneous and isotropic nature of the Universe is considered justified, if we ignore "details" at the scale of 500 million light years or less. In other words, we restrict ourselves to a very coarse description of the Universe by averaging over distances much larger than 100 million light years. Both observations and theory show that the Universe is expanding—space itself is expanding—at a rate $\dot{R} \equiv dR/dt$ which is proportional to the distance R between two points.[1]

---

[1]Clusters of galaxies containing hundreds of galaxies and smaller structures than that do not follow space in its expansion, because gravitational attraction keeps them together for the present era.

E.N. Economou, *From Quarks to the Universe*,
DOI 10.1007/978-3-319-20654-7_15

The proportionality coefficient is called Hubble's constant and is denoted by the letter $H \equiv \dot{R}/R$. The constancy of H means that it is the same no matter what the distance R is, as long as $R \gg 100$ million light years. On the other hand, H is time dependent and the basic equation determining this time evolution is the following:

$$\left(\frac{\dot{R}}{R}\right)^2 = \frac{8\pi}{3}\frac{G}{c^2}\varepsilon \tag{15.1}$$

where ε is the average energy density in the Universe. Notice that both H and ε are independent of R (the energy density ε because of the homogeneity assumption), but they do depend on time. This picture shows that in the past the Universe was denser and hotter; following this trip to the past, a time would be reached where the whole observable Universe together with its space was almost infinitely small, dense, and hot. From this zero point on, the Universe is expanding and cooling, a process termed *Big Bang Model*. By applying the known laws of physics, we can follow the history of the Universe (in coarse terms) and reach conclusions regarding its 'present' features, such as the value and the composition of the energy density, the value of Hubble's constant, the percentage of chemical elements, the radius of the observable Universe, the spectral and angular distribution of cosmic photons, as well as larger structures of the Universe such as galaxy-clusters, voids, filaments etc. Notice that time 'present' may include time 'past' as well, since by observing now an event at a distance d from us, we actually see what was happening there at a time $t \approx d/c$ before now. Comparison of the theoretical and numerical results with a stream of very impressive observational data has enabled the human civilization for the first time to provide some reliable answers to the age old, yet ever present question: How did the World start and how did it evolve?

In this chapter we start with a few historical remarks regarding the field of Cosmology and the recent impressive observational data. Then, we "derive" the basic (15.1) by employing Newtonian mechanics, the results of which can be trusted as long as they agree with those of the general theory of relativity (GTR). By taking into account some observational data regarding the composition of the energy density of the Universe, we analytically solve (15.1) (after making some simple approximations in order to avoid numerical solutions). We determine this way the size and the temperature of the Universe as a function of time elapsed since the time zero of the Big Bang. Having as input the temperature and the density as well as the known laws of Physics we reconstruct in coarse terms some of the basic events in the history of the Universe.

15.2 Measuring the Temperature and Other Parameters of the Universe

The modern study of the Universe really started in 1917 with Einstein finding a time independent, but unstable solution of the metric of the Universe within the framework of his general theory of relativity (GTR) in which he included the so-called *cosmological constant*. Alexander Friedmann, Georges Lemaitre and others found during the 1920s time dependent solutions of the metric of the Universe exhibiting Big Bang characteristics with the expansion either continuing for ever or terminating and reversing direction leading eventually to a Big Crunch. This type of solutions received a substantial boost by an extremely important observational fact, namely Hubble's discovery that distant galaxies go away from Earth with a velocity proportional to their distance from Earth. This discovery strongly indicated that the Universe is expanding as predicted by the time dependent solutions of the equations of the GTR.

The next big observational discovery occurred in 1964 sixteen years after the 1948 theoretical work of G. Gamow, R. Alpher, and R. Herman anticipating the existence of a relic electromagnetic radiation extending almost uniformly and isotropically over the whole Universe. This so-called cosmic microwave background (CMB) radiation was serendipitously discovered by A. Penzias and R. Wilson in 1964 and played a crucial role in transforming the field of Cosmology from what most physicists considered as a kind of metaphysics to a genuine physical subject. Following the initial discovery of the CMB radiation, a much more extensive and precise spectral analysis of the CMB radiation was performed by a team headed by G. Smoot and J. Mather through the instrument FIRAS mounted on the COBE satellite (see Fig. 15.1). Their analysis has shown that the frequency distribution of the CMB radiation was in excellent agreement with that of a black body of temperature 2.725 K. Moreover, the COBE mission detected a minute anisotropy in the temperature of 2.725 K (of the order of one part in a 100,000), indicative of very small inhomogeneity in the early Universe.

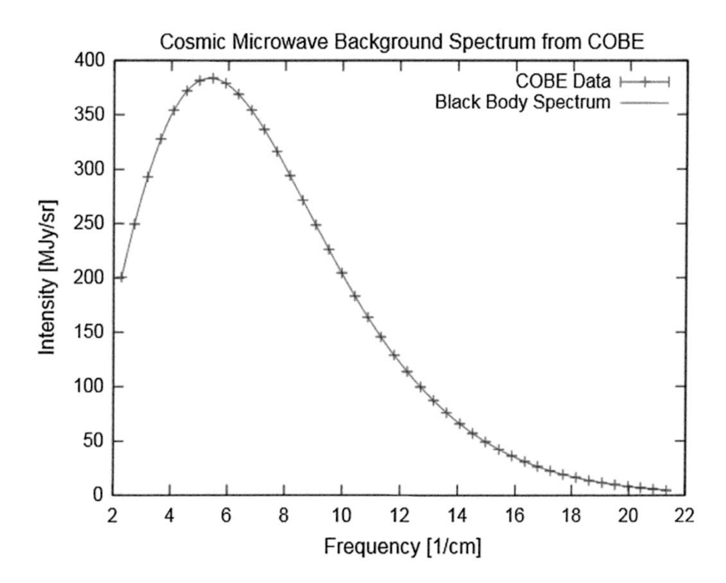

Fig. 15.1 The spectrum of the CMB radiation (crosses) and the black body spectrum; the latter depends on only one parameter, the temperature. The agreement is amazing

Inspired by the success of COBE many experiments were designed aiming at measuring this minute anisotropy as accurately as possible. Probably the most impressive among them is the one obtained through the WMAP mission which measured the temperature fluctuations, $\delta T(\theta, \phi)$, of CMB radiation as a function of the direction of observation, i.e., as a function of the angles θ and ϕ. This function can be expressed as a weighted sum of the so-called spherical harmonics $Y_{\ell m}(\theta, \phi)$ (we met these functions in Chap. 10, when we examined the angular dependence of the atomic orbitals). The average of the square of the absolute value of the coefficients of this expansion, denoted as C_l, is plotted in Fig. 15.2.

More recently (results announced in March 2013) the Planck space mission performed measurements of the cosmic microwave background (CMB) radiation at considerable higher resolution. The results were analyzed in terms of spherical harmonics as explained before and are presented in Fig. 15.3. Notice that in Fig. 15.3 the analysis reaches up to $l = 2500$ and the error bars have been dramatically reduced (especially for $l > 700$). These accurate measurements allowed us to estimate the

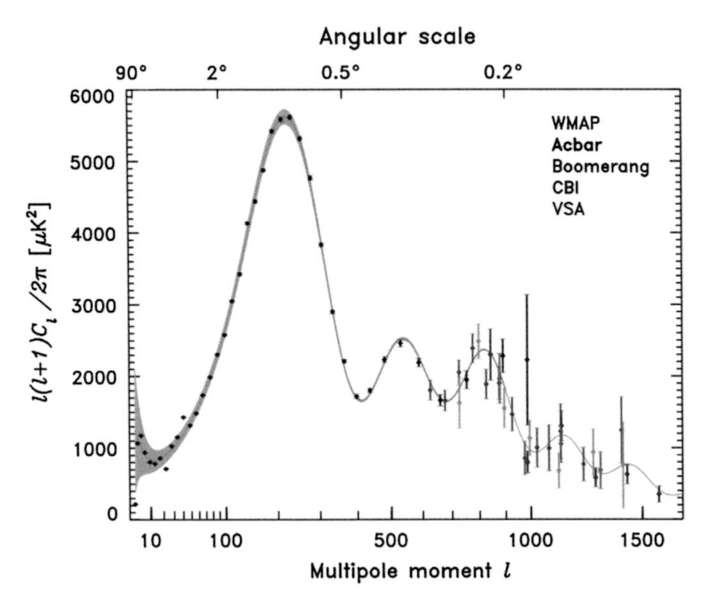

Fig. 15.2 The quantity C_l, defined in Sect. 15.2, multiplied by $l(l+1)/2\pi$ is plotted versus the index l (credit NASA). The measurements are denoted by points and *error bars*; different *colors indicate* measurements by different instruments and techniques. The continuous line shows the best fit to the data according to the prevailing theory which contains a minimum of six adjustable parameters, such as the various components of the total energy density, the Hubble constant, etc. The oscillatory character of the *above line* implies damped acoustic oscillations in the early Universe driven by the gravitational attraction and the photonic repulsion with the peaks representing the fundamental and the higher harmonics. These acoustic oscillations were evolved from quantum fluctuations in the very early Universe. The position of the first peak implies that the geometry of the Universe is Euclidean. The second and the third peaks determine the ratio of the baryonic mass density to the so-called dark mass density

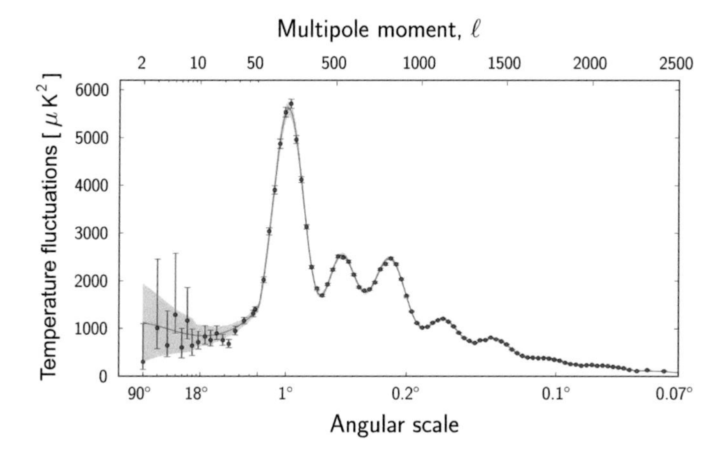

Fig. 15.3 The quantity C_l, defined above, multiplied by $l(l+1)/2\pi$ is plotted versus the index l(credit ESA). The measurements are denoted by points and *error bars*. The continuous *line* shows the best fit to the data according to the prevailing theory which contains a minimum of six adjustable parameters

values of the Hubble constant and the other parameters entering the theory with considerably smaller uncertainty.

The main conclusions resulting by the combination of theory and observational data are the following:

(1) The expansion of the Universe was decelerating in the distant past, but it begun accelerating nearly 6 Gyr ago. This implies that a cosmological constant-like term enters in the equations of general relativity or, equivalently, that the energy density contains a contribution almost time-independent in spite of the expansion of the Universe.
(2) The energy density of the Universe consists of several contributions:

$$\varepsilon = \varepsilon_b + \varepsilon_e + \varepsilon_v + \varepsilon_{dm} + \varepsilon_{ph} + \varepsilon_{de}. \tag{15.2}$$

The contribution ε_b is due to ordinary baryonic matter (protons, and neutrons bound, presently, in nuclei); ε_e is due to electrons and it is presently negligible in comparison to ε_b; ε_v comes from neutrinos and antineutrinos, otherwise called light dark matter; ε_{dm} is associated with the so-called dark matter, or more precisely heavy dark matter (several exotic particles predicted by some proposed theories, but not found yet experimentally, are candidates for making up the dark matter); ε_{ph} is due to photons; finally, ε_{de} is the so-called dark energy contribution which seems to be almost or exactly time-independent (in what follows we shall consider ε_{de} as exactly time-independent, which means that it plays a role equivalent to the cosmological constant). The nature of ε_{de} remains a mystery.

A clarification is needed regarding the number of electrons, protons, neutrons, neutrinos, etc. When the temperature of the Universe was much larger than

$T = m_e c^2/k_B = 5.93 \times 10^9$ K, the reaction $e + e^+ \rightleftarrows n\gamma$ was possible, where e^+ is the positron and n is the number of photons involved in the reaction (usually $n = 1$ or $n = 2$). As it was mentioned in Sect. 14.3, thermodynamic equilibrium implies a relation among the chemical potentials of the species involved in a reaction. For the above reaction the relation is $\mu_e + \mu_{e^+} = n\,\mu_\gamma$; since the chemical potential of photons is zero, it follows that $\mu_e = -\mu_{e^+}$. This last equation together with the equation $n_e - n_{e^+} = n_\ell$ (expressing the fact that the difference in the concentrations of electrons and positrons is equal to the concentration of electron lepton number n_ℓ) allows the calculation of n_{ph}, n_e and n_{e^+} as a function of the temperature and the electron lepton number concentration n_ℓ. Notice that both the lepton number and the baryon number $N_b = V\,n_b$ are conserved. Moreover, since eventually the neutron to proton ratio is 1:7 as it will be shown in Sect. 15.5, electrical neutrality implies that $n_\ell = (7/8)\,n_b$. We shall show below that the photon number is also approximately conserved and that the ratio $N_b/N_{ph} \approx 0.6 \times 10^{-9}$ remains almost constant throughout the history of the Universe. Such a small number of this ratio makes the following proposal very attractive: Initially, N_b, and N_ℓ were exactly zero, so that matter and antimatter was created out of electromagnetic energy through the reaction: photons $\rightleftarrows$ matter + antimatter; then a mechanism breaking to a minute degree the symmetry between matter and antimatter is assumed to intervene so that one part in a billion of matter remains after the annihilation. The electrically charged particle/antiparticle pairs includes the three types of leptons (electron/positron, $\mu/\bar{\mu}$, and $\tau/\bar{\tau}$) as well as the six types of quark/antiquark. The last pair to be annihilated is the electron/positron because it is the lightest charged pair. Before this last annihilation, i.e., for $T \gg m_e c^2/k_B = 5.93 \times 10^9$ K, the electron lepton number, $N_{e^-} - N_{e^+}$, is negligible in comparison with N_e or N_{e^+}; hence, $n_e \approx n_{e^+}$ and the result for the energy density is $\varepsilon_e \approx \varepsilon_{e^+} \approx \frac{7}{8}\varepsilon_{ph}$ (see unsolved problem 15.5). In the opposite limit, $k_B T \ll m_e c^2$, all the positrons have been annihilated and the left out electrons are equal to the electron lepton number. A similar to electron/positron process has taken place at higher temperatures with the heavier charged leptons and, in a much more complicated way, with the quark/antiquark and eventually with the proton/antiproton, neutron/antineutron pairs with the threshold temperature being, of course, associated with the rest energies of the proton, and neutron, $T = m_p c^2/k_B = 1.09 \times 10^{13}$ K. The three types of neutrino/antineutrino pairs, being electrically neutral, can participate in processes similar to the electron/positron pair but only through the weak interactions. As a result the neutrinos and antineutrinos became decoupled from the rest at an early stage ($T \approx 10^{10}$ K) and their number remains constant and comparable to the photon number since then.

Presently, on the basis of the data shown in Figs. 15.2 and 15.3, the various contributions to the total energy density of the Universe, ε, are as follows: $\varepsilon_b : \varepsilon_{dm} : \varepsilon_{ph} : \varepsilon_{de} : \varepsilon_\nu \approx 4.83 : 25.87 : 0.005 : 69.14 : 0.4$ (see Table 15.1). The present neutrino contribution is quite uncertain, since the neutrino mass is not known. The neutrino temperature is calculated to be 71.4 % of the photon temperature [1]. In Table 15.2 we show the best estimates of the cosmological parameters based on the data of Figs. 15.1 and 15.3.

Table 15.1 Composition of the Universe at the present epoch ($t = 13.7965 \pm 0.037$ billion years)

Kind of matter	Kind of particles	Particle mass (eV)	Number of particles in the observable Universe	Percentage in the energy of the Universe today	Percentage in pressure	Percentage in energy at $t = 380,000$ years
Baryonic	Protons, Neutrons (in nuclei)	$\approx 10^9$	0.86×10^{80}	$(4.825 \pm 0.05)\,\%$	≈ 0	12 %
Light dark matter	Neutrinos	$\ll 1$	10^{89}	$<1\,\%$	≈ 0	10 %
Heavy dark matter	Supersym-metric particles ?	10^{11}?	5×10^{78} ?	$(25.886 \pm 0.37)\,\%$	≈ 0	63 %
E/M radiation	Photons	6.34×10^{-4}	1.4×10^{89}	0.005%	≈ 0	15 %
Dark energy	? (Scalar particles?)	? (10^{-32}?)	? (10^{122}?)	$(69.14 \pm 1)\,\%$	100 %	$\approx 0\,\%$

The results shown in the 5th column of Table 15.1 are best fits using the measurements from Planck, WMAP polarization measurements, other CMB experiments focusing on high l C_ls, as well as baryonic acoustic oscillation (BAO) experiments

15.3 Derivation of the Equations Determining the Expansion Rate

We shall try first the dimensional analysis for Hubble's constant, whose dimension is one over time; we shall also follow the same approach for the reduced acceleration $\ddot{R}/R$. Obviously, the gravitational constant G must enter in the formulas for $\dot{R}/R$ and $\ddot{R}/R$; we expect that the total mass density $\rho = \varepsilon/c^2$ must also enter, since the gravitational field involves the product G times mass and hence the product $G\rho$. However, since the dimension of $G\rho$ is one over time squared, we can conclude that the most general relation for $\dot{R}/R$ and $\ddot{R}/R$ are of the form

$$\dot{R}/R = \sqrt{G\rho}f_1(x_1, x_2, \ldots) \quad \text{and} \quad \ddot{R}/R = G\rho f_2(x_1, x_2, \ldots)_2 \tag{15.3}$$

where $x_1, x_2, \ldots$ are dimensionless variables. One such variable is obviously the ratio of the pressure over the energy density, $x_1 = p/\varepsilon$; another is the ratio of $G\rho$ over c^2/R^2: $x_2 = G\rho/(c^2/R^2)$. This last variable x_2 will not appear in $\dot{R}/R$, if the latter is really constant, i.e., independent of R. Let us assume that there are no other dimensionless variables and let us proceed with the determination of f_1 and f_2.

Although the reliable way to obtain these functions is to employ the general theory of relativity (GTR), here we shall work within the Newtonian framework. The latter at closer scrutiny reveals inconsistencies in some of the formulae obtained; nevertheless, the main results are in agreement with the conclusions of the GTR. This allows us to employ the Newtonian approach, taking advantage of its simplicity and familiarity, and accept only those results which agree with these of the GTR.

Consider a far-away galaxy of mass m at a distance R from the observation point. Newton's equation, $m\ddot{R} = F$, becomes $\ddot{R} = -GM/R^2$, where M is the mass enclosed within the sphere of radius R centered at the observational point. By integrating this equation, *assuming constant mass M*, we obtain a relation showing that the sum of kinetic and potential energy is a constant, to be written as $\frac{1}{2}Cm$:

$$\frac{1}{2}\dot{R}^2 - \frac{GM}{R} = \frac{1}{2}C \tag{15.4}$$

Equation (15.4) can be brought to a form similar to (15.1) by the following operations: Substitute $\frac{4\pi}{3}R^3\rho$ for M; divide both sides of (15.4) by $R^2/2$; replace ρ by ε/c^2 and write $\varepsilon = \varepsilon' + \varepsilon_{de}$. We have then

$$\left(\frac{\dot{R}}{R}\right)^2 = \frac{8\pi G}{3c^2}\varepsilon - \frac{C}{R^2} = \frac{8\pi G}{3c^2}\varepsilon' + \frac{8\pi G}{3c^2}\varepsilon_{de} - \frac{C}{R^2} \tag{15.5}$$

Table 15.2 Current estimates of some cosmological parameters

Hubble constant	$H_o \simeq 0.21963 \times 10^{-17} s^{-1} = (67.77 \pm 0.036)\,\text{km}/(\text{s} \times Mpc)$
Age of the Universe	$t_o = 13.7965 \pm 0.037\,\text{Gyr}$
Age of decoupling	$t_{dec} = 380\,\text{kyr}$
Age of first star	$t_s \approx 108\,\text{Myr}$
Total density	$\Omega_{tot} \equiv \varepsilon/\varepsilon_{cr} = 1.02 \pm 0.01$
Critical density	$\rho_{cr} \equiv \varepsilon_{cr}/c^2 = 3H_o^2/8\pi G = 8.6274 \times 10^{-27}\,\text{kg/m}^3 \approx 5.158\,m_\text{p}/\text{m}^3$
Baryon density	$\rho_B \equiv \varepsilon_B/c^2 = 4.163 \times 10^{-28}\,\text{kg/m}^3 = 0.2489\,\text{m}_\text{p}/\text{m}^3$
Radius of observable Universe	$R_o \approx 4.359 \times 10^{26}\,\text{m} \approx 45.7$ Giga light years
Expansion since decoupling age	$R_o/R_{dec} = 1090$
Ratio of photons to baryons	$N_{ph}/N_B \approx 1.63 \times 10^9$
Temperature of CMBR	$T_\text{CMBR} \approx 2.7255 \pm 0.00057\,\text{K}$

The results of Table 15.2 are best fits using the measurements from Planck, WMAP polarization measurements, other CMB experiments focusing on high $l\,C_l$s, as well as baryonic acoustic oscillation (BAO) experiments

The separation of the total energy density ε in two terms ε' and ε_{de} makes sense, if the latter, as opposed to the former, is a universal constant. In this case we can define

$$\frac{8\pi G}{3c^2}\varepsilon_{de} \equiv \frac{\Lambda}{3}, \tag{15.6}$$

where Λ is the so-called cosmological constant. Equation (15.5), with the ε_{de} term assumed constant and written in terms of the cosmological constant, is exactly the outcome of Einstein's equations of GTR in combination with the assumptions of homogeneity and isotropy. In what follows we shall write (15.5) by setting $(8\pi G/3c^2)\varepsilon' + (8\pi G/3c^2)\varepsilon_{de} = (8\pi G/3c^2)\varepsilon$. It turns out that the constant C appearing in (15.5) determines the type of geometry of space: If C is positive, the geometry of the three-dimensional space is of positive curvature, the analogue of the two-dimensional geometry on the surface of a sphere; if C is negative, the geometry is of negative curvature, the analogue of the two-dimensional geometry on a saddle-like surface. Finally, if $C = 0$, the geometry is of zero curvature, i.e., it is the familiar Euclidean geometry; in this case we say that the three-dimensional space is flat. Notice that for $C = 0$, (15.5) reduces to (15.1). The $C = 0$ choice, which implies that the total energy as shown in (15.4) is zero, is supported by the *inflation* model. This model assures that just after time zero, more precisely during the period $t \approx 10^{-36}$ to $t \approx 10^{-32}$s, there was an exponentially fast expansion, a true 'Big Bang', which increased the distance between any two points by a factor of at least 10^{28}! There are proposed theories, of quantum nature, involving phase transitions among elementary particles, to account for this dramatic expansion process. But the main reason that inflation is tentatively accepted and incorporated in the "standard" cosmological model, is that it explains many puzzling data, such as (a) the same temperature of the CMB radiation coming from points of opposite directions relative to Earth and near the edges of the observable Universe,[2] (b) the so-called smoothness problem (i.e. the appropriate minute size of the quantum fluctuations during the inflation period acting as seeds for the eventual formation of galaxies of about the same size), and (c) the Euclidean geometry[3] as deduced from the data of Figs. 15.2 and 15.3, etc. Notice, however, that there is a minority of experts who attempt to explain these puzzling data by approaches other than those of the inflation model. Anyway, one must keep in mind that events at time $t \approx 10^{-35}$ s after the Big Bang involve energies as high as 10^{23} eV per particle, i.e., ten orders of magnitude higher than the upper limit achieved in the biggest accelerators on Earth; thus a direct experimental confirmation of such events is missing. Furthermore, a reliable theoretical study of events where both quantum and gravitational effects are involved requires a complete quantum theory of gravity, which is not yet available.

After the end of the inflation period it is expected that the evolution of the Universe obeyed the GTR, i.e., (15.1). From this equation and the first law of thermodynamics, $dU = -p\,dV$, one[4] can obtain an equation for the quantity $\ddot{R}/R$ which is in agreement with the corresponding one of the GTR. The proof follows: Taking into account that $U = V\varepsilon = \frac{4\pi}{3}R^3\varepsilon$, the first Law shows that

$$\frac{d\varepsilon}{dR} = -\frac{3}{R}(\varepsilon + p) \tag{15.7}$$

If we multiply (15.1) by R^2 and then differentiate it with respect to time we find $2\dot{R}\,\ddot{R} = (8\pi G/3c^2)[2R\dot{R}\varepsilon + R^2\dot{\varepsilon}]$. If in this last equation we replace the quantity $\dot{\varepsilon} \equiv d\varepsilon/dt$ by $(d\varepsilon/dR)\dot{R} = -(3/R)(\varepsilon + p)\dot{R}$, where (15.7) was taken into account, we obtain

$$\frac{\ddot{R}}{R} = -\frac{4\pi G}{3c^2}(\varepsilon + 3p) \tag{15.8}$$

[2] If the rate of expansion predicted by the GTR was the actual one at *all times*, the distance between these points would be at all times t much larger than ct; hence these points would never have the opportunity to communicate with each other and establish the same temperature; thus the coincidence of their temperatures would be an accident. In contrast, the existence of such a fast expansion rate as that of inflation, implies that these points were in close communication and interaction before the inflation period and, hence, they had the opportunity to have equal temperature.

[3] An expansion by thirty orders of magnitude in almost zero time will make the term C/R^2 sixty orders of magnitude smaller and , hence, absolutely negligible.

[4] The term TdS is practically zero, because the dissipation terms are negligible in comparison with the energy density which contains the rest energy of the particles.

Notice that (15.8) is not consistent with our initial Newtonian equation $\ddot{R} = -GM/R^2$, unless $p = 0$: This shows that results based on the Newtonian approach are to be trusted only if they agree with the ones based on the GTR. Another point to be noticed is that the time independence of any component of ε, such as the dark energy component, implies, according to (15.7), that $p_{de} = -\varepsilon_{de}$. Taking out of ε and p the corresponding dark energy components and recalling the relation $(8\pi G/3c^2)\varepsilon_{de} \equiv \Lambda/3$, we can rewrite (15.8) as follows:

$$\frac{\ddot{R}}{R} = -\frac{4\pi G}{3c^2}(\varepsilon' + 3p') + \frac{1}{3}\Lambda \qquad (15.8a)$$

15.4 Solutions of the Equation for the Expansion Rate

To solve either (15.1) or (15.8) we need the dependence on R of each component of ε or p, where from now on by R we shall denote the radius of the observable Universe.

For ε_{de}, following the observational evidence, we shall assume that it does not depend on R and, hence, that it is a constant equal to $-p_{de}$.

For ε_{ph} and p_{ph} all the information needed have been obtained in Sect. 6.2: $p_{ph} = \frac{1}{3}\varepsilon_{ph} = \frac{\pi^2}{45}(k_B T/\hbar c)^3 k_B T$ and $S_{ph} = \frac{4}{3}V\varepsilon_{ph}/T = \frac{16\pi^3}{135}k_B(R\, k_B T/\hbar c)^3$; the number of photons can be obtained by dividing the total energy $V\varepsilon_{ph}$ by the average energy of each photon which is $a\, k_B T$, where a turns out to be $a \approx 2.7$, $N_{ph} \approx 1.02(R\, k_B T/\hbar c)^3$. Since the entropy remains almost constant, it follows that $T_{ph} \propto 1/R$, $\varepsilon_{ph} \propto T_{ph}^4 \propto 1/R^4$ and the number of photons is also almost constant.

The R dependence of the hadronic contributions to ε and p is complicated because reactions due to strong interactions are taking place for temperatures higher than about 10^{13} K. Below this temperature the quark/gluon plasma has been transformed to baryons, the antibaryons have been annihilated, and the baryonic contributions to the energy density is, $\varepsilon_b \approx N_b m_B c^2/V \propto 1/R^3$, while the pressure, $p_b \approx N_b k_B T/V$, is negligible in comparison with ε_b, since $k_B T \ll m_B c^2$. Baryons will become important much later, at $t \geq 10^5$ years, when the temperature will be less than 6000 K.

For the joint electron and positron contribution we have

$$\begin{aligned} &\text{If } a' k_B T \gg m_e c^2, \quad \text{then} \quad 2p_e = \frac{2}{3}\varepsilon_e = \frac{2}{3}N_e a' k_B T/V = \frac{71}{43}\varepsilon_{ph} \propto 1/R^4 \\ &\text{If } m_e c^2 \gg a' k_B T, \quad \text{then} \quad p_e \approx N_\ell k_B T/V, \varepsilon_e \approx \varepsilon_\ell \approx N_\ell m_e c^2/V \propto 1/R^3 \end{aligned} \qquad (15.9)$$

Thus, for T much higher than 5.93×10^9 K the first line of (15.9) is valid, while for T much lower than 5.93×10^9 K the second line of (15.9) is valid. N_ℓ is the electron lepton number. We repeat here that observational data show that the number of photons is about a billion times larger than the number of baryons.

($N_{ph} \approx 1.65 \times 10^9 N_b$). It follows then from (15.9) that the electron/positron contribution to the energy density and the pressure is equal to 7/4 of the corresponding photon contribution for $T \gg 5.93 \times 10^9$ K and it is negligible for $T \ll 5.93 \times 10^9$ K.

For the unknown particles of the heavy dark matter it is not possible to make reliable predictions. Nevertheless, it is reasonable to expect that they will become decoupled from the rest at a much earlier time and that they exhibit a behavior of the form $\varepsilon \propto 1/R^3$ as a result of the absence of E/M and strong interactions. If the unknown particles of dark mass are of the supersymmetric type, their mass must be much larger than m_p. Anyway, there is strong evidence that the dark matter contribution to the energy density is about five times the baryonic contribution at least for $T \approx 10^{12}$ K or less. These assumptions are consistent with observational data. Based on the above relations and the corresponding data we find that the baryonic plus heavy dark matter contribution to the energy density becomes comparable to the photonic one around 10,000 K.

Finally, the light dark matter, the neutrinos and antineutrinos contribute to the energy density a term of the form $\varepsilon_\nu \approx N_\nu m_\nu c^2/V \propto 1/R^3$ at least for $T \leq 10^{10}$ K. The neutrino rest energy is not known exactly although it is extremely small, possibly less than 0.1 eV. The number of neutrinos is comparable to the number of photons, as we argued before. The above remarks imply that the evolution of the Universe can be separated, after the inflation period, into three distinct epochs depending on which component of the energy density is controlling its evolution:

(a) *Photon dominated*: Roughly, this epoch corresponds to $1000\,\text{s} < t < 20\,\text{kyr}$ for which $4.4 \times 10^8\,\text{K} > T > 16{,}000\,\text{K}$ and $10.7 \times 10^{18}\,\text{m} < R < 10.1 \times 10^{23}$ m.
(b) *Matter (baryonic and heavy dark matter) dominated*: Roughly, this epoch corresponds to $700\,\text{kyr} < t < 3\,\text{Gyr}$ for which $2000\,\text{K} > T > 7.4\,\text{K}$ and $6 \times 10^{23}\,\text{m} < R < 1.5 \times 10^{26}$ m.
(c) *Dark energy dominated*: It is expected to occur for t larger than about 20 Gyr corresponding to $T < 1\,K$, and $\varepsilon \approx \varepsilon_{de} = \text{const.}$

In each of those epochs as well as in the transition periods in between we shall determine the dependence of the total energy density ε on the scale R of the universe. Substituting then in the basic equation (15.1) we find differential equations which allow analytic solutions. More details follow:

(a) For the first epoch, the photon dominated one, where $\varepsilon \propto 1/R^4$, (15.1) becomes $R\dot{R} = \text{const.}$ or $d(R^2)/dt = \text{const.}$ or

$$R(t) = A\,t^{1/2}, \quad T(t) = \frac{C}{R(t)} = \frac{C}{A\,t^{1/2}}, \quad \varepsilon \propto t^{-2} \tag{15.10}$$

where the constant C (which remains almost unchanged throughout the history of the Universe) is approximately equal to

$$C \equiv RT = 1.186 \times 10^{27}\,\mathrm{K\,m} \tag{15.11}$$

The coefficient A is about equal to $A \approx 4.735 \times 10^{20}\,\mathrm{m\,year}^{-1/2}$ or $A \approx 8.429 \times 10^{16}\,\mathrm{m\,s}^{-1/2}$. With these choices of the coefficients A and C the radius corresponding to the temperature of 10,000 K, at which the matter contribution to the energy density becomes about equal to the photonic contribution, is $R = 1.2 \times 10^{23}$ m and the time at which this happens can be estimated to be 47,000 years. Up to this time, it is not unreasonable to still use the relation $R \propto \sqrt{t}$ for a rough estimate.

(b) Going now to the matter dominated epoch, where $\varepsilon \propto 1/R^3$, extending at least from 700,000 years to 3 Gyr, (15.1) becomes $R^{1/2}\dot{R} = \text{const.}$ or $d(R^{3/2})/dt = \text{const.}$ or

$$R \propto t^{2/3}, \quad T \propto t^{-2/3}, \quad \varepsilon \propto t^{-2} \tag{15.12}$$

(c) Finally, for the dark energy dominated epoch, extending beyond 20 Gyr, (15.1) becomes $\dot{R} = \text{const.}\,R$ or

$$R \propto \exp(Ht), \quad T \propto \exp(-Ht), \quad \varepsilon \approx \varepsilon_{de} = \text{const.} \tag{15.13}$$

where $H = \sqrt{b\,\varepsilon_{de}}$ is the Hubble constant at that time and $b \equiv 8\pi G/3c^2$; during this dark energy period the Hubble constant is constant not only in space but in time as well.

To obtain more accurate results for the time dependence of R, T, and ε we need to interpolate between the various epochs: from the inflation one to the photon (and matter/antimatter) epoch, from the latter to the matter epoch and from the matter era to the dark energy era. Although accurate results require more detailed considerations and numerical calculations, nevertheless, as we shall see, the transition periods from photon dominated to matter dominated and from matter dominated to dark energy dominated can be treated analytically. Let us start from the latter transition period. Indeed, from (15.1), we have that $\dot{R}/R = (b\varepsilon)^{1/2}$, where $b \equiv 8\pi G/3c^2$; $b\varepsilon = aR^{-3} + g$, $a \equiv 6.36\,b(3/4\pi)N_b m_b c^2$, $g \equiv b\,\varepsilon_{de}$. The factor 6.36 = 1 + 5.36 in a is there because we have included the heavy dark energy density which is taken to be equal to 5.36 times the baryon energy density $N_b m_b c^2/V$. (The best estimate for the present era of the ratio dark matter energy over baryonic energy is $(\varepsilon_{dm}/\varepsilon_b) = (25.87/4.83) = 5.36$). Thus, with this convenient notation, (15.1) takes the form,

$$\dot{R} = \left(\frac{a}{R} + g\,R^2\right)^{1/2} \tag{15.14}$$

The reader may verify that (15.14) is satisfied by the following function *R*(*t*):

$$R(t) = (a/g)^{1/3}[\sinh(\tfrac{3}{2}\sqrt{g}\,t)]^{2/3} \tag{15.15}$$

Notice that for $\frac{3}{2}\sqrt{g}\,t \ll 1$, (15.15) reduces to $R(t) \approx (\frac{3}{2})^{2/3} a^{1/3}\,t^{2/3}$ in agreement with (15.12). Similarly, for $\frac{3}{2}\sqrt{g}\,t \gg 1$, we have $R(t) \approx 2^{-2/3}(a/g)^{1/3}\exp(\sqrt{g}\,t)$, in agreement with (15.13).

We can obtain explicit numerical results by substituting in (15.15) the best estimates of the present era values for *a*/*g* and for $\sqrt{g}$ (see Tables 15.1 and 15.2). We have from $b\,\varepsilon = b\varepsilon_m + b\varepsilon_{dm} = a\,R^{-3} + g$ that $(a/g)^{1/3} = R(\varepsilon_m/\varepsilon_{de})^{1/3} = R_{pr}(\varepsilon_m/\varepsilon_{de})_{pr}^{1/3} \approx R_{pr}(0.30711/0.6914)^{1/3} = 0.763R_{pr}$. Taking into account that presently $\varepsilon_{de} \approx 0.6914\,\varepsilon$ and that $H_{pr} = \sqrt{b\,\varepsilon} \approx \sqrt{b\,\varepsilon_{de}}/\sqrt{0.6914}$ we have for $\frac{3}{2}\sqrt{g} \equiv \frac{3}{2}\sqrt{b\,\varepsilon_{de}} = \frac{3}{2}(0.6914)^{1/2}H_{pr} \approx 2.7393 \times 10^{-18}\mathrm{s}^{-1} = 0.086446\,\mathrm{Gyr}^{-1}$. Thus (15.15) becomes

$$R(t) \approx 0.763\,R_{pr}\left\{\sinh\left[0.086446\left(\frac{t}{1\,\mathrm{Gyr}}\right)\right]\right\}^{2/3} \tag{15.16}$$

Let us set our formula (15.16) under the scrutiny of a few tests: If we substitute for *t* the estimated present age of the Universe, *t* = 13.7965 Gyr we find $R(t_{pr}) \approx 0.998\,R_{pr}$, a discrepancy of less than 0.2 %. The time t_e at which $\varepsilon' \approx \varepsilon_{de}$ is obtained by noticing that the equality $\varepsilon' \approx \varepsilon_{de}$ implies that $g \approx a/R^3$ or $R \approx (a/g)^{1/3}$ which in combination with (15.15) leads to $\sinh(\frac{3}{2}\sqrt{g}\,t_e) \approx 1$ or $t_e \approx 10.19\,\mathrm{Gyr}$, which is 6 % larger than the more accurate numerical calculations. Let us find also the time t_o at which the acceleration vanishes, so that for $t < t_o$ the acceleration is negative, while for $t > t_o$ it is positive. Equation (15.8) shows that $\ddot{R} = 0$ is equivalent to $\varepsilon + 3p = 0$, or $\varepsilon' + \varepsilon_{de} - 3\varepsilon_{de} = 0$, or $\varepsilon' = 2\varepsilon_{de}$. We obtained these relations by taking into account that for *t* equal to a few billion years the only appreciable contribution to the pressure is the one coming from the dark energy. The relation $\varepsilon' = 2\varepsilon_{de}$ is equivalent to $a/R^3 \approx 2g$ or $R \approx (a/2g)^{1/3}$. Combining this last relation with (15.15) we obtain that $\sinh(\frac{3}{2}\sqrt{g}\,t_o) \approx 1/\sqrt{2}$ from which we get $t_o \approx 7.62\,\mathrm{Gyr}$. This estimate is very close to the best estimates for t_o, which is 7.83 Gyr. Finally, let us apply (15.16) to estimate the ratio of the present radius over the radius at *t* = 0.38 Myr, although (15.16) is not valid for *t* = 0.38 Myr (because calculations presented in solved problem 1, Sect. 15.8 indicate that at *t* = 0.38 Myr the neutrino and the photon contributions to ε are not negligible but they are 10 and 14 % respectively). The result, according to (15.16). is still a reasonable value of 1278, while the accepted value is 1090, a 13 % error. A better estimate will be given below.

The equation for the transition period between the photon domination and the matter domination is

$$\frac{\dot{R}^2}{R^2} = b\left[\frac{a_{ph}}{R^4} + \frac{a_m}{R^3}\right], \quad b \equiv \frac{8\pi G}{3\,c^2} \tag{15.17}$$

The reader may convince himself/herself that the solution of this differential equation is the following:

$$t = \frac{2a_{ph}^{3/2}}{b^{1/2}a_m^2}\left[\frac{1}{3}(1+a_0R)^{3/2} - (1+a_0R)^{1/2} + \frac{2}{3}\right], \quad a_0 \equiv a_m/a_{ph} \tag{15.18}$$

We take the limit of very small a_0R and by expanding the expressions $(1+a_0R)^{3/2}$ and $(1+a_0R)^{1/2}$ in powers of a_0R we obtain the following result

$$R \approx \left(2b^{1/2}a_{ph}^{1/2}t\right)^{1/2}, \quad a_0R \ll 1 \tag{15.19}$$

from which by comparing with (15.10) we find that

$$2b^{1/2}a_{ph}^{1/2} = 7.1065 \times 10^{33}\ \mathrm{m^2\,s^{-1}} \tag{15.20}$$

Similarly by going to the opposite limit of $a_0R \gg 1$ we find that

$$R \approx \left(\frac{3}{2}b^{1/2}a_m^{1/2}t\right)^{2/3}, \quad a_0R \gg 1 \tag{15.21}$$

from which by comparing with (15.16) in the limit $t \ll 1$ Gyr we obtain

$$\left(\frac{3}{2}\right)^{2/3} b^{1/3}a_m^{1/3}t^{2/3} = 1.4938 \times 10^{-12}\left(\frac{t}{1\ \mathrm{s}}\right)^{2/3} R_{pr} \tag{15.22}$$

The numerical values of a_{ph} and a_m are determined from (15.20) and (15.22) so that (15.18) becomes

$$\frac{t}{1\ \mathrm{kyr}} = 191.2\left\{\frac{1}{3}\left[1+\left(0.9658\frac{R}{10^{23}\ \mathrm{m}}\right)\right]^{3/2} - \left[1+\left(0.9658\frac{R}{10^{23}\ \mathrm{m}}\right)\right]^{1/2} + \frac{2}{3}\right\} \tag{15.23}$$

To test (15.23) we choose for R the value $R = R_{pr}/1090 = 4 \times 10^{23}$ m for which the established value of the corresponding time is 380 kyr. Equation (15.23) gives 389 kyr, a discrepancy of only 2.3 %. The value $R = 10^{23}$ m corresponds to $t = 35$ kyr. In Fig. 15.4 we plot the dependence of the radius of the observable Universe vs. time and in Fig. 15.5 we show the evolution of the fractional contribution of the various components to the energy density.

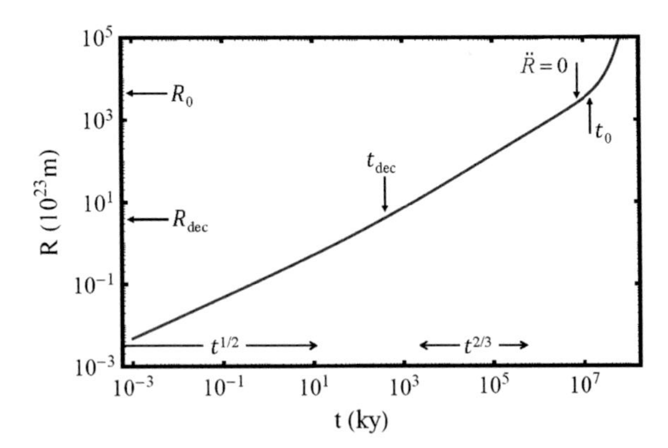

Fig. 15.4 The dependence of the radius of the observable Universe on time since the big bang; t_0 is the present age of 13.8 Gyr and t_{dec} is the time of decoupling at 380 kyr

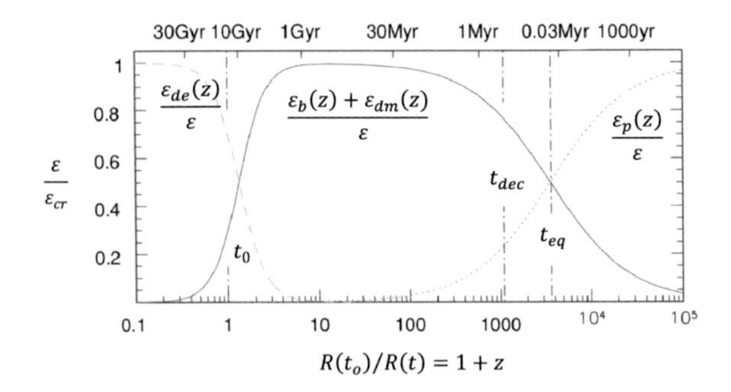

Fig. 15.5 The evolution of the fractional contribution of the various energy density components (ε_{ph}: radiation, $\varepsilon_b + \varepsilon_{dm} = \varepsilon_m$: ordinary baryonic and dark matter, ε_{de}: dark energy) to the total energy density ε of the universe as a function of "redshift", z. The total energy density is normalized to the critical density, ε_{cr}, (see Table 15.2). Note that the redshift of a distant object is defined by the ratio $R(t_0)/R(t)$ of the current radius of the Universe $R(t_o)$, to the radius of the Universe, $R(t)$, at time t, when the light from this object was emitted towards us (figure adapted from Fig. 8.7 "Galaxies in the Universe", L.S. Sparke and J.S. Gallagher, 2nd Edition.)

15.5 Formation of the Structures of the Universe

Having the temperature and the density of the Universe as a function of time, we can employ known laws of physics to examine the formation of various structures of matter as time goes on. For the very early period of the Universe, $t \leq 10^{-12}$ s, when the energy per particle was more than 10 TeV, rather detailed scenarios have been proposed including the inflation model; the latter was proven very effective in explaining several puzzling observations. These scenarios are based on theoretical schemes for which there is no direct experimental support. Here we shall restrict ourselves to briefly mention some crucial events in the history of the Universe starting from about 10^{-12} s after the time zero of the big bang corresponding to energies achievable experimentally.

Already in powerful accelerators on Earth energies approaching the 10 TeV per particle have been achieved and experiments are under way where fully ionized heavy nuclei moving with almost the velocity of light are forced to collide head-on with each other. Just after such a collision, the matter is in a state similar to that prevailing in the Universe during the period of about 10^{-12}–10^{-6} s, when the energy per particle was in the range from about 1.2 TeV to 1.2 GeV. For such energies quarks and gluons form a dense and very hot "soup" called *quark gluon plasma*.

As the energy per particle was dropping into the range 1–0.1 GeV corresponding to time in the range 10^{-6}–10^{-4} s the process of *baryogenesis* was taking place, during which quarks were combined, with the intervention of gluons, to form protons, neutrons and other short-lived baryons. During baryogenesis a thermodynamic equilibrium between protons and neutrons was established through reactions of the form $p + e^- \rightleftarrows n + \nu_e$ which led to a ratio of neutrons to protons equal to $n/p = \exp(-\Delta\varepsilon/k_B T)$, where $\Delta\varepsilon = (m_n - m_p)c^2 = 1.2934\,\text{MeV}$, is the rest energy difference between neutron and proton. As the temperature was dropping this ratio was changing; at $t \approx 3\,\text{s}$, when $k_B T \approx 0.7\,\text{MeV}$, $n/p \approx 1/6$. From this point on the rate of the expansion of the Universe was faster than the rate of the reactions changing neutrons to protons and vice versa; thus the ratio n/p "froze" temporarily at the value 1/6. Then the neutrons start decaying irreversibly through the reaction $n \to p + e^- + \bar{\nu}_e$ until the nucleosynthesis started and the neutrons were trapped and stabilized within the He^4 nucleus, at about $t \approx 500\,\text{s}$; at this point $n/p \approx 1/7$.

The *nucleosynthesis* started at about 120 s when $k_B T \approx 0.11\,\text{MeV}$ was low enough for the process to start with the reaction $n + p \to d + \gamma$; then it proceeded through a series of reactions which were concluded at about 1200s with almost all the neutrons bound in He-4 and in traces of deuterium. Thus, out of two neutrons and fourteen protons (ratio 1/7) two neutrons and two protons made one He-4 nucleus and 12 protons remained free; the result is one He-4 nucleus for each dozen of protons or a mass ratio of 25–75 % between He-4 and hydrogen, a prediction consistent with observational data.

As the Universe was expanding and cooling continuously, electrons started to bind around the He-4 nuclei and around the protons to form *neutral atoms* of helium and hydrogen. This process was concluded when the temperature was about 3000 K; at this temperature one can show that thermodynamic equilibrium dictates that practically all matter consists of neutral atoms of helium and hydrogen. The temperature of 3000 K was reached when the radius of the observable Universe was about 1090 smaller than the present radius and the time was about

t = 380,000 years. Equation (15.23) gave t = 389,000 years for $R_{pr}/R \approx 1090$). The radius of the observable Universe at t = 380,000 years was $R = 4 \times 10^{23}$ m and the concentration of neutral hydrogen atoms was $2.72 \times 10^8\,\mathrm{m}^{-3}$ i.e. 1090^3 times larger than the present one which is about 0.21 m^{-3}.

At T = 3000 K the maximum of the spectral distribution of the CMB radiation appears at $\lambda \approx 9650$ A. From equation (6.17) of Chap. 6 we can find that the total scattering cross-section of photons of wavelength 9650 A by neutral hydrogen atoms is equal to $6.58 \times 10^{-33}\,\mathrm{m}^2$. The scattering cross-section by He-4 neutral atoms is 9 % of that of hydrogen. The mean free path of photons at t = 380,000 years can be calculated from (6.20) and it is equal to $\ell \approx 5.6 \times 10^{23}$ m, i.e., slightly larger than the radius of the observable Universe at that time. The conclusion is that at $t \approx 380,000$ years the Universe became transparent to E/M radiation or, equivalently, matter and light were decoupled. If, at this time, the matter were fully ionized the scattering by electrons would dominate, as shown by (6.10), and the mean free path would be 4.8×10^{19}, i.e., four orders of magnitude smaller than the radius of the then observable Universe. Thus the matter by becoming neutral practically ceased to absorb and reemit photons and thus produce any light and, hence, became invisible. The dark age of the Universe started. This dark age lasted for about 100 Myr or so, according to the latest estimates, until the first protostar was formed and reionized the matter around and inside it. The formation of galaxies as well as the first stars within the galaxies followed; stars during their lifetime fuse hydrogen and He to heavier elements all the way to iron. Beyond iron the fusion of heavier elements requires an external supply of energy as shown in Fig. 9.1. This extra energy is provided during the explosive death of big stars through the supernova II process, which is credited for the creation of the nuclei (and hence atoms) heavier than iron. The material dispersed by the supernova explosions-containing all the chemical elements—is recycled by gravity to form new stars and planets such as ours, acting as possible cradles of life and eventually of intelligent life wondering about this World, "this small world, the great".

15.6 Summary of Important Relations

The basic equation is (15.1) connecting the square of the Hubble constant with the average energy density in the Universe. The various contributions to the average energy density including the dark matter and the dark energy are given in (15.2). The acceleration of the expansion of the Universe can be obtained (see (15.8)) from (15.1) and the first law of thermodynamics. It is worth noticing (15.11) which shows that the product RT remains almost constant over the history of the Universe. Finally, (15.15)/(15.16) and (15.18)/(15.23) give reasonably well the dependence of the radius of the observable Universe on time for at least after the first few seconds

15.7 Multiple-Choice Questions/Statements

1. The basic equation of Cosmology for a homogeneous, isotropic and without boundaries Universe having an energy density $\varepsilon = \rho c^2$ can be derived within the framework of Newtonian mechanics as follows: Consider the acceleration of a galaxy of mass m at a distance R from the observer: $m\ddot{R} = -GMm/R^2$. Integrating this equation under constant mass we obtain the basic equation, which is:
(a) $(\dot{R}/R)^2 = -(4\pi/3c^2)G\varepsilon - C/R^2$ (b) $(\dot{R}/R)^2 = (8\pi/3c^2)G\varepsilon - C/R^2$
(c) $(\dot{R}/R)^2 = (8\pi/3c^2)G\rho - C/R^2$ (d) $(\dot{R}/R)^2 = -(4\pi/3c^2)G\rho - C/R^2$

2. Starting from the basic equation (connecting the squared Hubble constant to the average energy density) and the first law $dU = -pdV$, the equation giving the acceleration of the expansion of the Universe is derived, which is:
(a) $(\ddot{R}/R) = -(4\pi G/3c^2)(\varepsilon' + 3p') + (\Lambda/3)$, $(\varepsilon_{d\varepsilon} = \Lambda c^2/8\pi G)$
(b) $(\ddot{R}/R) = -(4\pi G/3c^2)(\varepsilon' - 3p') + (\Lambda/3)$
(c) $(\ddot{R}/R) = (4\pi G/3c^2)(\varepsilon' - 3p') + (\Lambda/3)$
(d) $(\ddot{R}/R) = (4\pi G/3c^2)(\varepsilon' + 3p') + (\Lambda/3)$

3. If a component ε_i contributing to the energy density of the Universe remains constant in spite of the expansion of the Universe, as we assume it happens with ε_{de}, then from the conservation of energy it follows that:
(a) $p_i = \varepsilon_i$ (b) $p_i = 3\varepsilon_i$ (c) $p_i = -\varepsilon_i$ (d) $p_i = -3\varepsilon_i$

4. In the photon dominated era the energy density of the Universe satisfies one of the following relations:
(a) $\varepsilon \propto R^{-2}$ (b) $\varepsilon \propto R^{-3}$ (c) $\varepsilon \propto R^{-4}$ (d) $\varepsilon \propto R^{-5}$

5. In the matter dominated era the energy density of the Universe satisfies one of the following relations:
(a) $\varepsilon \propto R^{-5}$ (b) $\varepsilon \propto R^{-4}$ (c) $\varepsilon \propto R^{-3}$ (d) $\varepsilon \propto R^{-2}$

6. In the epoch where the dark energy will dominate the Universe the time dependence of R will be one of the following relations:
(a) $R \propto t^{1/2}$ (b) $R \propto t^{3/2}$ (c) $R \propto t^{9/2}$ (d) $R \propto \exp(Ht)$

7. In the matter dominated era the time dependence of R is one of the following:
(a) $R \propto t^{1/2}$ (b) $R \propto t^{2/3}$ (c) $R \propto t^{3/2}$ (d) $R \propto \exp(Ht)$

8. In the photon dominated era the time dependence of R is one of the following:
(a) $R \propto t^{1/2}$ (b) $R \propto t^{2/3}$ (c) $R \propto t^{3/2}$ (d) $R \propto \exp(H_0 t)$

9. In the transition period from the matter to the dark energy dominated era the time dependence of R is one of the following:

(a) $R(t) = 0.763 R_{pr} \{\sinh(0.0864464t)\}^{2/3}$, t in Gyr

(b) $R(t) = 0.763 R_{pr} \{\sinh(0.0864464t)\}^{1/3}$

(c) $R(t) = 0.763 R_{pr} \{\sinh(0.0864464t^{3/2})\}^{2/3}$

(d) $R(t) = 0.763 R_{pr} \{\sinh(0.0864464t^{1/3})\}^{2/3}$

10. The expansion of the Universe changed from decelerating to accelerating during one of the following cosmic epochs:
(a) the photon dominated era
(b) the matter dominated era
(c) the transition period from matter to dark energy dominated era
(δ) the transition period from photon to matter dominated era

11. The composition of the energy density of the Universe during the present era is according to the Planck data of March 2013:
(a) $\varepsilon_{de} : \varepsilon_{dm} : \varepsilon_b = 68.25 : 26.71 : 4.89\%$ (b) $\varepsilon_{de} : \varepsilon_{dm} : \varepsilon_b = 75.25 : 20.71 : 3.89\%$
(c) $\varepsilon_{de} : \varepsilon_{dm} : \varepsilon_b = 26.71 : 68.25 : 4.89\%$ (d) $\varepsilon_{de} : \varepsilon_{dm} : \varepsilon_b = 3.89 : 20.71 : 75.25\%$

12. Based on the Planck data of March 2013 the present value of the Hubble constant is:
(a) $H_0 \approx 85.5\,\text{km/s}\cdot\text{Mpc}$ (b) $H_0 \approx 67.1\,\text{km/s}\cdot\text{Mpc}$; $1\text{Mpc} = 10^6\,\text{pc}$ (see p. 193)
(c) $H_0 \approx 60.5\,\text{km/s}\cdot\text{Mpc}$ (d) $H_0 \approx 40.5\,\text{km/s}\cdot\text{Mpc}$

13. Based on the Planck data of March 2013 the present baryon energy density in the Universe is the equivalent of:
(a) 5 protons/ m^3 (b) 1 proton/ m^3 (c) 1/2 protons/ m^3 (d) 1/4 protons/ m^3

14. The baryogenesis took place during the period $10^{-5}\,\text{s} \leq t \leq 150\,\text{s}$, when the temperature was around the value:
(a) $T \approx 10^8\,\text{K}$ (b) $T \approx 10^{12}\,\text{K}$ (c) $T \approx 10^{16}\,\text{K}$ (d) $T \approx 10^{20}\,\text{K}$

15. The nucleosynthesis took place during the period $120\text{s} \leq t \leq 1200\text{s}$, when the temperature was around the value:
(a) $T \approx 10^9\text{K}$ (b) $T \approx 10^7\text{K}$ (c) $T \approx 10^5\text{K}$ (d) $T \approx 10^3\text{K}$

16. The present era temperature of the CMB radiation is:
(a) 273.15 K (b) 68.21 K (c) 23.65K (d) 2.725 K

17. The "age" of the Universe is:
(a) 13.8 Myr (b) 138 Myr (c) 1.38 Gyr (d) 13.8 Gyr

18. The present era temperature of the CMB radiation is 2.7255 K. The radius of the Universe since the decoupling time has increased by a factor of 1090. What was the temperature of the CMB radiation at the time of decoupling?
(a) 2.725 K (b) 0.2971K (c) 297.1 K (d) 2971 K

19. The CMB radiation as a function of angular frequency ω exhibits a maximum at $\hbar\omega = 0.665\,\text{meV}$. Based on this information we can deduce that the temperature of the CMB radiation is:
(a) 2.724 K (b) 0.297K (c) 297 K (d) 2970 K

15.8 Solved Problem

1. *Estimate the composition of the average energy density of the Universe at the decoupling time* $t = 380$ kyr. *It is given that at this time the energy density of the neutrinos was about 0.705 of the energy density of photons.*

Solution We shall use a subscript d to denote quantities at $t = 380$ kyr and no subscript for the quantities at time present. We also define $x \equiv T_d/T = R/R_d$, where $x = 1090$. We also have

$$\varepsilon_{ph,d} = x^4\varepsilon_{ph},\ \ \varepsilon_{\nu,d} = 0.705\varepsilon_{ph,d},\ \ \varepsilon_{b,d} = x^3\varepsilon_b,\ \ \varepsilon_{dm,d} = x^3\varepsilon_{dm},\ \ \varepsilon_{de,d} = \varepsilon_{de} \quad (15.24)$$

Because of the large factor x it is obvious that the percentage of the dark energy is quite negligible at the time of decoupling. Thus the total energy density at that time was

$$\varepsilon_{t,d} = \varepsilon_{ph,d} + \varepsilon_{\nu,d} + \varepsilon_{b,d} + \varepsilon_{dm,d} = \varepsilon_{dm,d}\left\{1 + \left(\varepsilon_{b,d}/\varepsilon_{dm,d}\right) + \left(1.705\varepsilon_{ph,d}/\varepsilon_{dm,d}\right)\right\} \quad (15.25)$$

Taking into account (15.24) and the fifth column of Table 15.1 we have

$$\varepsilon_{ph,d}/\varepsilon_{dm,d} = x(\varepsilon_{ph}/\varepsilon_{dm}) = 1090 \times (0.005/25.886) = 0.21 \quad (15.26a)$$

$$\varepsilon_{b,d}/\varepsilon_{dm,d} = \varepsilon_b/\varepsilon_{dm} = 4.825/25.886 = 0.186 \quad (15.26b)$$

$$\varepsilon_{t,d} = \varepsilon_{dm,d}\{1 + (0.186) + (1.705 \times 0.21)\} = 1.554\varepsilon_{dm,d} \quad (15.26c)$$

Therefore the percentages at the time of decoupling are

$$\begin{aligned}\varepsilon_{ph,d}/\varepsilon_{t,d} &= 0.21/1.554 \approx 14\,\%,\ \ \varepsilon_{\nu,d}/\varepsilon_{t,d} = 0.705 \times 14\,\% \approx 10\,\%,\\ \varepsilon_{b,d}/\varepsilon_{t,d} &= 0.186/1{,}554 \approx 12\,\%,\ \ \varepsilon_{dm,d}/\varepsilon_{t,d} = 1/1.554 \approx 64\,\%\end{aligned} \quad (15.27)$$

15.9 Unsolved Problems

1. Prove that the function (15.15) satisfies the differential equation (15.14)
2. Prove that the function (15.18) satisfies the differential equation (15.17)
3. Prove the relation (15.19)
4. Prove the relation (15.21)

5. When the range of temperatures is such that the reactions $e^- + e^+ \to 2\gamma$, $\gamma \to e^- + e^+$ are in equilibrium, show that the energy densities of electrons, positrons, and photons satisfy the relation $\varepsilon_{e^-} = \varepsilon_{e^+} = (7/8)\varepsilon_{ph}$
6. Use the present era results of Table 15.2 to calculate the quantity $C \equiv RT$ and compare with the value given in (15.11). Is the constancy of C verified?

Reference

1. S. Weinberg, *The First Three Minutes* (Fontana Paperbacks, London, 1983)

Epilogue
The Anthropic Principle

This World, he said, is your commandment
andin your bowels is written.
O. Elytis

Abstract The laws of physics and the values of the universal physical constants are such as to make the phenomenon of life possible.
Summary The biological fitness of the laws of Nature and the biological fitness of the values of the universal physical constants constitute a part of one version of the anthropic principle. A stronger version claims that the statement 'Life exists' is not only the shortest but the most comprehensive statement about the World. Most versions of the anthropic principle include also an 'explanation' of either the fitness or the assumed necessity of the physical laws and the values of the physical constants for the phenomenon of life as we know it on Earth.

In our short journey over the structures of the World we bypassed the most interesting ones, namely the living structures. The reason was that living structures do not seem to fit with what was the purpose of this book: *To derive some select but important quantitative (or semi-quantitative) results regarding structures of matter in a simple way starting from a few established general laws of Nature and the values of the universal physical constants.* The out of thermodynamic equilibrium living matter seems to be too complex and diverse to fit within this limited purpose and thus to allow a simple derivation of quantitative results from a few fundamental laws of physics.

However, the opposite direction, from the fact that "life exists" to the "laws of nature and to the values of the physical constants" is not without some merit [1, 2]. It must be admitted that this subject known under the generic name *anthropic principle* is rather controversial, because, on the one hand it is ideologically loaded, and on the other, it does not usually provide concrete quantitative conclusions. Nevertheless, in this short epilogue we shall mention a few points showing that one can deduce some fundamental laws of Nature or at least demonstrate their necessity from the fact that life exists. In what follows we shall refer mostly to the system Earth/Sun where carbon based life exists. However, most of our arguments are applicable to any other planet/star or satellite/star system sustaining Earth-like life.

E.N. Economou, *From Quarks to the Universe*,
DOI 10.1007/978-3-319-20654-7

The admirably precise reproducibility from generation to generation of living structures strongly suggests that matter is "digital" not "analog", meaning that matter is of discrete not of continuous character. It is very difficult to imagine a continuous complex medium in contact with an ever changing environment to emerge again and again with exactly the same form and the same features. For this impressive phenomenon to occur, the huge amount of information stored in the "central bank" of the DNA molecule has to be kept intact. This in turn requires the double discretization shown in Fig. 3.3d or (d'), because, otherwise, the DNA molecule would undergo a continuous change and it would never be the same. In other words, from the precise reproducibility of the living structures one is led to the discretization along both the vertical and the horizontal axis of Fig. 3.3(d or d'). The discretization along the vertical axis suggests, if not implies, the atomic idea. Similarly, the discretization along the horizontal axis is consistent with the wave-particle duality [2]. Whether one can reverse the argument and deduce from the double discretization, the atomic structure of matter (and forces) as well as the particle/wave duality is not clear.

The existence of an extremely weak, attractive, long range force, such as gravity, seems to be necessary in order to keep the atmosphere, the lakes, the oceans, and the living matter on the surface of the Earth; as well as to keep the Earth going around the Sun and drawing from it both the energy and the information required for the emergence and maintenance of life. This leads to the necessity of another force, the electromagnetic one, which has a dual responsibility: To transmit the energy and the information from the Sun to the Earth; and to bring and keep together all matter including the one from which the living structures consist. The strength of the E/M force has to be enormous in comparison with the gravitational one. This becomes apparent by considering what the size of a single-cell organism would be, if it would be held together by gravitational forces instead of the electromagnetic ones: its diameter would be an absurd size of 10^{21} m, instead of 10^{-5} m which is its actual size! Moreover, the E/M force by being both attractive and repulsive manages to be both long range in transmitting energy and information from the Sun and short range in bringing and keeping together the constituents of living matter.

Life on Earth requires the continuous flow of energy from the Sun over a long period (of the order of a few billion years). Such a supply of energy cannot be chemical, because it would be exhausted in a period of thousands of years. There must be another source of energy about a million times larger than the chemical one. Assuming that the available energy is of the same order of magnitude as the kinetic energy and taking into account Heisenberg's principle, we conclude that the required supply of energy E per particle must come from a region of size r and from particles of mass m such that $E/E_{chem} \approx (m_e a_B^2/m\,r^2) \approx 10^6$. The nuclear size of $r = 1$ fm and the proton mass of $1836\,m_e$ produces the necessary boost of 1.5×10^6. Hence, it is clear that the nucleus is the source of the energy supply in the Sun (and in any other active star). This implies that nuclei are composite particles held together by a third type of force-the strong one- capable to overcome the kinetic energy of the nucleons and the Coulomb repulsion among the protons. Finally, a fourth interaction, the weak one, is necessary in order to allow the transformation from proton to neutron and vice versa. Without this, the fusion of two protons to a deuteron (with the emission of a positron and a neutrino) would not be possible, neither would be possible the eventual fusion of four protons to a helium nuclei, the main reaction providing the energy of the stars.

Let us consider now the values of some universal physical constants. The mass of the electron is less than the mass difference between a neutron and a proton. This allows the neutron, if free, to decay to a proton, an electron, and an antineutrino. If this were not the case, electrons and protons would combine to free neutrons in the early Universe leaving very few protons to act as fuel in the stars. Thus, a neutron has to be heavier than the sum of an electron and a proton but not much heavier, because in such a case very few neutrons would be produced during the baryogenesis or within the nuclei (see Sect. 15.5). As a result of the greater percentage of protons heavier elements necessary for life would not be stable. Elements heavier than He and up to Fe are made by fusion within active stars. This stellar nucleosynthesis passes through a bottle-neck to reach the carbon nucleus. The reason is that the Be^8 nucleus created by the fusion of two He^4 nuclei and hence consisting of 4 protons and 4 neutrons is extremely metastable with a lifetime of the order of 10^{-15}s, because it is energetically favorable to break into two He nuclei. Thus it is extremely improbable for a third He atom to be incorporated in the Be^8 to produce the crucial for life carbon atom, unless a resonance effect is present. Indeed, such an effect occurs in the sense that the energy of Be^8 plus the energy of the third He nucleus slightly exceeds an excited energy level of carbon at 7.65 MeV above its ground state. This almost coincidence greatly enhances the probability of the incorporation of the third He nucleus and makes the carbon nucleus possible. It is worth mentioning that Fred Hoyle, by simply relying on the observed abundance of carbon in the Universe, *predicted* the occurrence of such an excited level in the carbon nucleus before it was found experimentally.

The existence of life implies that the dimensionless gravitational constant must have a value close to the measured one. Indeed, it has been estimated that about 10 billion years are needed for life to emerge in the Universe, since life is the final stage and the crowning of the matter dominated epoch. On the other hand, life ought to appear and evolve before the dark energy dominates, since the latter tends to dissolve the structures of matter. These two requirements imply that $\sqrt{g}\, t \approx 1$ for $t \approx 10^{10}$ys (see (15.15)). This in turn means that the dimensionless gravitational coupling constant must satisfy the relation $a_G \approx 4.7 \times 10^{-39}$ i.e. about 20 % smaller than its measured value. (The estimated value of $\rho_{de} = \varepsilon_{de}/c^2 \approx 0.6914 \times 8.627 \times 10^{-27}$kg/m^3 was used in obtaining this value of a_G).

As it was mentioned before, supernova explosions, in particular core-collapse type II supernova explosions (see below), are important for the creation and the recycling of the elements in general and for the emergence of life in particular. The heavier than iron elements are created during these events. Moreover, all elements are dispersed as a result of supernova explosions very efficiently in the interstellar medium to be recycled by gravity for the formation of the next generation of stars, of planetary systems and, eventually, of humans. In rough terms, a type II supernova explosion occurs as follows: The central part of a big star, upon the exhaustion of its nuclear fuel, collapses to a core so dense that even neutrinos are trapped in it. This superdense core attracts and accelerates the outer part of the star, which hits the core at high speed. This internal collision forces the outer part to rebound and at the same time releases the neutrinos which push the outer part with explosive force leading to the supernova II. For the neutrino flux to be capable of catapulting the outer part of the star against the gravitational attraction, the following relation connecting the weak dimensionless coupling constant a_W with the gravitational one a_G, must be satisfied:

$$a_W^4 \approx a_G (m_p/m_e)^6 \quad \Rightarrow \quad a_W \approx 2.18 \times 10^{-5}$$

The resulting value of a_W is in reasonable agreement with its theoretical and measured value which is about 10^{-5}.

We shall conclude this epilogue by drawing attention to the mystery of the cosmological constant Λ, or equivalently the density of the dark energy, ε_{de}. If we accept the established value of the gravitational constant, G, the measured value of Λ has the right order of magnitude for the transition from matter domination to dark energy domination to occur at $t \approx 10$ billion years, i.e., when the time was ripe for life to emerge. This looks like an extremely fine tuning of the value of Λ, in view of theoretical considerations producing values of Λ 10^{120} larger than the actual one!

References

1. S.W. Hawking, L. Mlodinow, *The Grand Design* (Bantam books, NY, 2010)
2. E. Schrödinger, *What is Life*? (Gambridge University Press, Cambridge, 1967)

Appendix A
Oscillations and Waves

Oscillations occur locally in systems where their energy can change back and forth from the form of potential energy E_P to the form of kinetic energy E_K. A typical mechanical example is the pendulum shown in Fig. A.1a. A typical electrical example is the LC circuit shown in Fig. A.1b; in this case the potential energy is equal to $\frac{1}{2}Q^2/C$, while the kinetic-type energy is $\frac{1}{2}LI^2/c^2$ (which includes both the systematic kinetic energy $\frac{1}{2}N_e m_e \upsilon^2$ of the electrons and the magnetic energy associated with the current $I = (N_e/\ell)\, e\, \upsilon$). C is the capacitance, L is the self-inductance$I = dQ/dt$, and ℓ is the overall length of the wire making up the circuit. In cases of macroscopic oscillations, as those in Fig. A.1a, b macroscopic kinetic energy is transferred through collisions with microscopic particles to the internal energyof both the system and the environment. This process is described macroscopically as the action of frictional "forces", which eventually lead to the "loss" of the macroscopic oscillating energy and the eventual termination of the macroscopic oscillation. If there is no transfer of energy from the oscillating system, then conservation of energy, $E_t = E_K + E_P$, implies that $E_{KM} - E_{Km} = E_{PM} - E_{Pm}$, where the subscripts M and m indicate maximum and minimum values respectively. This observation allows us to obtain the angular frequency$\omega = 2\pi f$ (f is the frequency) of the oscillation. E.g., in the case of Fig. A.1b by setting $LI_M^2/c^2 = Q_M^2/C$, and taking into account that $I^2 = \omega^2 Q^2$, we obtain $\omega^2 = c^2/LC$ (see also Sect. 5.1).

Waves are oscillations which migrate from local subsystems to neighboring subsystems within an extended medium (and, consequently, are spread and are delocalized). The medium can be discrete, as the coupled pendulums shown in Fig. A.1c, or continuous, as in the case of sea-waves shown schematically in Fig. A.1d. A wave in addition to the angular frequency ω of the propagating oscillation is usually characterized by its wavelengthλ (or equivalently by its wavenumber$k \equiv 2\pi/\lambda$) and its velocity υ of propagation. The three quantities ω, k, υ are related as follows:

$$\omega = \upsilon k \tag{A.1}$$

E.N. Economou, *From Quarks to the Universe*,
DOI 10.1007/978-3-319-20654-7

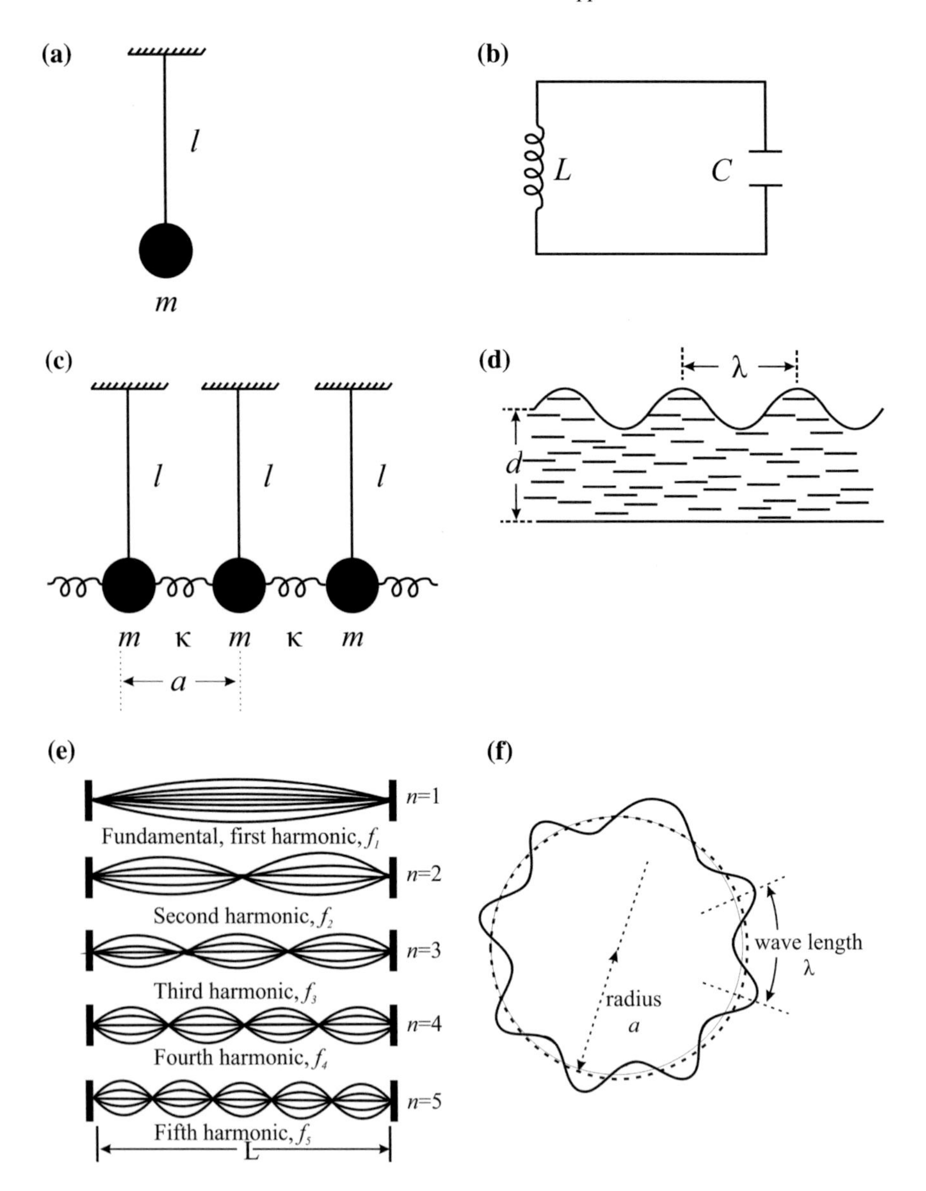

Fig. A.1 **a** A pendulum. **b** An LC circuit. **c** A medium of coupled pendulums. **d** Sea waves. **e** Standing or stationary waves in a string of fixed ends. **f** Standing or stationary wave around a *circle*

The velocity υ depends on the properties of the medium and possibly of the wavenumber k (see also Sect. 5.5).

Waves confined in a finite region of space have to satisfy certain boundary conditions, which restrict the allowed values of the angular frequency to a discrete set, ω_o, ω_1, $\omega_2, \ldots$ (see Fig. 3.3). To each value ω_i correspond one or more waves, called standing waves which, having spread over the available space, are not traveling anymore. Examples of standing waves are the ones appearing in a string of fixed ends (see Fig. A.1e) or the ones going around in a circle (see Fig. A.1f).

Appendix B
Elements of Electrodynamics of Continuous Media

B.1 Field Vectors, Potentials and Maxwell's Equations

Within a medium we define the following quantities by averaging them over a volume ΔV: *The total macroscopic current density,* $\mathbf{j}_t$, *the total macroscopic polarization,* $\mathbf{P}$, *and the total macroscopic magnetization,* $\mathbf{M}$. The term *total* means that all charges of the medium are included both "bound" and 'free'. The averaging volume ΔV is centered at the point $\mathbf{r}$ and is such that $a^3 \ll \Delta V \ll \lambda^3$, where the length a is of atomic scale and λ is the length characterizing the variation of the averaged fields.

$$\mathbf{j}_t(\boldsymbol{r}) \equiv \frac{1}{\Delta V}\sum_{i,\Delta V} q_i \boldsymbol{\upsilon}_i = \frac{1}{\Delta V}\int_{\Delta V} \rho(\mathbf{r}')\boldsymbol{\upsilon}(\mathbf{r}')\mathrm{d}^3 r' \tag{B.1}$$

$$\mathbf{P}(\mathbf{r}) \equiv \frac{1}{\Delta V}\sum_{i,\Delta V} q_i \mathbf{r}_i \equiv \frac{1}{\Delta V}\int_{\Delta V} \mathbf{r}'\rho(\mathbf{r}')\mathrm{d}^3 r' \tag{B.2}$$

$$\mathbf{M} \equiv \frac{1}{2c}\frac{1}{\Delta V}\sum_{i,\Delta V} q_i\, \mathbf{r}_i \times \boldsymbol{\upsilon}_i = \frac{1}{2c}\frac{1}{\Delta V}\int_{\Delta V} \mathbf{r}' \times \rho(\mathbf{r}')\boldsymbol{\upsilon}(\mathbf{r}')\mathrm{d}^3 r, \text{G-CGS} \tag{B.3}$$

Obviously, outside the material, in the vacuum, the quantities, $\mathbf{j}_t$, $\mathbf{P}$, $\mathbf{M}$, and the charge density ρ are zero.[1] For the connection of the Gauss(G)-CGS system of units with the SI one, see Footnote 1 and the last two pages of this appendix.

From (B.1) and (B.2) follows that

$$\mathbf{j} = \frac{\partial \mathbf{P}}{\partial t} \tag{B.4}$$

[1] In SI set $c = 1$. To change formulae from SI to G-CGS (and vice versa) see the table at the end. The relation $\varepsilon_0\mu_0 = 1/c^2$ may help in transforming formulas from SI to G-CGS.

E.N. Economou, *From Quarks to the Universe*,
DOI 10.1007/978-3-319-20654-7

Usually we combine **E** and **P** to define the so-called *electric induction* **D**; we also combine **B** and **M** to define the so-called "auxiliary" magnetic field **H**:

$$\begin{aligned}\mathbf{D} &\equiv \varepsilon_0\mathbf{E} + \mathbf{P}, \quad \text{SI},\\ &\equiv \mathbf{E} + 4\pi\mathbf{P}, \quad \text{G-CGS},\end{aligned} \tag{B.5}$$

$$\begin{aligned}\mathbf{H} &\equiv \frac{1}{\mu_0}\mathbf{B} - \mathbf{M}, \quad \text{SI},\\ &\equiv \mathbf{B} - 4\pi\mathbf{M}, \quad \text{G-CGS},\end{aligned} \tag{B.6}$$

where ε_o, the permittivity of vacuum, and μ_o, the magnetic permeability of the vacuum, are universal constants (see Table I.1 of Appendix I) whose product equals to the inverse of the square of the velocity of light in vacuum.

$$c^2 = \frac{1}{\varepsilon_0\mu_0}$$

In terms of the vectors **E**, **D**, **B**, and **H** Maxwell's equations are:

$$\boxed{\begin{aligned}\nabla \cdot \mathbf{D} &= \rho_e, \quad \text{SI},\\ &= 4\pi\rho_e, \quad \text{G-CGS},\end{aligned}} \tag{B.7}$$

$$\boxed{\begin{aligned}\nabla \times \mathbf{E} &= -\frac{\partial \mathbf{B}}{\partial t}, \quad \text{SI},\\ &= -\frac{1}{c}\frac{\partial \mathbf{B}}{\partial t}, \quad \text{G-CGS},\end{aligned}} \tag{B.8}$$

$$\boxed{\nabla \cdot \mathbf{B} = 0,} \tag{B.9}$$

$$\boxed{\begin{aligned}\nabla \times \mathbf{H} &= \mathbf{j}_e + \frac{\partial \mathbf{D}}{\partial t}, \quad \text{SI},\\ &= \frac{4\pi}{c}\mathbf{j}_e + \frac{1}{c}\frac{\partial \mathbf{D}}{\partial t}, \quad \text{G-CGS},\end{aligned}} \tag{B.10}$$

where ρ_e and $\boldsymbol{j}_e$ are external charges and currents not belonging to the material body. The advantage of employing **D** and **H** is due to the fact that their sources are only the external ones, which usually are assumed known.

B.2 Relations Among the Fields

To solve Maxwell' s equations we have to express two of the four vectors **E**, **B**, **D**, **H** in terms of the other two, e.g.

$$\mathbf{D} = f_1(\mathbf{E}, \mathbf{H}; x_i), \tag{B.11}$$

$$\mathbf{B} = f_2(\mathbf{E}, \mathbf{H}; x_i), \tag{B.12}$$

where x_i stand for other variables which influence how **D** and **B** depend on **E** and **H**. E.g., x_i may represent mechanical stresses (which lead to the phenomenon of piezoelectricity) or temperature gradient (which lead to thermoelectric effects) or chemical potential gradients, or rotations (which leads to gyromagnetic phenomena), etc. In general the dependence of **D** and **B** on **E** and **H** is complicated and non-linear especially in the presence of powerful lasers where several very interesting phenomena with technological applications are due exactly to this non-linearity. Non-linear effects appear also in the presence of strong static magnetic fields (e.g., in the phenomena of magnetoresistance and the Hall effect) or strong electric field (e.g. dielectric breakdown in insulators).

For weak fields the rhs of (B.11) and (B.12) can be expanded in powers of **E** and **H**. Keeping up to the first power we have[2]

$$\mathbf{D} = \mathbf{D}_o + \hat{\varepsilon}\,\mathbf{E}, \tag{B.13}$$

$$\boldsymbol{B} = \boldsymbol{B}_o + \hat{\mu}\boldsymbol{H}, \tag{B.14}$$

where $\hat{\varepsilon}$ and $\hat{\mu}$ are in general *linear tensor operators*. The quantity $\hat{\varepsilon}$ is called *permittivity* or *electrical permeability* or *dielectric constant* (or function).[3] The quantity $\hat{\mu}$ is called *magnetic permeability* or simply *permeability*. In the unusual case where $\boldsymbol{D}_o \neq 0$, the material is called *pyroelectric*. In the particular case where the change from $\mathbf{D}_o \neq 0$ to $\mathbf{D}_o = 0$ is taking place through a second order phase transition the material is called *ferroelectric*. For most materials, $\mathbf{D}_o = 0$; even in the cases where $\mathbf{D}_o \neq 0$, its value is very small. Materials for which $\mathbf{B}_o$ is different from zero are called *ferromagnets*. In this book we assume that $\mathbf{D}_o = \mathbf{B}_o = 0$

Employing the definitions (B.5) and (B.6) and the relations (B.13) and (B.14) respectively we can express **P** and **M** in terms of **E** and **H**, respectively:

$$\begin{aligned} \mathbf{P} &= (\hat{\varepsilon} - \varepsilon_0)\mathbf{E}, \quad \text{SI}, \\ &= \frac{\hat{\varepsilon} - 1}{4\pi}\mathbf{E}, \quad \text{G-CGS}, \end{aligned} \tag{B.15}$$

---

[2]Because **D** and **E** are polar vectors, while **B** and **H** are axial vectors, no linear relation between **D** and **H** or between **B** and **E** is possible, unless, e.g., there is another polar vector $\boldsymbol{g}$ to form the cross product $\mathbf{g} \times \mathbf{H}$or the cross product $\mathbf{g} \times \mathbf{E}$.

[3]In the SI system it is customary to call $\hat{\varepsilon}$ permittivity and the ratio $\hat{\varepsilon}/\varepsilon_0$ dielectric constant, (or dielectric function) where ε_0 is the value of ε in vacuum. We shall adopt this convention in this book.

$$\begin{aligned}\mathbf{M} &= \frac{\hat{\mu} - \mu_0}{\mu_0}\mathbf{H}, \quad \text{SI}, \\ &= \frac{\hat{\mu} - 1}{4\pi}\mathbf{H}, \qquad \text{G-CGS},\end{aligned} \tag{B.16}$$

The dimensionless quantity

$$\begin{aligned}\hat{\chi}_e &\equiv \frac{\hat{\varepsilon} - \varepsilon_0}{\varepsilon_0}, \qquad \text{SI}, \\ &\equiv \frac{\hat{\varepsilon} - 1}{4\pi}, \quad \text{G-CGS},\end{aligned} \tag{B.17}$$

is called *electric susceptibility*, while the corresponding dimensionless quantity

$$\begin{aligned}\hat{\chi}_m &\equiv \frac{\hat{\mu} - \mu_0}{\mu_0}, \quad \text{SI}, \\ &\equiv \frac{\hat{\mu} - 1}{4\pi}, \quad \text{G-CGS},\end{aligned} \tag{B.18}$$

is called *magnetic susceptibility*.[4] It should be pointed out that the values of χ_e in SI and G-CGS are not the same. Actually

$$\hat{\chi}_{e,\text{SI}} = 4\pi\hat{\chi}_{e,\text{G-CGS}}. \tag{B.19}$$

Similarly

$$\hat{\chi}_{m,\text{SI}} = 4\pi\hat{\chi}_{m,\text{G-CGS}}. \tag{B.20}$$

From (B.17) and (B.18) we can express $\hat{\varepsilon}$ and $\hat{\mu}$ in terms of $\hat{\chi}_e$ and $\hat{\chi}_m$ respectively

$$\begin{aligned}\hat{\varepsilon}/\varepsilon_0 &= (1 + \hat{\chi}_e), \qquad \hat{\mu}/\mu_0 = (1 + \hat{\chi}_m), \qquad \text{SI}, \\ \hat{\varepsilon} &= 1 + 4\pi\,\hat{\chi}_e, \qquad \hat{\mu} = 1 + 4\pi\,\hat{\chi}_m, \qquad \text{G-CGS}.\end{aligned} \tag{B.21}$$

In terms of $\hat{\chi}_e$ and $\hat{\chi}_m$ the polarization and the magnetization become

$$\begin{aligned}\mathbf{P} &= \varepsilon_0\hat{\chi}_e\mathbf{E}, \;\; \text{SI}, \\ &= \hat{\chi}_e\mathbf{E}, \text{G-CGS},\end{aligned} \tag{B.22}$$

$$\mathbf{M} = \hat{\chi}_m\mathbf{H}. \tag{B.23}$$

[4]The magnetic susceptibility as defined here is equal to $\hat{\chi}_m \equiv (\partial\mathbf{M}/\partial\mathbf{H})_T$ in the limit $H \to 0$. This last condition can be removed; then, $\hat{\chi}_m$ may become a function of **H**, or **B**.

Let us assume now that the time dependence of the various fields is of the form $\exp(-i\omega t)$. Then (B.4) becomes

$$\begin{aligned} \mathbf{j} = -i\omega\,\mathbf{P} = -i\omega\varepsilon_0\hat{\chi}_e\mathbf{E}, \quad & \text{SI}, \\ = -i\omega\,\hat{\chi}_e\mathbf{E}, \quad & \text{G-CGS}. \end{aligned} \tag{B.24}$$

The coefficient in front of $\mathbf{E}$ in the rhs of (B.24) is by definition the AC electrical conductivity $\hat{\sigma}(\omega)$ (which, according to our definition includes the contributions of both the "free" and the "bound" charges). Hence, $\hat{\sigma}(\omega)$ is directly related with the permittivity and the electrical susceptibility:

$$\begin{aligned} \hat{\sigma}(\omega) = -i\,\omega(\hat{\varepsilon} - \varepsilon_0) = -i\,\omega\varepsilon_0\hat{\chi}_e, \quad & \text{SI}, \\ = -i\,\omega\frac{\hat{\varepsilon} - 1}{4\pi} = -i\,\omega\hat{\chi}_e, \quad & \text{G-CGS}. \end{aligned} \tag{B.25}$$

Inverting (B.25) we can express $\hat{\varepsilon}(\omega)$ in terms of $\hat{\sigma}(\omega)$

$$\boxed{\begin{aligned} \frac{\hat{\varepsilon}(\omega)}{\varepsilon_0} &= 1 + \frac{i\hat{\sigma}(\omega)}{\varepsilon_0\omega}, \quad \text{SI}, \\ \hat{\varepsilon}(\omega) &= 1 + \frac{4\pi i\hat{\sigma}(\omega)}{\omega}, \quad \text{G-CGS}. \end{aligned}} \tag{B.26}$$

Relations (B.18), (B.25), and (B.26) allow us to obtain immediately two of the three quantities, $\hat{\varepsilon}$, $\hat{\chi}_e$, $\hat{\sigma}$, if we know the third. Usually -especially for conductors- it is easier to calculate first the AC conductivity, $\hat{\sigma}(\omega)$, and then, using (B.26) and (B.25), to obtain $\hat{\varepsilon}(\omega)$ and $\hat{\chi}_e(\omega)$ respectively. In the case of atomic or molecular insulators we calculate first each atomic (or molecular) dipole moment$\boldsymbol{p}_\alpha$ which is assumed to be proportional to the *local*electric field,

$$\mathbf{p}_a = \hat{\alpha}_p\mathbf{E}_{\text{loc}}, \tag{B.27}$$

and then, taking into account that $\Delta\mathbf{p} = \sum_a \mathbf{p}_a$ (where summation is over the particles within the volume ΔV), we obtain $\mathbf{P}$ from (B.2). We need to express $\mathbf{E}_{\text{loc}}$ in terms of the average field $\mathbf{E}$ to end up with $\hat{\chi}_e(\omega)$. Under the assumption of cubic symmetry we have that $\mathbf{E}_{\text{loc}} = \mathbf{E} + (\mathbf{P}/3\,\varepsilon_0) = [1 - (\sum a_p/3\,\varepsilon_0\Delta V)]\mathbf{E}$. The quantity $\hat{\alpha}_p$ is called atomic (or molecular or ionic) *electricpolarizability*.

We shall conclude this appendix by giving the formula for the EM energy flux $\mathbf{S}$ in vacuum (i.e. the EM energy crossing at normal incidence a surface per unit time and area)

$$\boxed{\begin{aligned} \mathbf{S} &= \frac{1}{\mu_0}\mathbf{E} \times \mathbf{B}, \quad \text{SI} \\ \mathbf{S} &= \frac{c}{4\pi}\mathbf{E} \times \mathbf{B}, \quad \text{G-CGS} \end{aligned}} \tag{B.28}$$

The flux of energy is equal to the energy densityu_V of the EM field in vacuum times the velocity of light. It follows then that

$$\boxed{\begin{aligned} u_V &= \frac{\varepsilon_0}{2}\mathbf{E}^2 + \frac{1}{2\mu_0}\mathbf{B}^2, && \text{SI} \\ u_V &= \frac{1}{8\pi}\mathbf{E}^2 + \frac{1}{8\pi}\mathbf{B}^2, && \text{G-CGS} \end{aligned}} \tag{B.29}$$

In the following two columns we summarize some of the main formulae of electromagnetism both in SI and in G-CGS as well as the correspondence of various quantities in the two systems. We remind the reader that $\hat{\varepsilon}(\omega), \hat{\chi}_e, \hat{a}_e\, \hat{\sigma}, \hat{\mu}, \hat{\chi}_m, \hat{a}_m$ are in general linear, integral, tensor operators.

SI	G-CGS
$\mathbf{F} = q(\mathbf{E}+\boldsymbol{v}\times\mathbf{B})$	$\mathbf{F} = q\left(\mathbf{E}+\frac{1}{c}\boldsymbol{v}\times\mathbf{B}\right)$
$\nabla\cdot\mathbf{P} = -\bar{\rho}$	$\nabla\cdot\mathbf{P} = -\bar{\rho}$
$\nabla\times\mathbf{M} = \mathbf{j}_M$	$\nabla\times\mathbf{M} = \frac{1}{c}\mathbf{j}_M$
$\mathbf{D} = \varepsilon_0\mathbf{E}+\mathbf{P}$	$\mathbf{D} = \mathbf{E}+4\pi\mathbf{P}$
$\mathbf{H} = \frac{1}{\mu_0}\mathbf{B}-\mathbf{M}$	$\mathbf{H} = \mathbf{B}-4\pi\mathbf{M}$
$\nabla\times\mathbf{E} = -\frac{\partial\mathbf{B}}{\partial t}$	$\nabla\times\mathbf{E} = -\frac{1}{c}\frac{\partial\mathbf{B}}{\partial t}$
$\nabla\cdot\mathbf{D} = \rho_e$	$\nabla\cdot\mathbf{D} = 4\pi\rho_e$
$\nabla\times\mathbf{H} = \mathbf{j}_e+\frac{\partial\mathbf{D}}{\partial t}$	$\nabla\times\mathbf{H} = \frac{4\pi}{c}\mathbf{j}_e+\frac{1}{c}\frac{\partial\mathbf{D}}{\partial t}$
$\nabla\cdot\mathbf{B} = 0$	$\nabla\cdot\mathbf{B} = 0$
$\mathbf{D} = \hat{\varepsilon}\mathbf{E}=\varepsilon_0\mathbf{E}+\mathbf{P}$	$\mathbf{D} = \hat{\varepsilon}\mathbf{E}=\mathbf{E}+4\pi\mathbf{P}$
$\mathbf{B} = \hat{\boldsymbol{\mu}}\mathbf{H} = \mu\mathbf{H}+\mathbf{M}$	$\mathbf{B} = \hat{\boldsymbol{\mu}}\mathbf{H} = \mathbf{H}+4\pi\mathbf{M}$
$\mathbf{P} = \varepsilon_0\hat{\chi}_e\mathbf{E}$	$\mathbf{P} = \hat{\chi}_e\mathrm{E}$
$\mathbf{P} = \frac{1}{V}\sum_i\mathbf{p}_i$	$\mathbf{P} = \frac{1}{V}\sum_i\mathbf{p}_i$
$\mathbf{p} = \hat{\alpha}_p\mathbf{E}_{loc}$	$\mathbf{p} = \hat{a}_p\mathbf{E}_{loc}$
$\hat{\varepsilon} = \varepsilon_0+\varepsilon_0\hat{\chi}_e$	$\hat{\varepsilon} = 1+4\pi\hat{\chi}_e$
$\mathbf{M} = \hat{\chi}_m\mathbf{H}$	$\mathbf{M} = \hat{\chi}_m\mathbf{H}$
$\mathbf{M} = \frac{1}{V}\sum_i\mathbf{m}_i$	$\mathbf{M} = \frac{1}{V}\sum_i\mathbf{m}_i$
$\mathbf{m} = \hat{a}_m\mathbf{H}$	$\mathbf{m} = \hat{a}_m\mathbf{H}$
$\hat{\mu}/\mu_0 = 1+\hat{\chi}_m$	$\hat{\mu} = 1+4\pi\hat{\chi}_m$
$(\hat{\varepsilon}/\varepsilon_0) = 1+\frac{i\hat{\sigma}}{\varepsilon_0\omega}$	$\varepsilon = 1+\frac{4\pi i\hat{\sigma}}{\omega}$
$\hat{\sigma} = -i\omega(\hat{\varepsilon}-\varepsilon_0)$	$\hat{\sigma} = -\frac{i\omega}{4\pi}(\hat{\varepsilon}-1)$
$\delta U = \int\mathbf{E}\cdot\delta\mathbf{D}dV$	$\delta U = \frac{1}{4\pi}\int\mathbf{E}\cdot\delta\mathbf{D}dV$
$\delta U' = -\int\mathbf{P}\cdot\delta\mathbf{E}_o dV$	$\delta U' = -\int\mathbf{P}\cdot\delta\mathbf{E}_o dV$
$\delta U_t = \int\mathbf{H}\cdot\delta\mathbf{B}dV$	$\delta U_t = \frac{1}{4\pi}\int\mathbf{H}\cdot\delta\mathbf{B}dV$

(continued)

SI	G-CGS
$\delta\tilde{U}' = -\int \mathbf{M}\cdot\delta\mathbf{B}_o dV$	$\delta\tilde{U}' = -\int \mathbf{M}\cdot\delta\mathbf{B}_o dV$
$\mathbf{P} = -\frac{1}{V}\left(\frac{\partial F'}{\partial \mathbf{E}_o}\right)_T$	$\mathbf{P} = -\frac{1}{V}\left(\frac{\partial F'}{\partial \mathbf{E}_o}\right)_T$
$\mathbf{M} = -\frac{1}{V}\left(\frac{\partial F'}{\partial \mathbf{B}_o}\right)_T$	$\mathbf{M} = -\frac{1}{V}\left(\frac{\partial F'}{\partial \mathbf{B}_o}\right)_T$
$\mathbf{B}$	$\mathbf{B}/c$
ε_0	$1/4\pi$
μ_0	$4\pi/c^2$
$\mathbf{M}$	$c\mathbf{M}$
$\mathbf{D}$	$\mathbf{D}/4\pi$
$\mathbf{H}$	$c\mathbf{H}/4\pi$
L	L/c^2
χ_e	$4\pi\chi_e$
χ_m	$4\pi\chi_m$

The range of the visible spectrum

$$400\,\text{nm} < \lambda < 750\,\text{nm}$$
$$750\{\text{THz}\} > f > 400\{\text{THz}\}$$
$$4.71\times 10^{15}\,\text{rad/s} > \omega > 2.51\times 10^{15}\,\text{rad/s}$$
$$3.1\,\text{eV} > \hbar\omega > 1.66\,\text{eV}$$

References for Electromagnetism

1. D.J. Griffiths, *Introduction to Electrodynamics*, 3rd edn. (Benjamin, New York, 2008)
2. L.D. Landau, E.M. Lifshitz, *The Classical Theory of Fields*, 2nd edn. (Pergamon Press, Oxford, 1965)
3. L.D. Landau, E.M. Lifshitz, *Electrodynamics of Continuous Media*, 2nd edn. (Pergamon Press, Oxford, 1984)

Appendix C
Elements of Thermodynamics and Statistical Mechanics

C.1 Thermodynamic Relations

We remind the reader here some basic thermodynamic definitions:

$$\text{The Internal Energy of simply Energy} \qquad U \text{ or } E \tag{C.1}$$

$$\text{The Helmholtz Free Energy} \qquad F \equiv U - TS \tag{C.2}$$

$$\text{The Gibbs Free Energy} \qquad G \equiv F + PV = \mu N \tag{C.3}$$

$$\text{The Enthalpy } H \equiv U + PV \tag{C.4}$$

$$\text{The Grand Thermodynamic Potential } \Omega = F - \mu N = -PV \tag{C.5}$$

and the corresponding relations for their differentials

$$dU \leq T\,dS - P\,dV + \mu\,dN \tag{C.6}$$

$$dF \leq -SdT - PdV + \mu dN \tag{C.7}$$

$$dG \leq -SdT + VdP + \mu dN \tag{C.8}$$

$$dH \leq T\,dS + V\,dP + \mu\,dN \tag{C.9}$$

$$d\Omega \leq -SdT - PdV - Nd\mu \tag{C.10}$$

In the above expressions T is the absolute temperature, S is the entropy and μ is the so-called *chemical potential*, a quantity of special interest. The expression of the Gibbs free energy is based on the assumption that the work done by the system on the environment is given exclusively by PdV; if there are other contributions to the work, as, e.g., the electrostatic contribution, $V\boldsymbol{P} \cdot d\mathbf{E}_o$, then corresponding terms such as $-V\boldsymbol{P} \cdot d\mathbf{E}_o$ should be included to the right-hand side of the differentials dU,

E.N. Economou, *From Quarks to the Universe*,
DOI 10.1007/978-3-319-20654-7

dF, $d\Omega$. The expression of Ω is based on the assumption that the system consists of one type of particles; if there are several types of particles the term μN is replaced by $\sum_\alpha \mu_\alpha N_\alpha$.

Various derivatives of the thermodynamic potentials are also of interest. Among them are the specific heats

$$C_V \equiv T\left(\frac{\partial S}{\partial T}\right)_{V,N} = \left(\frac{\partial U}{\partial T}\right)_{V,N}, \tag{C.11}$$

$$C_P \equiv T\left(\frac{\partial S}{\partial T}\right)_{P,N} = \left(\frac{\partial H}{\partial T}\right)_{P,N}, \tag{C.12}$$

the chemical potential

$$\mu \equiv \left(\frac{\partial U}{\partial N}\right)_{S,V} = \left(\frac{\partial F}{\partial N}\right)_{T,V} = \frac{G}{N}, \tag{C.13}$$

the bulk[5] thermal expansion coefficient

$$\alpha_b \equiv \frac{1}{V}\left(\frac{\partial V}{\partial T}\right)_{P,N}, \tag{C.14}$$

and the bulk moduli

$$B_T \equiv -V\left(\frac{\partial P}{\partial V}\right)_{T,N}, \tag{C.15}$$

$$B_S \equiv -V\left(\frac{\partial P}{\partial V}\right)_{S,N}. \tag{C.16}$$

Under equilibrium conditions for which the equality sign holds in (C.6)–(C.10), several new relations can be produced. E.g.:

$$S = -\left(\frac{\partial F}{\partial T}\right)_{V,N} = -\left(\frac{\partial \Omega}{\partial T}\right)_{V,\mu}, \tag{C.17}$$

$$P = -\left(\frac{\partial F}{\partial V}\right)_{T,N} = -\left(\frac{\partial \Omega}{\partial V}\right)_{T,\mu} = -\frac{\Omega}{V}, \tag{C.18}$$

[5]Quite often the linear thermal expansion coefficient is used, defined by the relation $a_L \equiv (\partial L/\partial T)_{P,N}/L$ and connected to a_b by $a_b = 3a_L$.

$$\left(\frac{\partial V}{\partial T}\right)_{P,N} = -\left(\frac{\partial S}{\partial P}\right)_{T,N}, \tag{C.19}$$

etc. With more complicated manipulations we can prove the following useful relations

$$a_b \equiv \frac{1}{B_T}\left(\frac{\partial P}{\partial T}\right)_{V,N}, \tag{C.20}$$

$$B_S = \frac{C_p}{C_V} B_T, \tag{C.21}$$

$$C_P - C_V = TVB_T \alpha_b^2. \tag{C.22}$$

C.2 Basic Relations of Statistical Mechanics

In this subsection we consider only *statistically independent*[6] systems in *thermodynamic equilibrium*. Let Ψ_{I} be the eigenfunctions of the Hamiltonian of such a system with eigenenergies E_I and number of particles N_I. The quantity of central importance in Statistical Mechanics is the *probability*, P_I, *to find such a statistically independent system*[6]*in theeigenstate* Ψ_{I}. For such an *open* system exchanging both energy and particles with the environment and, hence, having no fixed energy and number of particles, P_I depends only on the eigenvalues E_I, and N_I with the temperature T and the chemical potentialμ as parameters

$$\boxed{P_I = \frac{1}{Z_G} e^{-\beta(E_I - \mu N_I)};} \tag{C.23}$$

$\beta \equiv (k_B T)^{-1}$ is the inverse of the product of Boltzmann's constant, k_B, times T and (as a result of $\sum_I P_I = 1$)

$$Z_G = \sum_{\mathrm{I}} e^{-\beta(\mathrm{E}_I - \mu N_I)}. \tag{C.24}$$

The basic equation (C.23) follows from the fact that the quantity $\ln P_I$ is additive (due to statistical independence) and conserved for a closed system (a consequence of a theorem known as Liouville's theorem). As such $\ln P_I$ is a linear combination of

[6]Statistically independent means that the probability P_I to find the system in its state $|\Psi_I\rangle$ does not depend on what is the probability Q_J for any neighboring system to be in its state Φ_J: thus the joint probability $\Pi_{IJ} = P_I \cdot Q_J$ and $\ln \Pi_{IJ} = \ln P_I + \ln Q_J$)

the independent, additive, and conserved quantities (for a closed system) such as the energy and the number of particles. (The latter are conserved in the absence of "chemical" reactions).

The connection with thermodynamics is obtained through the relation between the P_Is (as given by (C.23)) and the entropy:

$$\boxed{S \equiv -k_B \sum_I P_I \ln P_I,} \tag{C.25}$$

This more general definition of the entropy coincides with that of (4.3) in the case where all P_Is are equal to each other (see below). From (C.25) and (C.23) follows (using (C.2) and (C.5), $U \equiv \sum_I P_I E_I$, and $N \equiv \sum_I P_I N_I$) that

$$\Omega = -k_B \mathrm{T} \ln Z_G. \tag{C.26}$$

If there is no exchange of particles with the environment, N_I is fixed, $N_I = N$, and the term $\exp(\beta\,\mu N)$ in (C.23) can be incorporated in the normalization factor. Thus

$$P_I = \frac{1}{Z} e^{-\beta E_I}, \tag{C.27}$$

where

$$Z = Z_G e^{-\beta\mu N} = \sum_I e^{-\beta E_I}, \tag{C.28}$$

and

$$-k_B T \ln Z = -k_B T \ln Z_G + k_B T \beta\mu N = \Omega + \mu N \equiv F. \tag{C.29}$$

Finally, if the system exchanges neither particles nor energy with the environment, both E_I and N_I are fixed, $E_I = U$ and $N_I = N$ and, consequently,

$$P_I = \frac{1}{\Delta\Gamma}, \tag{C.30}$$

where

$$\Delta\Gamma = Z e^{\beta U}, \tag{C.31}$$

and

$$k_B \ln \Delta\Gamma = k_B \ln Z + k_B \beta U = -\frac{F}{T} + \frac{U}{T} = \frac{TS}{T} = S. \tag{C.32}$$

Equation (C.30) means that all eigenstates of equal energy and equal number of particles are of equal probability to occur (for a statistically independent system in thermodynamic equilibrium). Furthermore, (C.32) implies that their total number, $\Delta\Gamma$, is directly connected to the entropy through the famous relation $S = k_B \ln \Delta\Gamma$. This relation follows also by combining (C.25) and (C.30).

C.3 Non-interacting Particles

When the system consists of non-interacting particles, the general formulae of Sect. C.2 are drastically simplified because they can be expressed in terms of the eigenenergies, ε_i, of a *single*-particle instead of the eigenenergies, E_I, of the *wholesystem.*

C.3.1 Non-interacting Electrons

We consider the single particle eigenstate ψ_i (including the spin variable) with eigenenergy ε_i. According to (C.23) the probability of having zero electrons in this state is

$$P_0 = \frac{1}{Z_{Gi}}, \tag{C.33}$$

since in this case $n_i = 0$ and consequently $n_i\varepsilon_i = 0$. The probability of having one electron in this state is according to (C.23).

$$P_1 = \frac{1}{Z_{Gi}} e^{-\beta(\varepsilon_i - \mu)}. \tag{C.34}$$

Because of Pauli's principle the probability of having n electrons in the state ψ_i, where $n \geq 2$, is zero. Hence, $P_0 + P_1 = 1$ and consequently

$$Z_{Gi} = 1 + e^{-\beta(\varepsilon_i - \mu)}. \tag{C.35}$$

Combining (C.34) and (C.35) we find the average number of particles, $\bar{n}_i = 0 \cdot P_0 + 1 \cdot P_1 = P_1$, in the eigenstate ψ_i (the so called *Fermi distribution* valid for any fermionic particle):

$$\boxed{\bar{n}_i = \frac{1}{e^{\beta(\varepsilon_i - \mu)} + 1}}. \tag{C.36}$$

It follows that the total average number of electrons N is

$$N = \sum_i \bar{n}_i = \sum_i \frac{1}{e^{\beta(\varepsilon_i - \mu)} + 1}, \tag{C.37}$$

and the total average energy U is

$$U = \sum_i \bar{n}_i \varepsilon_i = \sum_i \frac{\varepsilon_i}{e^{\beta(\varepsilon_i - \mu)} + 1}. \tag{C.38}$$

Combining (C.35), (C.26), and (C.5) we have

$$-k_B \mathrm{T} \sum_i \ln\left(1 + e^{-\beta(\varepsilon_i - \mu)}\right) = \Omega = -PV. \tag{C.39}$$

The summations in (C.37), (C.38), and (C.39), which are over all *single* particle eigenstates, can be transformed into integrals by introducing the single-particle density of states (DOS) *per spin*, $\rho(\varepsilon)$ and the number, $R(\varepsilon)$, of eigenstates *per spin* with eigenenergy less than ε (see Appendix D, Sect. D.3). (Obviously, $dR(\varepsilon) \equiv R(\varepsilon + d\varepsilon) - R(\varepsilon) = \rho(\varepsilon)d\varepsilon$).

$$N = 2 \int_{\varepsilon_{\min}}^{\infty} \frac{d\varepsilon\, \rho(\varepsilon)}{e^{\beta(\varepsilon - \mu)} + 1}, \tag{C.40}$$

$$U = 2 \int_{\varepsilon_{\min}}^{\infty} \frac{d\varepsilon\, \varepsilon\, \rho(\varepsilon)}{e^{\beta(\varepsilon - \mu)} + 1}, \tag{C.41}$$

and

$$\begin{aligned} -\Omega = PV = 2k_B T \int_{\varepsilon_{\min}}^{\infty} d\varepsilon\, \rho(\varepsilon) \ln\left[1 + e^{-\beta(\varepsilon - \mu)}\right] = \\ = 2 \int_{\varepsilon_{\min}}^{\infty} \frac{d\varepsilon\, R(\varepsilon)}{e^{\beta(\varepsilon - \mu)} + 1}. \end{aligned} \tag{C.42}$$

The last term in (C.42) was obtained by an integration by parts. The relation (C.37) or (C.40) is very important because it allows the determination of the chemical potentialμ in terms of the (average) number of particles, N, the parameters (such as the volume) in $\rho(\varepsilon)$, and the temperature T. In particular for $T = 0$ (i.e. $\beta = \infty$)

$$\begin{aligned}\bar{n}_i &= 1 \quad \text{for } \varepsilon_i < \mu(0), \\ &= 0 \quad \text{for } \varepsilon_i > \mu(0),\end{aligned} \tag{C.43}$$

which, in view of Fig. 3.1, implies that

$$E_F = \lim_{T \to 0} \mu(\mathrm{T}). \tag{C.44}$$

C.3.2 Ionic Vibrations and Phonons

For an independent one-dimensional harmonic oscillator of eigenenergies $\varepsilon_i(n + \frac{1}{2})$ where $\varepsilon_i = \hbar\omega_i$ and $n = 0, 1, 2, 3$ and in equilibrium with a heat bath of temperature $T = 1/k_B\beta$ we have according to (C.28):

$$Z_i = \sum_{n=0}^{\infty} e^{-\beta\varepsilon_i(n+\frac{1}{2})} = e^{-\beta\varepsilon_i/2} \sum_{n=0}^{\infty} (e^{-\beta\varepsilon_i})^n = \frac{e^{-\beta\varepsilon_i/2}}{1 - e^{-\beta\varepsilon_i}}. \tag{C.45}$$

Hence, the corresponding free energy, F_i, is according to (C.29):

$$F_i = \frac{\varepsilon_i}{2} + k_B \mathrm{T} \ln\left(1 - e^{-\beta\varepsilon_i}\right). \tag{C.46}$$

Since the system of harmonically vibrating ions is equivalent to $3N_\alpha - 6 \approx 3N_\alpha$ independent 1-D harmonic oscillators we have for the total free energy, F, of the ions

$$F = \sum_{i=1}^{3N_\alpha} F_i = \frac{1}{2}\sum_{i=1}^{3N_\alpha} \varepsilon_i + k_B T \sum_{i=1}^{3N_\alpha} \ln\left(1 - e^{-\beta\varepsilon_i}\right). \tag{C.47}$$

The first term of the rhs of (C.47) is the zero point contribution to the ionic energy. The second term, which vanishes at $\mathrm{T} = 0\,\mathrm{K}$, is the so-called phonon contribution to the ionic free energy

$$F_{ph} \equiv k_B T \sum_{i=1}^{3N_\alpha} \ln\left(1 - e^{-\beta\varepsilon_i}\right) = 3k_B T \sum_{\mathbf{k}}^{k_D} \ln\left(1 - e^{-\beta\hbar\omega_k}\right), \qquad \omega_k = \bar{c}\,k. \tag{C.48}$$

In (C.48) we took into account that the vibrating ions (or atoms) are acoustic waves of frequency $\omega_k = \bar{c}\,k$, $\varepsilon_k \equiv \hbar\omega_k$ and that for each $\mathbf{k}$ there are two transverse waves and one longitudinal wave; $\bar{c}$ is an appropriately averaged velocity of sound in the solid. The summation over $\mathbf{k}$ can be transformed to an integration over $d^3k = 4\pi k^2 dk$ times the standard factor $V/(2\pi)^3$. The integration over k can be changed to an integration over ω since $4\pi k^2 dk = (4\pi/\bar{c}^3)\omega^2 d\omega$. Thus

$$F_{ph} = \frac{3k_B T\, V}{(2\pi)^3} \frac{4\pi}{\bar{c}^3} \int_0^{\omega_D} d\omega\, \omega^2 \ln\left(1 - e^{-\beta\hbar\omega_k}\right) = \frac{9k_B T\, N_a}{\varepsilon_D^3} \int_0^{\varepsilon_D} d\varepsilon\, \varepsilon^2 \ln\left(1 - e^{-\beta\varepsilon}\right) \tag{C.49}$$

In (C.49) we took into account that the quantity $\left(V \times \frac{4\pi}{3} k_D^3\right)/(2\pi)^3$, which gives all the eigenmodes, ought to be equal to $3N_a$; we also change variables from ω to ε, $\varepsilon \equiv \hbar\,\omega$, $\varepsilon_D \equiv \hbar\,\omega_D$. By performing an integration by parts we ended up with the following relation:

$$F_{ph} = -\frac{3N_a \hbar}{\omega_D^3} \int_0^{\omega_D} \frac{d\omega\, \omega^3}{e^{\beta\hbar\omega} - 1} = -\frac{3N_a}{\varepsilon_D^3} \int_0^{\varepsilon_D} \frac{d\varepsilon\, \varepsilon^3}{e^{\beta\varepsilon} - 1}, \tag{C.50}$$

From (C.49) and (C.17) we obtain the phononic contribution to the entropy

$$S_{ph} = -\frac{F_{ph}}{T} + \frac{9N_a}{T\, \varepsilon_D^3} \int_0^{\varepsilon_D} \frac{d\varepsilon\, \varepsilon^3}{e^{\beta\varepsilon} - 1} \tag{C.51}$$

Then the phononic contribution to the energy is according to (C.2)

$$U_{ph} = F_{ph} + TS_{ph} = \int_0^{\varepsilon_D} \frac{d\varepsilon\, \varepsilon\, \phi(\varepsilon)}{e^{\beta\varepsilon} - 1}, \quad \phi(\varepsilon) \equiv \frac{9\, N_a\, \varepsilon^2}{\varepsilon_D^3}. \tag{C.52}$$

The phononic contribution to the pressure is obtained by combining (C.48) and (C.18)

$$P_{ph} = -\left(\frac{\partial F_{ph}}{\partial V}\right)_{T,N_\alpha} = -\sum_{i=1}^{3N_\alpha} \frac{\partial \varepsilon_i}{\partial V} \frac{1}{e^{\beta\varepsilon_i} - 1} = \int_0^{\varepsilon_D} d\varepsilon\, \frac{\partial \Phi}{\partial V} \frac{1}{e^{\beta\varepsilon} - 1}, \quad \Phi(\varepsilon) = \frac{3N_a \varepsilon^3}{\varepsilon_D^3} \tag{C.53}$$

The phononic contribution to the specific heatC_V is calculated by differentiating U_{ph} with respect to the temperature

$$C_{Vph} = \left(\frac{\partial U_{ph}}{\partial T}\right)_{V,N_\alpha} = k_B \int_0^{\varepsilon_D} d\varepsilon \phi(\varepsilon) \frac{(\beta\varepsilon)^2 e^{\beta\varepsilon}}{\left(e^{\beta\varepsilon} - 1\right)^2}. \tag{C.54}$$

The expression

$$U_{ph} = \int_0^{\varepsilon_D} \frac{d\varepsilon\varepsilon\phi(\varepsilon)}{e^{\beta\varepsilon} - 1} = \sum_{i=1}^{3N_\alpha} \frac{\varepsilon_i}{e^{\beta\varepsilon_i} - 1}, \tag{C.55}$$

compared with the expression $U_{ph} = \sum_{i=1}^{3N_\alpha} \bar{n}_i \varepsilon_i$ allows us to determine the average number $\bar{n}_i$ of vibrational quanta each of energy $\varepsilon_i = \hbar\omega_i$:

$$\bar{n}_i = \frac{1}{e^{\beta\varepsilon_i} - 1}. \tag{C.56}$$

In analogy with the quantum of the EM field called photon, each quantum of the eigenmode i of the acoustic field is called *phonon*; then (C.56) gives the average number of phonons in the phononic state i.

Phonons, being the quanta of a classical field, have two similarities with photons: First they are bosons and second their number is not conserved; the consequence of the latter is that the chemical potential of a phonon gas is zero.

In general the average number of non-interacting identical bosons whose number *is conserved* in a single particle eigenstate ψ_i (including the spin variable) is given by a formula similar to (C.56) with ε_i replaced by $\varepsilon_i - \mu$:

$$\boxed{\bar{n}_i = \frac{1}{e^{\beta(\varepsilon_i - \mu)} - 1}} \tag{C.57}$$

References

1. E. Fermi, *Thermodynamics*, (Dover Publications, NY, 1956)
2. M.W. Zemansky, R.H. Dittman, *Heat and Thermodynamics*, 7th edn. (McGraw-Hill, New York, 1997)
3. L. Landau, E. Lifshitz, *Statistical Physics*, Part 1, 3rd edn. (Pergamon Press, Oxford, Butteworth-Heinemann, 2000)

Appendix D
Semiconductors Revisited

D.1 Holes

A fully occupied band is inert in the sense that it cannot respond to the application of external fields or temperature gradients. The reason is that any response requires a redistribution of the electrons which in a fully occupied band is prevented by the Pauli principle. Hence, a fully occupied VB gives no contribution to the conductivity. It must be stressed that this zero contribution is not due to the electrons being trapped in the bond regions; on the contrary the electrons in the VB of a semiconductor are almost as mobile as in metals. What causes the zero conductivity is the inability of the electrons to readjust their state in response to the electric field; this is due to the absence of nearby empty states in combination with Pauli's principle.

Let us now consider a band which is *almost* fully occupied. We denote by the index **k**, o the occupied states (characterized by **k** and the spin orientation) and by **k**, e the rest of the band states which are empty. Then the electrical current density $\boldsymbol{j}$ can be written as follows:

$$\boldsymbol{j} = \frac{-|e|}{V}\sum_{\mathbf{k},o} \boldsymbol{\upsilon}_{\mathbf{k}} = \frac{-|e|}{V}\sum_{\mathbf{k},o} \boldsymbol{\upsilon}_{\mathbf{k}} + \frac{-|e|}{V}\sum_{\mathbf{k},e} \boldsymbol{\upsilon}_{\mathbf{k}} - \frac{-|e|}{V}\sum_{\mathbf{k},e} \boldsymbol{\upsilon}_{\mathbf{k}} = \frac{|e|}{V}\sum_{\mathbf{k},e} \boldsymbol{\upsilon}_{\mathbf{k}} \tag{D.1}$$

In the above expression we have added and subtracted the summation over the empty states. This way we have the summation over all states (which gives zero) minus the summation over the empty states. Thus we are left with the summations over the empty states of the quantity $\boldsymbol{\upsilon}_{\mathbf{k}}$. The conclusion is that the quantity $\boldsymbol{j}$ (and others) in an almost fully occupied band can be obtained in two equivalent ways: Either by the electrons in the occupied states or by fictitious entities called *holes* occupying the empty states. The properties of the holes denoted by the subscript h are:

$$e_h = |e|,\ \boldsymbol{k}_h = -\boldsymbol{k}_e,\ \boldsymbol{s}_h = -\boldsymbol{s}_e,\ \boldsymbol{\upsilon}_h(\boldsymbol{k}_h) = \boldsymbol{\upsilon}_e(\boldsymbol{k}_e),\ \varepsilon_h(\mathbf{k}_h) = -\varepsilon_e(\boldsymbol{k}_e) \tag{D.2}$$

E.N. Economou, *From Quarks to the Universe*,
DOI 10.1007/978-3-319-20654-7

The subscript e refers to the missing electrons. Thus the hole can be thought of as a particle of positive charge $|e|$, of velocity equal to the one that the missing electron would have, and of energy opposite to the one that the missing electron would have. In essence the energy of the hole is the energy required for its creation, i.e. the energy needed to transfer an electron from the state $\boldsymbol{k}_e$ to the level chosen as the zero of energy.

D.2 Effective Masses

Let us consider the model shown in Fig. 12.5a. It is clear from (12.17) and from Fig. 12.6 that the only allowed eigenenergies (as $N \to \infty$) lie in a finite band of width $4|V_2|$ centered at the atomic level ε. Near the lower band edge we can approximate $E(k)$ as follows:

$$E(k) - \varepsilon + 2|V_2| \approx |V_2|a^2k^2, \quad |k|a \ll \pi, \tag{D.3}$$

while near the upper band edge by a similar expansion we have

$$E(k) - \varepsilon - 2|V_2| \approx -|V_2|a^2\delta k^2, \quad 0 < a|\delta k| \ll \pi, \tag{D.4}$$

where $\delta k = (\pi/a) - k$ or $-(\pi/a) + |k|$

This quadratic dependence on k or δk allows us to define an *effective mass* for the lower and the upper band edges respectively

$$|V_2|a^2k^2 \equiv \frac{\hbar^2k^2}{2m^*} \quad \Rightarrow \quad m^* \equiv \frac{\hbar^2}{2|V_2|a^2}, \quad k \approx 0, \tag{D.5}$$

$$-|V_2|a^2(\delta k)^2 \equiv \frac{\hbar^2(\delta k)^2}{2m^*} \quad \Rightarrow \quad m_e^* \equiv -\frac{\hbar^2}{2|V_2|a^2}, \quad k \approx \pm\pi/a. \tag{D.6}$$

It must be stressed that the effective mass at the upper band edge is *negative*! This unusual feature is connected with the appearance of gaps and consequently the existence of upper limits in the spectrum as a result of wave motion in a periodic potential, in contrast to the case of a constant potential which possesses only a single lower band limit. In the present simple model the two effective masses shown in (D.5) and (D.6) are of equal magnitude; in general the lower and upper band effective masses are different both in sign and in magnitude. The concept of the effective mass near the lower or the upper band edge makes, as we shall see, the electronic motion in a periodic potential equivalent in many respects to free of force motion but with the electronic mass being replaced by m^*.

More generally the effective mass is defined in terms of the *band structure*, i.e. the $E(k)$ versus k relation, as follows:

$$\boxed{\frac{1}{m^*} \equiv \frac{\partial^2 E(k)}{\hbar^2 \partial k^2}} \tag{D.7}$$

Notice that, according to this more general definition, the effective electron mass near the upper band edge is necessarily negative. We will be on more familiar ground by employing the concept of hole (i.e., of missing electron), introduced in the previous section, having an effective hole mass m_h^* equal to minus the effective electronic mass, $m_h^* = -m_e^*$. Thus m_h^* is a positive number at the upper band edge.

There is a direct generalization of (D.7) to the 3-D case

$$\boxed{\left(m^{-1}\right)_{ij} \equiv \frac{\partial^2 E(\mathbf{k})}{\hbar^2 \partial k_i \partial k_j}.} \tag{D.8}$$

Thus the inverse of the effective mass is a symmetric tensor of rank two. With a proper orientation of the Cartesian axes at each point **k**, the effective mass tensor at this point can become diagonal

$$\frac{1}{m_i} = \frac{\partial^2 E(\mathbf{k})}{\hbar^2 \partial k_i^2}, \quad i \text{ along, each one of the three principal axes.} \tag{D.9}$$

Taking into account (D.2) we have for the effective mass of a hole

$$\frac{1}{m_{hi}} = -\frac{1}{m_{ei}}, \; i \text{ along, each one of the three principal axes} \tag{D.10}$$

Since the electronic effective mass at the top of the valence band (or at the top of any band) is negative, it follows that the hole effective mass at the top of the valence band is positive. This justifies the introduction of the hole concept as a convenient tool, since the almost full valence band can then be described as occupied by a few holes of positive effective mass at its top, instead of a huge number of electrons of negative effective mass.

D.3 Density of States

It is quite common to encounter expressions of the form $\sum_{\mathbf{k}} f_{\mathbf{k}}$ where the summation is over all values of **k** associated with a band. This summation in the limit of a macroscopic system can be transformed to an integral as follows:

$$\sum_{\mathbf{k}} f_{\mathbf{k}} = \frac{V}{(2\pi)^3} \int d^3k f_{\mathbf{k}} \tag{D.11}$$

Equation (D.11) is valid for the 3-D case. In the 2-D case the integration is over d^2k and the prefactor is $A/(2\pi)^2$ where A is the area of the 2-D system. In 1-D the integration is over k from $-(\pi/a)$ to (π/a) and the prefactor is $L/(2\pi)$.

In the case where the function $f_{\mathbf{k}}$ has the form $f(E(\boldsymbol{k}))$ we can transform (D.11) to an integral over energy by introducing a quantity called *density of states* (DOS). We shall show how this is achieved in the very common case where $E = \hbar^2 k^2 / 2\, m^*$ in 3-D. In this case $k = (2m^*/\hbar^2)^{1/2} E^{1/2}$ and $dk = (m^*/2\hbar^2)^{1/2} E^{-1/2}$. Hence

$$d^3k = 4\pi k^2 dk = 4\pi \frac{2m^*}{\hbar^2} \left(\frac{m^*}{2\hbar^2}\right)^{1/2} E^{1/2} = 4\pi\sqrt{2} \left(\frac{m^*}{\hbar^2}\right)^{3/2} E^{1/2} \tag{D.12}$$

Substituting in (D.11) we obtain

$$\frac{V}{(2\pi)^3} \int d^3k f(E(\mathbf{k})) = \int dE\, \rho(E) f(E) \tag{D.13}$$

where

$$\boxed{\rho(E) = \frac{V}{\sqrt{2}\,\pi^2} \left(\frac{m^*}{\hbar^2}\right)^{3/2} E^{1/2}} \tag{D.14}$$

The integral of $\rho(E)$ from the lowest value of E up to E gives by definition the number of states $R(E)$ with energy less than E

$$R(E) \equiv \int_{E_m}^{E} dE' \rho(\mathrm{E}') = \frac{2}{3} \frac{V}{\sqrt{2}\,\pi^2} \left(\frac{m^*}{\hbar^2}\right)^{3/2} \left(E^{3/2} - E_m^{3/2}\right) = \frac{1}{(2\pi)^3} V \left(\frac{4\pi}{3} k^3\right) \tag{D.15}$$

D.4 Effective Hamiltonian

As we have seen in the previous section, the concept of the effective mass is very useful because it allows us to obtain the electronic DOS near the band edges simply by replacing the electronic mass m_e in the free electron formula by the effective mass.[7] However, the effective mass concept shows its full power in the very

[7]If the effective mass is anisotropic (see (D.9)), m_e must be replaced by $m^* = (m_1 m_2 m_3)^{1/3}$.

common and important case where there are deviations from the periodic crystal potential due to unavoidable impurities and defects or to the presence of external electromagnetic fields. In those cases the electronic Hamiltonian $\hat{H}$ has the form

$$\hat{H} = \frac{\hat{p}^2}{2m_e} + \mathcal{V}_p + \mathcal{V}_a, \tag{D.16}$$

where $\mathcal{V}_p$ is the periodic crystal potential experienced by the electron, and $\mathcal{V}_a$ is the non-periodic perturbation due to impurities, defects, external fields, etc. If $\mathcal{V}_a$ is such that it produces negligible transitions from a given band to any other band, and if the relation between the energy and the wavevector$\boldsymbol{k}$, is of the form $\varepsilon(\boldsymbol{k}) = \varepsilon_o + \left(\hbar^2\boldsymbol{k}^2/2m^*\right)$, then

$$\hat{H} = \frac{\hat{p}^2}{2m_e} + \mathcal{V}_p + \mathcal{V}_a \quad \approx \quad \hat{H}_e = \varepsilon_0 + \frac{\hat{p}^2}{2m^*} + \mathcal{V}_a \tag{D.17}$$

Equation (D.17) is valid only for those eigenfunctions of $\hat{H}_e$ for which their extent is much larger than the lattice constant.

Equation (D.17) means that the periodic potential $\mathcal{V}_p$, under the said conditions, can be replaced by a constant potential (i.e. zero force), if at the same time the electronic mass is replaced by the effective mass,[8] m^*. The analytical and calculational advantages of using $\hat{H}_e$ instead of $\hat{H}$ shown in (D.16) are quite obvious.

D.5 Impurity Levels and Doping

D.5.1 Impurity Levels: The General Picture

Consider the case of an electron moving under the influence of a periodic potential responsible for the creation of bands and gaps and a perturbing non-periodic potential $\mathcal{V}_a$. If the latter is local and attractive, its effect is as in the case of a potential well examined in Chap. 3, Sect. 3.8, Problem 3: This means that the local attractive potential deforms the local unperturbed DOS by increasing it near the lower edge of each band (and by decreasing it near the upper edge). As the strength of this local attractive potential exceeds a critical, band-dependent value, a level is pulled down out of the band and into the gap below it; the corresponding eigenstate is bound in the vicinity of the local potential $\mathcal{V}_a$. As the strength of the local potential increases further, the bound level is moving away from the band edge

---

[8] If the band structure is of the more general form $\varepsilon(k) = \varepsilon_o + \left(\hbar^2k_1^2/2m_1\right) + \left(\hbar^2k_2^2/2m_2\right) + \left(\hbar^2k_3^2/2m_3\right)$ then the effective hamiltonian $\hat{H}_e$ will be assumed for simplicity to be as in (D.17) with $m^* = (m_1m_2m_3)^{1/3}$.

towards the center of the gap and the corresponding state becomes more localized. Impurity levels located well within the interior of the gap are called *deep levels*; they may be detrimental to the function of semiconductor-based devices, since they act as electron-hole recombination centers. If the local potential $\mathcal{V}_a$ is repulsive, a behavior is exhibited which has no analog in the case of the potential well of Sect. 3.8: The local DOS at each band is deformed as to enhance its values near the upper band edge (at the expense of the values near the lower band edge); as the strength of this repulsive potential exceeds a critical, band-dependent value, a level is pushed up out of the band and into the gap above it. Thus, in the presence of gaps, a repulsive potential may sustain bound eigenstates! In other words, as an attractive local potential may create a bound state by pushing down out of the bottom of the band a level in the gap, in an analogous way, a repulsive local potential may create a bound state by pulling up out of the top of the band a level in the gap.

D.5.2 Impurity Levels: Doping

An important consequence of the preceding comments is the possibility of controlling the levels in the gap by introducing appropriate impurities. This is of crucial importance for semiconductor devices, such as diodes, solar cells, transistors, integrated circuits, etc, i.e., of crucial importance for our technological civilization. Of particular interest are substitutional impurities whose valence ζ_i is equal to $\zeta_o \pm 1$, where ζ_o is the valence of the host atom. If $\zeta_i = \zeta_o + 1$ the impurity is called *n-type* or *donor*, because out of its $\zeta_i$electrons the ζ_o participate in the bonding, as before the substitution, while the extra one is free to be donated to the CB by thermal excitation (see Sect. D.6 below); thus, this substitution provides an extra electron to populate the CB (besides the ζ_o electrons per atom which populate the VB and contribute to the bonding), i.e., an extra negative carrier is now available. If $\zeta_i = \zeta_o - 1$ the impurity is called *p-type* or *acceptor*, because it can accept through thermal excitation an electron from the VB. Thus a hole is created in the VB, i.e. a positively charged carrier. In Table D.1 we give a list of n-type and p-type impurities in various host material. The process of introducing n-type (donors) or p-type (acceptors) impurities is called *doping* and the semiconductors with such dopants are called *doped* or *extrinsic* (see next section).

An n-type substitutional impurity attracts the extra electron (the one not tetrahedrally bonded) by a potential of the form

$$\mathcal{V}_a = -\frac{e^2}{\varepsilon r}, \tag{D.18}$$

Table D.1 Host atoms and n-type (donor) and p-type (acceptor) substitutional impurities in various semiconductors

Impurities	Host		GaAs	
	Si	Ge	Ga	As
n-type, donors	P, As, Sb, Bi	P, As, Sb, Bi	Si, Ge	S, Se
p-type, acceptors	B, Al, Ga, In, Tl	B, Al, Ga, In, Tl	Be, Mg, Zn, Cd, Mn[a]	Si, Ge

[a]Mn is important as a dopant in connection with the so-called diluted magnetic semiconductors (DMS)

where r is the distance between the impurity ion and the extra electron and ε is the dimensionless static dielectric function[9] of the host material. Combining (D.17) and (D.18) we see that the problem has been reduced to that of the hydrogen atom with the replacements

$$m_e \to m_c^*, \quad e^2 \to e^2/\varepsilon, \quad \varepsilon_o = E_g, \tag{D.19}$$

where m_c^* is a properly averaged effective mass at the bottom of the CB, and E_g is the energy at the bottom of the CB (The top of the VB is chosen as the zero of energy). Hence

$$a_{\mathrm{B}} \to a_{\mathrm{B}}^* = \frac{\hbar^2}{m_c^*(e^2/\varepsilon)} = a_{\mathrm{B}} \frac{m_e}{m_c^*} \varepsilon, \tag{D.20}$$

$$E_n \to E_n^* = E_g - \frac{1}{2n^2} \frac{e^2}{\varepsilon a_B^*} = E_g - E_n \frac{m_c^*}{m_e \varepsilon^2}. \tag{D.21}$$

Thus the ground state binding energy is $\varepsilon_d \equiv E_g - E_1^* = \left(13,6\, m_c^*/m_e\varepsilon^2\right)$ eV. For the results (D.20) and (D.21) to be even approximately valid the effective Bohr radiusa_B^* must be much larger than the lattice constant a: $a_B^* \gg a$. The reader is urged to check if this inequality is satisfied for Si ($\varepsilon \approx 12$, $m_1 = m_2 = 0.1905\, m_e$, $m_3 = 0.9163\, m_e$ for Ge ($\varepsilon \approx 16.2$, $m_1 = m_2 = 0.0807\, m_e$, $m_3 = 1.57\, m_e$), and for GaAs ($\varepsilon \approx 12.85$, $m_1 = m_2 = m_3 = 0.067\, m_e$).

For p-type impurities which push up states out of the VB and into the gap we can treat the problem either in terms of electrons or, equivalently, in terms of holes. Let us use the electron language. The impurity ion has now a proton less and thus it creates a minus e charge. Hence, an electron feels a potential $\mathcal{V}_a$ of the form

$$\mathcal{V}_a = \frac{e^2}{\varepsilon r}, \tag{D.22}$$

[9]The employment of the static dielectric function is justified only if the average distance r is much larger than the interatomic distance d or the lattice constant, a.

and its effective Hamiltonian in its simplest version is as follows

$$\hat{H}_e = -\frac{\hbar^2}{2m_e^*}\nabla^2 + \frac{e^2}{\varepsilon r},$$

where $m_e^* = -m_h$ is a negative number. Hence, Schrödinger's equation$\hat{H}_e\psi = E\psi$ multiplied by -1, becomes

$$-\frac{\hbar^2}{2m_h}\nabla^2\psi - \frac{e^2}{\varepsilon r}\psi = -E\psi, \tag{D.23}$$

where m_h is the positive effective hole mass. Notice that if we had used the hole option, we would have obtained (D.23) with $-E$ replaced by E_h. Equation (D.23) becomes equivalent to the hydrogen atom case with the following replacements:

$$m_e \to m_h,\ e^2 \to e^2/\varepsilon, \tag{D.24}$$

$$a_B \to a_B^* = a_B\frac{m_e\varepsilon}{m_h}, \tag{D.25}$$

$$E_n \to E_n^* = -E_n\frac{m_h}{\varepsilon^2 m_e} = |E_n|\frac{m_h}{m_e\varepsilon^2}. \tag{D.26}$$

Thus the lowest lying acceptor impurity level is in the gap above the top of the VB by an amount equal to $13.6(m_h/m_e\varepsilon^2)$ eV.

D.6 Concentration of Carriers as a Function of Temperature *T*

The great importance of the donor or acceptor impurities stems from their potential to control dramatically the concentration of carriers (of electrons in the CB and of holes in the VB) and, consequently, the conductivity of the semiconductors. In this section we shall demonstrate this point.

In Fig. D.1 we plot schematically the DOS in and around the gap in a doped semiconductor with both p-type (or acceptor) and n-type (or donor) substitutional impurities.

Let us assume that the concentration of acceptor impurities N_a is larger than the concentration of donor atomsN_d. Then, at absolute zero temperature all electrons from the donor atoms would go to the lower acceptor level; as a result, the number of holes at ε_a (per unit volume) would be equal to $N_a - N_d$, the number of holes at the VB would be zero and the number of electrons at the level $E_g - \varepsilon_d$ and at the CB would be zero. At a non-zero temperature let p and p_a be the number of holes per unit volume at the VB and the acceptor level respectively, while n_d and n be the

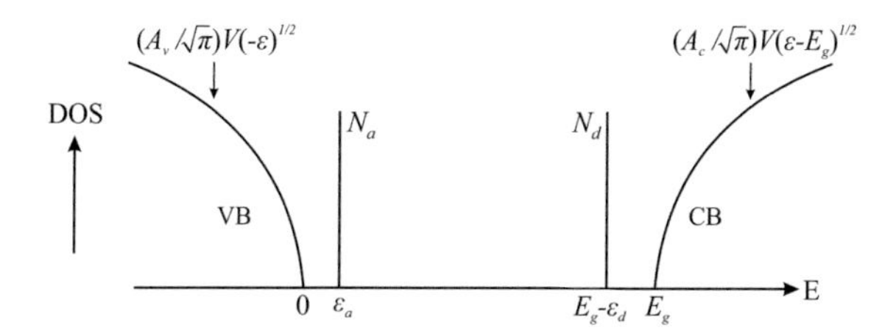

Fig. D.1 DOS for a doped semiconductor with both acceptor and donor impurities, the levels of which are at ε_a and $E_g - \varepsilon_d$ respectively. N_a, N_d are their respective concentrations. The constants A_c and A_υ for the CB and the VB DOS are: $A_j = \left(\sqrt{2}/2\right)\left(m_j/\pi\hbar^2\right)^{3/2}, j = c, \upsilon$

number of electrons per unit volume at the donor level and the CB respectively. Obviously the number of electrons is equal to the number of additional holes (beyond those, $N_a - N_d$, already existing at $T = 0\,\text{K}$).

Hence

$$\boxed{n + n_d = p + p_a - (N_a - N_d)} \quad \text{or} \quad \boxed{n + n_d - N_d = p + p_a - N_a} \tag{D.27}$$

The reader may convince himself/herself that (D.27) remains valid in the opposite case where N_d is larger than N_a.

The concentration of electrons in the CB $n = N/V$, is given by

$$n = \frac{2}{V}\int_{E_g}^{\infty} \frac{d\varepsilon\,\rho_c(\varepsilon)}{e^{\beta(\varepsilon-\mu)} + 1}. \tag{D.28}$$

where the factor 2 is for the two orientations of the electronic spin, $\left(e^{\beta(\varepsilon-\mu)} + 1\right)^{-1}$ is the average number of electrons in a state of energy ε, and μ is the chemical potential.

In a similar way we find that the concentration of holes in the VB is

$$p = \frac{2}{V}\int_{-\infty}^{0} d\varepsilon\,\rho_v(\varepsilon)\left[1 - \frac{1}{e^{\beta(\varepsilon-\mu)} + 1}\right] = \frac{2}{V}\int_{-\infty}^{0} d\varepsilon\,\rho_v(\varepsilon)\frac{1}{e^{-\beta(\varepsilon-\mu)} + 1} \tag{D.29}$$

Usually, but not always,[10] $\beta(E_g - \mu) \gg 1$ and $\beta\mu \gg 1$, where $\beta = 1/k_BT$. Under these circumstances the Fermi distributions are reduced to the Boltzmann distributions by omitting the 1 in the denominators. Then, given the DOS shown in Fig. D.1, the integrals in (D.28) and (D.29) can be performed analytically with the following results:

[10] If these inequalities are violated, the semiconductor is called *degenerate*.

$$n = A_c (k_B T)^{3/2} e^{-\beta(E_g - \mu)}, \tag{D.30}$$

$$p = A_\upsilon (k_B T)^{3/2} e^{-\beta\mu}, \tag{D.31}$$

where $A_c = \left(\sqrt{2}/2\right)\left(m_c/\pi\hbar^2\right)^{3/2}$ and $A_\upsilon = \left(\sqrt{2}/2\right)\left(m_\upsilon/\pi\hbar^2\right)^{3/2}$. The product np is independent of μ and is given by

$$np = A_c A_\upsilon \left(k_\beta \mathrm{T}\right)^3 e^{-\beta E_g}. \tag{D.32}$$

Intrinsic case.
This is the case where the concentrationsN_d and N_a of both donor and acceptor impurities are negligible (which means that $N_d \ll \sqrt{np}$ and $N_a \ll \sqrt{np}$). Then, (D.27) implies that $n = p$ which in combination with (D.32) gives the *intrinsic concentration*, n_i and p_i of electrons in the CB and holes in the VB

$$n_i = p_i = (A_c A_\upsilon)^{1/2} (k_B T)^{3/2} e^{-\beta E_g/2}. \tag{D.33}$$

Notice that the exponent in (D.33) is $E_g/2k_B T$ and not $E_g/k_B T$. If we choose the effective masses to be equal to m_e, we have for the intrinsic concentrations

$$n_i = p_i = 4.84 \times 10^{15}\, T^{3/2} \exp\left[-5802\, E_g/T\right] \mathrm{cm}^{-3}, \tag{D.34}$$

where T is in degrees Kelvin and E_g in eV. The chemical potentialμ_i for the intrinsic case is obtained by equating (D.31) with (D.33)

$$\mu_i = \frac{E_g}{2} + \frac{3k_B T}{4} \ln \frac{m_\upsilon}{m_c} \to \frac{E_g}{2}, \quad T \to 0\ \mathrm{K}. \tag{D.35}$$

Extrinsic case.
This is the case where the concentration N_d or N_a of either donor or acceptor impurities or both is much larger than the intrinsic concentration. If $N_d = N_a$, the doped semiconductor is called *compensated*. If the chemical potential happens to be in between the acceptor and the donor levels but several times $k_B T$ away from both, then $n_d \ll N_d$and $p_a \ll N_a$ so that, for uncompensated samples, we have $|n_d - p_a| \ll |N_a - N_d|$. In this case (D.27) becomes

$$n - p \approx N_d - N_a \equiv \delta N. \tag{D.36}$$

Knowing both the difference $n - p = \delta N$ and the product np (see (D.32) we obtain both n and p

$$n \approx \frac{1}{2}\left[\left(\delta N^2 + 4n_i^2\right)^{1/2} + \delta N\right], \tag{D.37}$$

$$p \approx \frac{1}{2}\left[\left(\delta N^2 + 4n_i^2\right)^{1/2} - \delta N\right]. \tag{D.38}$$

As it was mentioned before, for (D.37) and (D.38) to be valid we must have $E_g - \varepsilon_d - \mu \gg k_B T$ and $\mu - \varepsilon_a \gg k_B T$; thus we have to calculate μ to see whether these inequalities are indeed satisfied. If in (D.30) and (D.31) we replace μ by μ_i we shall obtain n_i and p_i respectively. Hence

$$\frac{n}{n_i} = e^{\beta(\mu-\mu_i)} \text{ and } \frac{p}{p_i} = e^{-\beta(\mu-\mu_i)},$$

or

$$e^{\beta(\mu-\mu_i)} - e^{-\beta(\mu-\mu_i)} = 2\sinh\beta(\mu-\mu_i) = \frac{n-p}{n_i} \approx \frac{\delta N}{n_i},$$

or

$$\mu = \mu_i + k_B T \sinh^{-1}\frac{\delta N}{2n_i}. \tag{D.39}$$

If μ, as calculated by (D.39), does not satisfy the inequalities $\mu - \varepsilon_a \gg k_B T$ and $E_g - \varepsilon_d - \mu \gg k_B T$, use this μ in (D.30) and (D.31) to obtain approximately n and p.

We remind the reader that the conductivity of a semiconductor is obtained in terms of the concentrations of carriers in the VB and the CB by the following formula:

$$\sigma = e\, n\, \mu_c + e\, p\, \mu_h \tag{D.40}$$

where the so-called mobility is defined as follows:

$$\mu_j = \left| e\, \tau_j / m_j^* \right|, \quad j = c,\ h \tag{D.41}$$

Problem: For *Si* and at $T = 295\,\text{K}$ calculate:

a. The intrinsic concentration n_i.
b. The chemical potentialμ_i for the intrinsic case
c. The *mobility* μ_c for electrons in the CB. (For *Si* at $T = 295\,\text{K}\ \tau = \ell/\upsilon_{th}$, $\upsilon_{th} \simeq (3k_B T/m^*)^{1/2}$ and $\ell \approx 530\,\text{A}$)
d. The resistivity, ρ, in the intrinsic case ($\rho_e = \sigma^{-1}$, $\sigma = en\mu_e + ep\mu_h$) (the hole mobility in *Si* is $\mu_h \approx 0.36\,\mu_e$)

e. The resistivity of *Si* doped with *P* atoms of concentration $n_P = 10^{-6}\, n_{Si}$, where n_{Si} is the concentration of *Si* atoms.

References

1. P.Y. Yu, M. Cardona, *Fundamentals of Semiconductors*, 3rd edn. (Springer, Berlin, 2005)
2. S.M. Sze, *Physics of Semiconductor Devices*, 3rd edn. (Wiley-Interscience, Hoboken, 2007)

Appendix E
Useful Concepts and Definitions

Length, ℓ Time, t Mass, m	Dimensions
Angular frequency: $\omega \equiv 2\pi/t = 2\pi f$	$[\omega] = 1/[t]$
Velocity: $\mathbf{\upsilon} \equiv d\mathbf{r}/dt$	$[\upsilon] = [\ell]/[t]$
Acceleration: $\mathbf{a} \equiv d\mathbf{\upsilon}/dt$	$[a] = [\ell]\Big/[t]^2$
Angular acceleration: $\dot{\boldsymbol{\omega}} \equiv d\boldsymbol{\omega}/dt$	$[\dot{\omega}] = 1\Big/[t]^2$
Mass density: $\rho_M \equiv dm/dV$	$[\rho_M] = [m]\Big/[\ell]^3$
Momentum: $\mathbf{p} = m_r\mathbf{\upsilon}$	$[p] = [m][\ell]/[t]$
Angular momentum: $\mathbf{L} = \mathbf{r} \times \mathbf{p}$	$[L] = [m][\ell]^2\Big/[t]$
Force: $\mathbf{F}$	$[F] = [E]/[\ell] = [m][\ell]\Big/[t]^2$
Energy: $E = E_o + E_K + E_P$	$[E] = [m][\ell]^2\Big/[t]^2$
Internal energy: $\langle E\rangle \equiv U = E_o + E_K + E_P,$	$[E]$, $E_o + E_K = \sum_i \sqrt{m_{oi}^2 c^4 + c^2 p_i^2}$
Kinetic energy: $\varepsilon_K \approx p^2/2m$ if $cp \ll m_o c^2$,	$[E]$, $\varepsilon_K \approx cp$ if $cp \gg m_o c^2$
Potential energy: ε_P	$[E]$
Work: $W = \int \mathbf{F} \cdot d\ell$	$[E]$
Heat: Q	$[E]$
Temperature: T It appears physically as $k_B T$	$[k_B T] = [E]$
Entropy: $S = k_B \ell n\, \Delta\Gamma_N(U,V) = -k_B \sum_I P_I \ell n P_I$	$[S] = [E]/[T] = [k_B]$
Chemical potential: $\mu = (\partial U/\partial N)_{S,V}$	$[\mu] = [E]$
Pressure: $P = F/A = E/V$	$[P] = [F]\Big/[\ell]^2 = [E]\Big/[\ell]^3$
Moment of force: $\mathbf{M} \equiv \mathbf{r} \times \mathbf{F}$	$[M] = [E]$
Electric charge: q	$\boxed{[q] = *}$
Electric charge density: $\rho_q = dq/dV$	$[q]/[\ell]^3$
Electric current: $I \equiv dq/dt$	$[I] = [q]/[t]$
Density of electric current: $j \equiv I/A$	$[j] = [q]\Big/[t][\ell]^2$

(continued)

E.N. Economou, *From Quarks to the Universe*,
DOI 10.1007/978-3-319-20654-7

Electric field: $\mathbf{E} \equiv \mathbf{F}_e/q$	$[\mathbf{E}] = [E]/[\ell][q]$
EMF $\equiv \oint \mathbf{E} \cdot d\ell$	$[\text{HE}\Delta] = [E]/[q]$
Magnetic field: $q\mathbf{v} \times \mathbf{B} \equiv \mathbf{F}_m^*$	$[\mathbf{B}] = [m]/[t][q]$
Flux of a vector field **A**	$\Phi_A = \iint \mathbf{dS} \cdot \mathbf{A}$
Circulation of a vector field **A**	$K = \oint \boldsymbol{d\ell} \cdot \mathbf{A}$
Density of flux of a quantity A	$\mathbf{j}_A = \rho_A \cdot \mathbf{v}_A$
Derived quantities	
Moment of inertia: $I = \sum_i m_i \rho_i^2$	$[I] = [m][l^2]$
Specific heat: $C = \frac{dQ}{dT}$	$[C] = [E]/[T] = [k_B]$
Bulk and shear moduli:	$[B] = [\mu_s] = [P]$
Dipole moment: $\mathbf{p}_e = q\,\mathbf{r}$,	$[p_e] = [q][\ell]$
Electric resistance: $R = V_e/I$,	$[R] = [E][t]/[q]^2$
Resistivity: ρ_e: $R = \rho_e \ell/A$;	Conductivity: $\sigma_e = 1/\rho_e$, $[\rho_e] = [R][\ell]$
Capacitance: $C_e = q/V_e$	$[C_e] = [q]^2/[E]$
Self-induction: $L_e = \Phi_B/I$, Φ_B is the flux of **B**	$[L_e] = [m][l]^2/[q]^2$
Permittivity or dielectric function: ε	$[\varepsilon] = [q]^2[t]^2/[m][l]^3$
Permeability: μ	$[\mu] = [m][l]/[q]^2$
Surface tension: σ,	$[\sigma] = [\text{E}]/[\ell]^2$
Viscosity: η	$[\eta] = [P][t]$
Index of refraction: $n = \sqrt{\varepsilon\,\mu}$,	dimensionless
Polarizability: $a_p = \mathbf{p}_e/\mathbf{E}$	$[a_p] = [\ell]^3$, G-CGS
Susceptibility χ: $\varepsilon = 1 + 4\pi\chi$, G-CGS	dimensionless

*The system SI has much more familiar EM units, while the G-CGS has simpler EM formulae, since, instead of ε_0, μ_0, it contains the velocity of light c. See the Appendix B, particularly its last two pages

Less familiar quantities are written in italics

Appendix F
25 Important Relations

1.	Conservation of angular momentum $\mathbf{L}$ of an isolated system, $(\mathbf{L} = \sum_i \mathbf{r}_i \times \mathbf{p}_i = \mathbf{I}\boldsymbol{\omega})$ $I = \sum_i \Delta m_i \rho_i^2 = \int dV \rho_M \rho^2$
2.	Conservation of momentum $\mathbf{P}$ of an isolated system, $\{\mathbf{P} = \sum_i \mathbf{p}_i\}$, $\mathbf{p}_i = m_i \boldsymbol{\upsilon}_i$, $m_i = m_{oi}/\sqrt{1-(\upsilon_i^2/c^2)}\}$
3.	Conservation of energy E of an isolated system, $E = E_o + E_K + E_P$, $E_o + E_K = \sum_i \sqrt{m_{oi}^2 c^4 + c^2 p_i^2} = \sum_i m_i c^2$
First Law: $dU = \bar{d}Q - \bar{d}W + \bar{d}E_m$	
4.	Conservation of internal characteristic quantities of an isolated system, (charge, baryon number, lepton number, etc) as well as parity, time reversal etc for some interactions
5.	Entropy and the Second Law, $S \equiv k_B \ell n \Gamma_N(U, V, \ldots) \equiv -k_B \sum_I P_I \ell n P_I$
6.	$dS = \bar{d}S_e + \bar{d}S_i$ (όταν $\bar{d}E_m = 0$), όπου $\bar{d}S_e = \frac{\bar{d}Q}{T}$ and $\boxed{\bar{d}S_i \geq 0}$
7.	$\bar{d}W = PdV$ etc, $\bar{d}E_m = \mu dN$ and generalizations
8.	$P_I = \frac{e^{-E_I/k_B T}}{Z}$, $Z = \sum_I e^{-E_I/k_B T}$
9.	$n_i = \frac{1}{e^{(\varepsilon_i - \mu)/k_B T} \pm 1}$, valid for non-interacting particles
10.	$G \equiv U + PV - TS \equiv H - TS$, $dG \leq -SdT + VdP + dE_m$
11.	$\left.\begin{array}{l} x' = x,\ y' = y \\ z' = \gamma(z - \upsilon_o t),\ ct' = \gamma(ct - \upsilon_o z/c) \\ \gamma = 1/\sqrt{1-(\upsilon_o/c)^2} \end{array}\right\}$ Lorentz transformations
12.	$\mathbf{F}_m = -\frac{GmM}{r^2}\mathbf{r}_o$, outline of GTR
13.	$\mathbf{F}_{HM} = q(\mathbf{E} + \boldsymbol{\upsilon} \times \mathbf{B})$, SI $\mathbf{F}_{HM} = q(\mathbf{E} + \frac{\boldsymbol{\upsilon}}{c} \times \mathbf{B})$, G-CGS
14.	Maxwell's equations in vacuum, SI $\left.\begin{array}{l} \Phi_\mathbf{E} = \frac{1}{\varepsilon_0} Q \Leftrightarrow \nabla \cdot \mathbf{E} = \frac{1}{\varepsilon_0}\rho_e \\ K_\mathbf{E} = -\frac{\partial \phi_\mathbf{B}}{\partial t} \Leftrightarrow \nabla \times \mathbf{E} = -\frac{\partial \mathbf{B}}{\partial t} \\ \Phi_\mathbf{B} = 0 \Leftrightarrow \nabla \cdot \mathbf{B} = 0 \\ K_\mathbf{B} = \mu_0 I + \mu_0 \varepsilon_0 \frac{\partial \phi_\mathbf{E}}{\partial t} \Leftrightarrow \nabla \times \mathbf{B} = \mu_0 \mathbf{j} + \mu_0 \varepsilon_0 \frac{\partial \mathbf{E}}{\partial t} \end{array}\right\}$ (SI)
15.	$c = \frac{1}{\sqrt{\varepsilon_0 \mu_0}}$, $\mathbf{S} = \frac{1}{\mu_0}\mathbf{E} \times \mathbf{B}$, $u = \frac{1}{2}\varepsilon_0 \mathbf{E}^2 + \frac{1}{2\mu_0}\mathbf{B}^2$, $\mathbf{p} = \varepsilon_0 \mathbf{E} \times \mathbf{B} = \frac{\mathbf{S}}{c^2}$, SI
16.	$U = \int u dV = n\hbar\omega$, $\mathbf{P} = \int \mathbf{p} dV = n\hbar\mathbf{k}$, $L = \frac{U}{\omega} = n\hbar$

(continued)

E.N. Economou, *From Quarks to the Universe*,
DOI 10.1007/978-3-319-20654-7

17.	Equation of motion: $\frac{d\mathbf{p}}{dt} = \mathbf{F}$ (non-quantum) , $\frac{d\mathbf{L}}{dt} = \mathbf{M}$
18.	$\varepsilon = \hbar\omega$, $\mathbf{p} = \hbar\mathbf{k}$, $\lvert\mathbf{k}\rvert = \frac{2\pi}{\lambda}$
19.	Equation of motion: $-\frac{\hbar^2}{2m}\left(\frac{\partial^2\psi}{\partial x^2} + \frac{\partial^2\psi}{\partial y^2} + \frac{\partial^2\psi}{\partial z^2}\right) + \mathcal{V}(\mathbf{r})\psi = i\hbar\frac{\partial\psi}{\partial t}$ (quantum)
20.	$\Delta x \cdot \Delta p_x \geq \frac{\hbar}{2}$
21.	$\varepsilon_K \geq \sqrt{m_o^2 c^4 + 9,12\frac{\hbar^2 c^2}{V^{2/3}}} - m_o c^2 \underset{c\to\infty}{\rightarrow} 4,56\frac{\hbar^2}{m_o V^{2/3}} \underset{m_o c^2 \ll cp}{\rightarrow} 3\hbar c/V^{1/3}$
22.	Pauli's exclusion principle (for half integer identical particles sharing the same volume)
23.	$E_K \geq 2,87N\frac{\hbar^2}{m}\left(\frac{N}{V}\right)^{2/3}$ ή $E_K \geq 2,32\frac{c\hbar N^{4/3}}{V^{1/3}}$ για $m_o c^2 \ll cp$
24.	$\delta E = c_1\hbar^2/mV^{2/3}$: Comparison of δE with $k_B T$ or $\hbar\omega$
25.	Number of single particle states: $\frac{VV_k}{(2\pi)^3}$,where V_k is the volume in $\boldsymbol{k}$space
The properties of macroscopic systems are also of quantum nature	

Appendix G
21 Numbers to be Remembered

1. Velocity of light, $c \approx 3 \times 10^8$ m/s
2. Electron mass, $m_e c^2 \approx 0.511\,\text{MeV}$, $m_e = 9.11 \times 10^{-31}\,\text{kg}$
3. Proton mass, $m_p \approx 1836,15\,\text{m}_\text{e}$
4. Neutron mass, $m_n \approx 1838.68\,m_e$
5. Atomic mass u (≡1/12 of carbon-12 mass), $u \approx 1822,89\,m_e$
6. Acceleration of gravity at Earth's surface, $g \approx 9.81\,\text{m/s}^2$
7. Avogadro's constant, $N_A \approx 6 \times 10^{23}$
8. Fine-structure constant, $\alpha \equiv e^2/\hbar c = \hbar/m_e a_B c = \upsilon_0/c \approx 1/137$
9. Gas constant, $R \equiv N_A k_B \approx 8.312\,\text{J/K}$
10. Bohr radius, $a_B \equiv \hbar^2/m_e e^2 \approx 0.53\,\mathring{\text{A}}$
11. Unit of energy in atomic system, $E_0 \equiv e^2/a_B = \hbar^2/m_e a_B^2 \approx 27.2\,\text{eV}$
12. Unit of time in atomic system, $\hbar/E_0 \approx 2.42 \times 10^{-17}\,\text{s}$
13. Unit of pressure in atomic system, $P_0 \equiv E_0/a_B^3 \approx 294\,\text{Mbar}$
14. Dimensionless strengths of the basic interactions
    Gravity, $G m_p^2/\hbar\, c \approx 5.9 \times 10^{-39}$
    EM, $e^2/\hbar\, c = \alpha \approx 1/137$
    Weak, $\alpha_w \approx 10^{-5}$
    Strong, $\alpha_s \approx 1$
15. Circumference of Earth, $\Pi \approx 4 \times 10^7$ m
16. Strong interaction per pair of nucleons, $\varepsilon \approx 11.4\,\text{MeV}$
17. Mass of the Sun, $M_S \approx 1.99 \times 10^{30}\,\text{kg}$
18. Mass of the Earth, $M_E = 5.97 \times 10^{24}\,\text{kg}$
19. $N_A u = 1\,\text{g}$
20. 1 eV/atom ≈ 96.48 kJ/mol = 23.06 kcal/mol

$$1\,\text{eV} \leftrightarrow 12400\,\text{A} \leftrightarrow 8065\,\text{cm}^{-1} \leftrightarrow 2.42 \times 10^{14}\text{Hz};\, 290^\circ\,\text{K} \Leftrightarrow (1/40)\,\text{eV}$$

Light year: 1 ly = 9.4607305×10^{15} m
Parsec: 1 pc = 3.0856776×10^{16} m = 3.2616 ly
Astronomical unit: 1 AU = 1.4959787×10^{11} m

E.N. Economou, *From Quarks to the Universe*,
DOI 10.1007/978-3-319-20654-7

Appendix H
Answers to Multi-Choice Questions/Statements

Chapter 2, Section 2.6
1,d; 2,c; 3,b; 4,a; 5,c; 6,c; 7,d; 8,b; 9,a; 10,d; 11,b; 12,c

Chapter 3, Section 3.6
1,c; 2,b; 3,c; 4,c; 5,b; 6,a; 7,c; 8,b; 9,b; 10,b

Chapter 3, Section 3.9
5, $\frac{5\hbar^2}{8m}\left\{\frac{1}{a^2}+\frac{1}{b^2}+\frac{1}{c^2}\right\} > \frac{15\hbar^2}{8m}\frac{1}{r^2}, \quad r^3 = abc$

Chapter 4, Section 4.8
1,c; 2,d; 3,a; 4,c; 5,c; 6,a; 7,b; 8,b; 9,c; 10,c

Chapter 5, Section 5.6
1.c; 2,b; 3,c; 4,c; 5,c; 6,b; 7,a; 8,a

Chapter 6, Section 6.9
1,c; 2,b; 3,a; 4,c; 5,b; 6,d; 7,c; 8,b; 9,d; 10,c; 11,d; 12,c; 13,a; 14,a; 15,c

Chapter 7, Section 7.6
1,a; 2,d; 3,b

Chapter 8, Section 8.4
1,a; 2,a
Section 8.5
3. 1st, Yes (strong interactions). 2nd, No (violation of baryon number and spin). 3rd, Yes (strong interactions). 4th, Yes (weak interactions, $\Delta S = 1$). 5th, Yes (weak interactions, $\Delta S = 1$). 6th, No (violation of lepton muon number). 7th, No (violation of lepton muon and electron number). 8th, No (violation of baryon number). 9th, No (violation of charge conservation).

Chapter 9, Section 9.6
1,c; 2,b; 3,c; 4,c; 5,c; 6,a; 7,d; 8,a; 9,d; 10,c;

Chapter 10, Section 10.6
1,b; 2,a; 3,d; 4,c; 5,a; 6,b; 7,b; 8,c; 9,a; 10,d; 11,a; 12,c; 13,b; 14,d; 15,b; 16,d; 17,b

E.N. Economou, *From Quarks to the Universe*,
DOI 10.1007/978-3-319-20654-7

Chapter 11, Section 11.7

1,a; 2,d; 3,b; 4,d; 5,c; 6,d; 7,c; 8,d; 9,b; 10,a; 11,c; 12,b; 13,a; 14,d; 15,a; 16,c; 17,d; 18,a; 19,c; 20,a; 21,b; 22,b; 23,d; 24,a; 25,c

Chapter 12, Section 12.11

1,d; 2,b; 3,a; 4,b; 5,a; 6,d; 7,d; 8,c; 9,c; 10,b; 11,c; 12,d; 13,b; 14,b; 15,d; 16,d; 17,a; 18,d; 19,c; 20,d; 21,a; 22,d; 23,d; 24,c; 25,b; 26,c; 27,a; 28,d; 29,d; 30,d; 31,c; 32,d; 33,a; 34,b; 35,c; 36,b; 37,c; 38,b; 39,b; 40,a

Chapter 13, Section 13.7

1,a; 2,b; 3,b; 4,c; 5,c; 6,a; 7,c; 8,b; 8,b; 9,c; 10,b; 11,c; 12,a; 13,d

Chapter 14, Section 14.8

1,b; 2,c; 3,b; 4,c; 5,c; 6,c; 7,b; 8,a; 9,b; 10,c; 11,c; 12,d; 13,d; 14,c; 15,b; 16,c; 17,d; 18,b; 19,a; 20,c; 21,b; 22,c; 23,a; 24,c; 25,b

Chapter 15, Section 15.7

1,b; 2,a; 3,c; 4,c; 5,c; 6,d; 7,b; 8,a; 9,a; 10,c; 11,a; 12,b; 13,d; 14,b; 15,a; 16,d; 17,d; 18,d; 19,a

Appendix I
Tables

See Tables I.1, I.2 and I.3

Table I.1 Physical constants

Quantity	Symbol	Value (year 2008)	Units
Planck constant over 2π	$\hbar = h/2\pi$	$1.054\,571\,628\,(53) \times 10^{-34}$	$\mathrm{J \cdot s}$
	$\hbar$	$6.582\ 118\ 99(16) \times 10^{-16}$	$\mathrm{eV \cdot s}$
Velocity of light	c	299 792 458	$\mathrm{m \cdot s^{-1}}$
Gravitational constant	G	$6.674\ 28(67) \times 10^{-11}$	$\mathrm{m^3 kg^{-1} s^{-2}}$
Proton charge	e	$1.602\ 176\ 487(40) \times 10^{-19}$	C
Electron mass	m_e or m	$9.109\ 382\ 15(45) \times 10^{-31}$	kg
Proton mass	m_p	$1.672\ 621\ 637(83) \times 10^{-27}$	kg
Neutron mass	m_n	$1.674\ 927\ 211(84) \times 10^{-27}$	kg
Atomic mass constant $\frac{1}{12} m(C^{12})$	u (or m_u)	$1.660\ 538\ 782(83) \times 10^{-27}$	kg
Vacuum permittivity	ε_0	$8.854187817\ldots \times 10^{-12}$	$\mathrm{F \cdot m^{-1}}$
Vacuum permeability	μ_0	$4\pi \times 10^{-7}$	$\mathrm{N \cdot A^{-2}}$
Boltzmann constant	k_B	$1.380\ 6504(24) \times 10^{-23}$	$\mathrm{J \cdot K^{-1}}$
Avogadro constant	N_A	$6.022\ 141\ 79(30) \times 10^{23}$	$\mathrm{mol^{-1}}$
Fine-structure constant	α[a]	$(137.035\ 999\ 679(94))^{-1}$	
Magnetic flux quantum	Φ_0[a]	$2.067\ 833\ 667(52) \times 10^{-15}$	Wb
Quantum Hall resistance	R_H[a]	25 812.8075(80)	Ω
Bohr magneton	μ_B[a]		

(continued)

E.N. Economou, *From Quarks to the Universe*,
DOI 10.1007/978-3-319-20654-7

Table I.1 (continued)

Quantity	Symbol	Value (year 2008)	Units
		$927.400\,915(23) \times 10^{-26}$	$\mathrm{J \cdot T^{-1}}$
Nuclear magneton	μ_N [a]	$5.050\,783\,24(13) \times 10^{-27}$	$\mathrm{J \cdot T^{-1}}$
Electron magnetic moment	μ_e	$-1.001\,159\,652\,181\,11(74)$	μ_B
Proton magnetic moment	μ_p	$2.792\,847\,356(23)$	μ_N
Neutron magnetic moment	μ_n	$-1.913\,042\,73\,(45)$	μ_N
Gas constant	$R \equiv N_A k_B$	$8.314\,472(15)$	$\mathrm{J \cdot mol^{-1} \cdot K^{-1}}$
Bohr radius	$a_B \equiv 4\pi\mathbf{e}_0\hbar^2/m_e e^2$	$0.529\,177\,208\,59(36) \times 10^{-10}$	m

Numbers in parentheses give the standard deviation in the last two digits

[a] $\alpha = e^2/4\pi\varepsilon_0\hbar c$, $\Phi_0 = h/2e$, $R_H \equiv h/e^2$, $\mu_B \equiv e\hbar/2m_e$, $\mu_N \equiv (m_e/m_p)\mu_B$ (all in SI)

$\alpha = e^2/\hbar c$, $\Phi_0 = hc/2e$, $a_B = \hbar^2/m_e e^2$, $\mu_B = e\hbar/2m_e c$, $\mu_N \equiv (m_e/m_p)\mu_B$ (all in G-CGS)

Table I.2 Atomic system of units ($m_e = 1$; $e = 1$; $a_B = 1$; $k_B = 1$) $c = 1/\alpha$ in G-CGS; $\varepsilon_0 = 1/4\pi$, $\mu_0 = 4\pi\alpha^2$ in SI

Length	$l_0 = a_B$
Mass	$m_0 \equiv m_e$
Charge	$q_0 \equiv e$
Time	$t_0 \equiv m_e a_B^2/\hbar = 2.418\,884 \times 10^{-17}$ s
Energy	$E_0 \equiv \hbar^2/m_e a_B^2 = 4.359\,744 \times 10^{-18}$ J $= 27.211384$ eV
Angular frequency	$\omega_0 \equiv \hbar/m_e a_B^2 = 4.134\,137 \times 10^{16}$ rad/s
Velocity	$\upsilon_0 \equiv a_B/t_0 = \hbar/m_e a_B = \alpha c = 2\,187.691$ km/s
Mass density	$\rho_0 = m_e/a_B^3 = 6.147\,315$ kg/m^3
Temperature	$T_0 \equiv E_0/k_B = \hbar^2/m_e a_B^2 k_B = 315\,775$ K
Pressure	$P_0 \equiv E_0/a_B^3 = \hbar^2/m_e a_B^5 = 2.942\,101 \times 10^{13}$ N / m^2 $= 2.942\,101 \times 10^8$ bar
Electrical resistance	$R_0 \equiv \hbar/e^2 = R_H/2\pi = 4\,108.236\ \Omega$
Resistivity	$\rho_{\rho 0} = R_0 a_B = \hbar a_B/e^2 = 21.739\,848\ \mu\Omega \times$ cm
Conductivity	$\sigma_0 = 1/\rho_{c0} = e^2/\hbar a_B = 4.599\,848 \times 10^7\ \Omega^{-1}\mathrm{m}^{-1}$
Electric current	$i_0 \equiv e/t_0 = 6.623\,618 \times 10^{-3}$ A
Voltage	$V_0 \equiv E_0/e = 27.211\,384$ V
Electric field	$\lvert\mathbf{E}_0\rvert \equiv V_0/a_B = 5.142\,206 \times 10^{11}$ V/m
Magnetic field[a]	$B_0 \equiv c\hbar/e a_B^2 = 2.350\,517 \times 10^5$ T,
Electric polarizability[b]	$a_{e0} = 4\pi\varepsilon_0 a_B^3 = 1.648\,777 \times 10^{-41}$ Fm^2
Electric induction	$D_0 \equiv e/a_B^2 = 57.214\,762$ C/m^2

(continued)

Table I.2 (continued)

Magnetic moment	$\mu_0 \equiv 2\mu_B = 1.854\,802 \times 10^{-23}$ JT^{-1}
Magnetization	$M_0 \equiv \mu_0 / a_B^3 = 1.251\,682 \times 10^8$ A/m
Magnetic field	$H_0 \equiv M_0 = 1.251\,682 \times 10^8$ A/m

For any quantity X we define $\bar{X} = X/X_0$ (see (1.16)).

The value of G in atomic units is 2.4×10^{-43}

[a]In SI set $c = 1$

[b]In G-CGS set $4\pi\varepsilon_0 = 1$

Table I.3 Periodic table of the elements

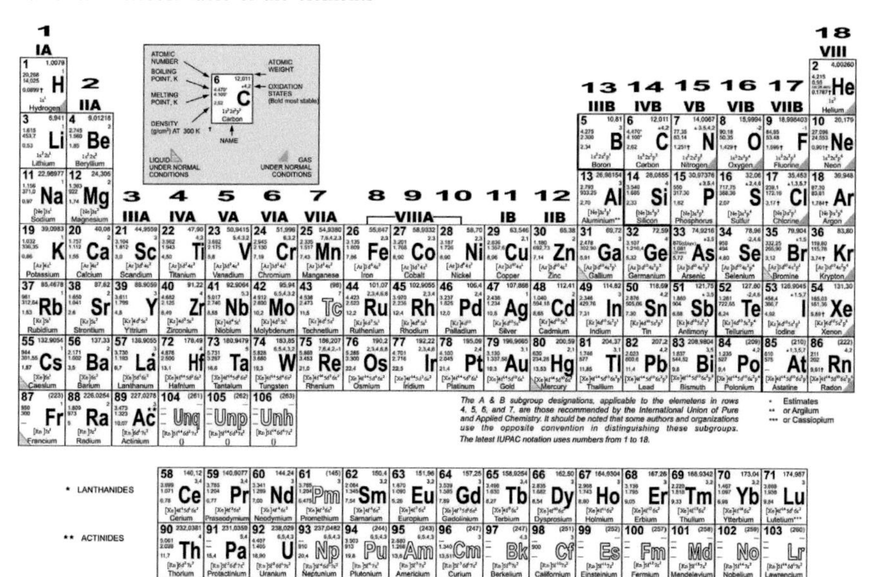

† For the gas phase (noted by the symbol) the units is g/l, for T=273K and P=1atm.

Index

E.N. Economou, *From Quarks to the Universe*,
DOI 10.1007/978-3-319-20654-7